Data in Science and Technology

Editor in Chief: R. Poerschke

Magnetic Properties of Metals

d-Elements, Alloys and Compounds

Editor: H. P. J. Wijn

Springer-Verlag Berlin Heidelberg New York
London Paris Tokyo Hong Kong
Barcelona Budapest

Cover picture: Spin density wave in Cr (see Fig. 3 on page 3).

ISBN 3-540-53485-7 Springer-Verlag Berlin Heidelberg NewYork
ISBN 0-387-53485-7 Springer-Verlag NewYork Berlin Heidelberg

Library of Congress Cataloging-in-Publication Data
Magnetic properties of metals : d-element, alloys, and compounds / editor, H. P. J. Wijn.
p. cm. -- (Data in science and technology)
Includes bibliographical references.
ISBN 0-387-53485-7
1. Semiconductors--Magnetic properties--Handbooks, manuals, etc.
2. Metals--Magnetic properties--Handbooks, manuals, etc. I. Wijn, H. P. J., 1922-. II. Series
QC611.8.M25M32 1991
620.1'1278--dc20 91-6299 CIP

Printed in Germany

Typesetting: Thomson Press, Ltd., New Delhi, India
Printing: Druckhaus Langenscheidt KG, Berlin
Bookbinding: Lüderitz & Bauer GmbH, Berlin

2163/3040-543210 – Printed on acid-free paper

Preface

During the last decades the knowledge of the magnetic properties of the d transition elements and of their metallic alloys and compounds has increased widely. The improvement of preparation techniques for well-defined substances, the development of sophisticated measuring methods and above all the drive to obtain more insight in the origin of magnetic interactions in solids have resulted in the publication of many specific magnetic properties for an abundance of all kinds of metallic materials.

The data assembled in this booklet are selected from the comprehensive compilation of magnetic and related properties of metals in the Landolt-Börnstein New Series Group III subvolumes 19a, b and c. It has been attempted to include preferentially those properties which are of a basic character and which therefore are most often needed by scientists active in the field of solid state magnetism.

In the field of magnetism, there is a gradual transition from the use of cgs/emu units to SI units. It was, however, not intended to represent all data in the units of one system, regardless of how nice this would have been from a systematic point of view. Instead, mostly preference was given to the system of units that was originally used by the authors whose work is quoted. Thus cgs/emu units occur most frequently. Of course the user of the tables and figures is helped in several ways to convert the data to the units which he is most familiar with, see, e.g., the list of definitions, units and conversion factors for the magnetic quantities occuring most frequently.

Aachen, April 1991 **The Editor**

Table of contents

List of symbols

Symbol	Unit	Quantity
A	$\mathrm{erg\,cm^{-3}}$	exchange stiffness constant
a, b, c	Å	lattice parameters
B	bar	bulk modulus
C_g	$\mathrm{cm^3\,Kg^{-1}}$	Curie constant per unit mass
C_m	$\mathrm{cm^3K\,mol^{-1}}$	Curie constant per mole
C_p, C_V	$\mathrm{cal\,K^{-1}}$	heat capacity at constant pressure/volume
D	$\mathrm{eV\,Å^2}$	spin wave stiffness constant
E_F	eV	Fermi energy
ΔE_Q	$\mathrm{mm\,s^{-1}}$	quadrupole splitting
f	MG	dHvA frequency
g		spectroscopic splitting factor
g'		magnetomechanical ratio
H	Oe, $\mathrm{Am^{-1}}$	magnetic field
H_{appl}	Oe, $\mathrm{Am^{-1}}$	applied magnetic field
H_c	Oe	coercive field
H_{hyp}	Oe	magnetic hyperfine field
h_i		magnetostriction coefficient
h'_i	$\mathrm{Oe^{-1}}$	forced magnetostriction coefficient
IS	$\mathrm{mm\,s^{-1}}$	isomer shift
J	meV	exchange integral
K		Knight shift
K_i	$\mathrm{erg\,cm^{-3}}$	magnetocrystalline anisotropy constant
k_F	$\mathrm{Å^{-1}}$	Fermi wavevector
k_B		Boltzmann constant
l	Å	interatomic distance
M_s	G	spontaneous (saturation) magnetization
n		number of electrons per atom
p	kbar	pressure
p	μ_B	atomic magnetic moment in paramagnetic phase
p_{eff}	μ_B	atomic magnetic moment in paramagnetic phase, derived from Curie-Weiss law
$\bar{p}$	μ_B	average magnetic moment
$p_{at}(\bar{p}_{at})$	μ_B	(average) magnetic moment per atom
$p_{ce}(\bar{p}_{ce})$	μ_B	(average) conduction electron magnetic moment per atom
$\bar{p}_{loc}$	μ_B	average localized magnetic moment per atom
$p_M(\bar{p}_M)$	μ_B	(average) magnetic moment of atom M
p_{orb}	μ_B	orbital magnetic moment per atom
p_s	μ_B	spontaneous (saturation) magnetic moment per atom
Q	$\mathrm{Å^{-1}}$	spin density wavevector
Q	$\mathrm{\mu V K^{-1}}$	thermoelectric power
R_H	$\mathrm{\Omega cm\,G^{-1}}$	Hall coefficient
S		atomic long-range order parameter
S		Stoner enhancement factor
T	K, °C	temperature
T_C	K	ferromagnetic Curie temperature
T_{ci}	K	commensurate-incommensurate transition temperature
T_M	K	martensitic transition temperature

Symbol	Unit	Quantity
T_m	K, °C	melting point temperature
T_N	K	Néel temperature
T_s	K	superconducting transition temperature; transition temperature between two types of magnetic order
T_{sf}	K	spin flip transition temperature
T_t	K	tetragonal phase transition temperature
t_a	s	annealing time
V	cm^3	volume
Z		atomic number
α	K^{-1}	linear thermal expansion coefficient
α	rad, deg	Kerr effect
β	rad, deg	angle
γ	$mJ\,mol^{-1}\,K^{-2}$	electronic specific heat coefficient
γ	$kHz\,G^{-1}$	gyromagnetic ratio
γ		fraction of 3d electrons in E_g state
δ		incommensurability parameter of spin density wave
δ		transverse (equatorial) Kerr effect
ε'_{ij}		real part of dielectric tensor element
ε''_{ij}		imaginary part of dielectric tensor element
Θ	K	paramagnetic Curie temperature
Θ_D	K	Debye temperature
θ	rad, deg	angle
λ	$W\,cm^{-1}\,K^{-1}$	thermal conductivity
λ		electron-phonon interaction constant
λ	s^{-1}	Landau-Lifshitz damping temperature
λ_s		linear saturation magnetostriction
λ_V		volume magnetostriction
μ_B		Bohr magneton
ν	Hz	frequency
ϱ	$g\,cm^{-3}$	density
ϱ	$\mu\Omega\,cm$	electrical resistivity
ϱ_x	$g\,cm^{-2}$	X-ray density
σ	$G\,cm^3\,g^{-1}$	magnetic moment per unit mass
σ_m	$G\,cm^3\,mol^{-1}$	magnetic moment per mole
σ_s	$G\,cm^3\,g^{-1}$	spontaneous (saturation) magnetic moment per unit mass
ϕ	rad, deg	angle
χ_{ac}		magnetic ac susceptibility
χ_g	$cm^3\,g^{-1}$	magnetic mass susceptibility
χ_{HF}	$cm^3\,g^{-1}$	high-field magnetic susceptibility
χ_m	$cm^3\,mol^{-1}$	magnetic molar susceptibility
χ_o		temperature-independent contribution to paramagnetic susceptibelity
χ_{orb}		orbital magnetic susceptibility
χ_{spin}		spin susceptibility
χ_{sp-orb}		spin-orbit interaction contribution to magnetic susceptibility
χ_V		magnetic volume susceptibility
ω		volume magnetostriction
$\partial\omega/\partial H$	Oe^{-1}	forced volume magnetostriction

Definitions, units and conversion factors

In the SI, units are given for both defining relations of the magnetization, $\boldsymbol{B}=\mu_0(\boldsymbol{H}+\boldsymbol{M})$ and $\boldsymbol{B}=\mu_0\boldsymbol{H}+\boldsymbol{M}$, respectively. $\mu_0=4\pi\cdot10^{-7}\,\mathrm{Vs\,A^{-1}\,m^{-1}}$, A: molar mass, ϱ: mass density.

Quantity	cgs/emu	SI	
$\boldsymbol{B}$	$\mathrm{G}=(\mathrm{erg\,cm^{-3}})^{1/2}$	$\mathrm{T}=\mathrm{Vs\,m^{-2}}$	
	$1\,\mathrm{G}\;\widehat{=}$	$10^{-4}\,\mathrm{T}$	
$\boldsymbol{H}$	$\mathrm{Oe}=(\mathrm{erg\,cm^{-3}})^{1/2}$	$\mathrm{Am^{-1}}$	
	$1\,\mathrm{Oe}\;\widehat{=}$	$10^3/4\pi\,\mathrm{Am^{-1}}$	
$\boldsymbol{M}$	$\boldsymbol{B}=\boldsymbol{H}+4\pi\boldsymbol{M}$	$\boldsymbol{B}=\mu_0(\boldsymbol{H}+\boldsymbol{M})$	$\boldsymbol{B}=\mu_0\boldsymbol{H}+\boldsymbol{M}$
	G	$\mathrm{Am^{-1}}$	T
	$1\,\mathrm{G}\;\widehat{=}$	$10^3\,\mathrm{Am^{-1}}$	$4\pi\cdot10^{-4}\,\mathrm{T}$
$\boldsymbol{P}$	$\boldsymbol{P}=\boldsymbol{M}V$	$\boldsymbol{P}=\boldsymbol{M}\mathrm{V}$	$\boldsymbol{P}=\boldsymbol{M}\mathrm{V}$
	$\mathrm{G\,cm^3}$	$\mathrm{Am^2}$	$\mathrm{Vs\,m}$
	$1\,\mathrm{G\,cm^3}\;\widehat{=}$	$10^{-3}\,\mathrm{Am^2}$	$4\pi\cdot10^{-10}\,\mathrm{Vs\,m}$
σ	$\sigma=M/\varrho$	$\sigma=M/\varrho$	$\sigma=M/\varrho$
	$\mathrm{G\,cm^3\,g^{-1}}$	$\mathrm{Am^2\,kg^{-1}}$	$\mathrm{Vs\,m\,kg^{-1}}$
	$1\,\mathrm{G\,cm^3\,g^{-1}}\;\widehat{=}$	$1\,\mathrm{Am^2\,kg^{-1}}$	$4\pi\cdot10^{-7}\,\mathrm{Vs\,m\,kg^{-1}}$
σ_m	$\sigma_\mathrm{m}=\sigma A$	$\sigma_\mathrm{m}=\sigma A$	$\sigma_\mathrm{m}=\sigma A$
	$\mathrm{G\,cm^3\,mol^{-1}}$	$\mathrm{Am^2\,mol^{-1}}$	$\mathrm{Vs\,m\,mol^{-1}}$
	$1\,\mathrm{G\,cm^3\,mol^{-1}}\;\widehat{=}$	$10^{-3}\,\mathrm{Am^2\,mol^{-1}}$	$4\pi\cdot10^{-10}\,\mathrm{Vs\,m\,mol^{-1}}$
χ	$\boldsymbol{P}=\chi\boldsymbol{H}$	$\boldsymbol{P}=\chi\boldsymbol{H}$	$\boldsymbol{P}=\chi\mu_0\boldsymbol{H}$
	$\mathrm{cm^3}$	$\mathrm{m^3}$	$\mathrm{m^3}$
	$1\,\mathrm{cm^3}\;\widehat{=}$	$4\pi\cdot10^{-6}\,\mathrm{m^3}$	$4\pi\cdot10^{-6}\,\mathrm{m^3}$
χ_V	$\chi_V=\chi/\mathrm{V}$	$\chi_V=\chi/V$	$\chi_V=\chi/V$
	$\mathrm{cm^3\,cm^{-3}}$	$\mathrm{m^3\,m^{-3}}$	$\mathrm{m^3\,m^{-3}}$
	$1\,\mathrm{cm^3\,cm^{-3}}\;\widehat{=}$	$4\pi\,\mathrm{m^3\,m^{-3}}$	$4\pi\,\mathrm{m^3\,m^{-3}}$
χ_g	$\chi_\mathrm{g}=\chi_V/\varrho$	$\chi_\mathrm{g}=\chi_V/\varrho$	$\chi_\mathrm{g}=\chi_V/\varrho$
	$\mathrm{cm^3\,g^{-1}}$	$\mathrm{m^3\,kg^{-1}}$	$\mathrm{m^3\,kg^{-1}}$
	$1\,\mathrm{cm^3\,g^{-1}}\;\widehat{=}$	$4\pi\cdot10^{-3}\,\mathrm{m^3\,kg^{-1}}$	$4\pi\cdot10^{-3}\,\mathrm{m^3\,kg^{-1}}$
χ_m	$\chi_\mathrm{m}=\chi_\mathrm{g}A$	$\chi_\mathrm{m}=\chi_\mathrm{g}A$	$\chi_\mathrm{m}=\chi_\mathrm{g}A$
	$\mathrm{cm^3\,mol^{-1}}$	$\mathrm{m^3\,mol^{-1}}$	$\mathrm{m^3\,mol^{-1}}$
	$1\,\mathrm{cm^3\,mol^{-1}}$	$4\pi\cdot10^{-6}\,\mathrm{m^3\,mol^{-1}}$	$4\pi\cdot10^{-6}\,\mathrm{m^3\,mol^{-1}}$

List of abbreviations

AF	antiferromagnetic
AF_0	commensurate spin density wave state
AF_1	transverse incommensurate spin density wave state
AF_2	longitudinal incommensurate spin density wave state
bcc	body-centered cubic
CAF	commensurate spin density wave state
dhcp	double hexagonal close-packed
dHvA	de Haas-van Alphen
F	ferromagnetic
fcc	face-centered cubic
GM	giant magnetic moment
hcp	hexagonal close-packed
IRM	isothermal remanent magnetization
I-SDW	incommensurate spin density wave
KS	Kohn-Sham potential
L	Lifshitz point
L-SDW	longitudinal spin density wave
MAG	magnetization
NMR	nuclear magnetic resonance
P	paramagnetic
RKKY	Rudermann-Kittel-Kasuya-Yosida
RT	room temperature
SDW	spin density wave
SG	spin glass
TQ	magnetic torque measurement method
T-SDW	transverse spin density wave
vBH	von Barth-Hedin exchange correlation potential

1 3d elements

1.1 Ti, V, Cr, Mn

Survey

Section	Metal	Property	Fig.	Table
1.1.1	Ti	$\chi_g(T)$	1	
1.1.2	V	$\chi_g(T)$	2	
1.1.3	Cr	SDW	3, 5	1
		magn. phase diagram	4	
		$\chi_g(T)$	6, 7	
1.1.4	Mn	magn. structure	8	2
		$\chi_g(T)$	9	

1.1.1 Ti

Titanium metal is a Pauli paramagnet; no localized magnetic moments have been observed. Since Ti becomes superconducting below 0.4 K, probably no magnetic ordering occurs.

The crystallographic structure of α-Ti, the most stable phase at room temperature, is hexagonal; in single crystals the magnetic susceptibility is therefore a function of the angle between the direction of the magnetic field and the *c* axis.

Next to α-Ti, there are two other phases of Ti known: β-Ti, with a body-centered cubic crystallographic structure, which is stable above 1155 K [56 M 1], and ω-Ti, with a hexagonal crystallographic structure, stable only under high pressure, but metastable at pressure zero [74 D 1].

1.1.2 V

Vanadium metal is a Pauli paramagnet. Neutron diffraction measurements revealed no localized magnetic moments; if they exist they are smaller than 0.01 μ_B [77 A 1]. Since vanadium becomes a superconductor at 5.265 K, probably no magnetic ordering occurs.

Vanadium has a body-centered cubic crystallographic structure; the magnetic susceptibility is isotropic, and a smooth function of temperature above the superconducting transition, see Fig. 2; it is possible that χ shows a very shallow maximum between 100 K and 200 K [65 K 1]. A good room temperature value is: $\chi_g = 7.3 \cdot 10^{-8}\ m^3/kg$.

Sometimes discontinuities have been observed in the susceptibility and other physical properties of V at temperatures between 120···240 K. Rostoker and Yamamoto [55 R 1] observed a crystallographic transition at −30 °C; this has not been confirmed by later investigations. Kostina [71 K 1] found a peak in the susceptibility at 240 K, and corresponding anomalies in the resistivity and the Hall effect. Kondorskii [73 K 1] found two peaks, respectively at 120 K and 190 K, and corresponding anomalies in the magnetostriction and thermal expansion.

The nature of these not very reproducible anomalies, which were sometimes interpreted as a hypothetical antiferromagnetic Néel point, is at the moment not understood.

The magnetomechanical ratio of vanadium is $g' = 1.18(10)$ [71 H 1].

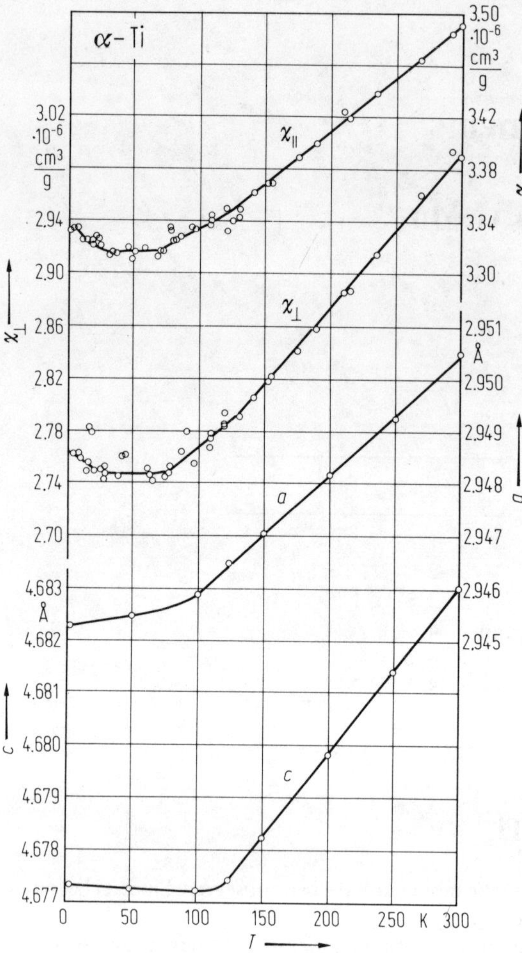

Fig. 1. Temperature dependence of the magnetic mass susceptibility χ_g of a single crystal of pure α-Ti compared with the temperature dependences of the lattice parameters a and c [67 E 1]. The susceptibility of Ti is seen to be practically temperature-independent up to about 70 K. The low-temperature upturn is accountable to a trace of dissolved Mn [71 C 1].

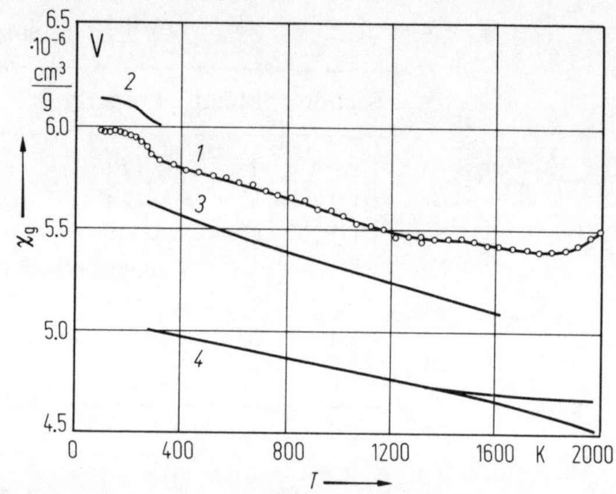

Fig. 2. Temperature dependence of the magnetic mass susceptibility χ_g for polycrystalline V. Curve *1*: [65 K 1], sample impurities, in [wt%]: 0.032 C; 0.07 O_2; 0.031 N_2; 0.001 H_2; others <0.095. *2*: [61 B 1]. *3*: [62 T 1]. *4*: [53 K 1].

1.1.3 Cr

The crystallographic structure of chromium metal is body-centered cubic, and its magnetic structure is very peculiar. Chromium is antiferromagnetic at temperatures below the Néel point (T_N) of about 312 K. This antiferromagnetism is, however, not caused by local magnetic moments aligning themselves antiparallel; Overhauser [62 O 1] showed that the antiferromagnetic ordering in chromium may be described by a *spin density wave* in the itinerant 3d-electrons, having a wavelength incommensurate with the lattice constant. Whereas in some other substances spin density waves only exist as excitations, in Cr metal at low temperatures the ground state is a spin density wave with a finite amplitude, see Fig. 3.

Analytically, the magnetic moment per atom in antiferromagnetic Cr, as a function of the position in space, $\boldsymbol{R}$, is given by:

$$\boldsymbol{S}(\boldsymbol{R}) = \boldsymbol{S}_1 \cos(\boldsymbol{Q} \cdot \boldsymbol{R}) + \boldsymbol{S}_3 \cos(3\boldsymbol{Q} \cdot \boldsymbol{R}) + \dots,$$

where the main amplitude of the magnetic moment, S_1, has a value of about 0.6 μ_B at low temperatures, and a value of about 0.2 μ_B just below T_N [65 A 1]. $\boldsymbol{S}_1$ is always directed along one of the cube axes of the body-centered cubic crystal lattice; $\boldsymbol{S}_3$ is always directed opposite to $\boldsymbol{S}_1$, and has an absolute value of a few percent of S_1, the "spin density wave" $\boldsymbol{S}(\boldsymbol{R})$ is therefore somewhat more "rectangular-like" than a pure sine function, see [81 I 1]

Accurate measurements show no detectable hysteresis of the Néel point T_N in well-annealed, pure Cr [80 W 2]; T_{sf} has a hysteresis of about 1 K [82 B 1].

Free energy expressions, dependent on the applied field and/or the strain, are given by [81 B 1] (near T_{sf}) and [80 W 1] (near T_N).

Surface magnetization of Cr has been mentioned by [82 S 1].

The magnetic anisotropy torque of Cr in the $3Q$-state, and of Cr in the $1Q$-state, was measured by [64 M 1].

It appears that both at T_N and T_{sf} chromium has a first-order phase transition [65 A 1]. The latent heat of transition at T_{sf} is 0.04(2) J/mol [82 B 1].

The relative change in the volume at T_{sf} is $-1.4(6)\cdot 10^{-6}$ [69 S 1]; at T_N the change in volume could not be measured due to a change in the thermal expansion, it can be calculated to be about $-2\cdot 10^{-5}$.

In the $1Q$-state, chromium shows a tetragonal or, depending on the previous treatment, an orthorhombic deformation (or magnetostriction) of the order of magnitude 10^{-5} [69 S 1].

The magnetomechanical ratio of chromium metal is $g' = 1.21(7)$ [71 H 1].

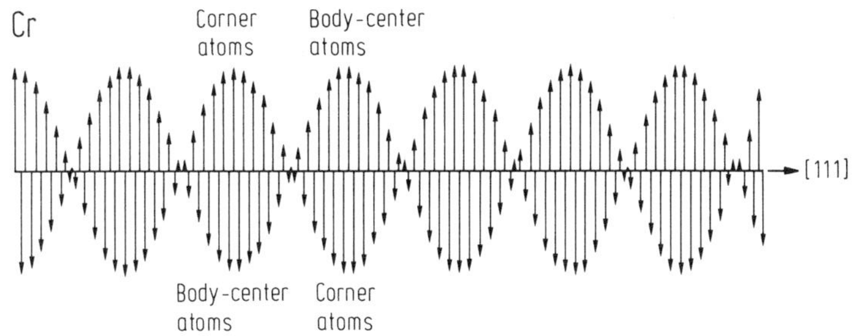

Fig. 3. Spin density wave in Cr [81 F 1]. The magnetic moments of two successive atoms on the body-diagonal of the cubic lattice are antiparallel. The magnitude of the atomic moments on each sublattice is given by a sinusoidal function of the position.

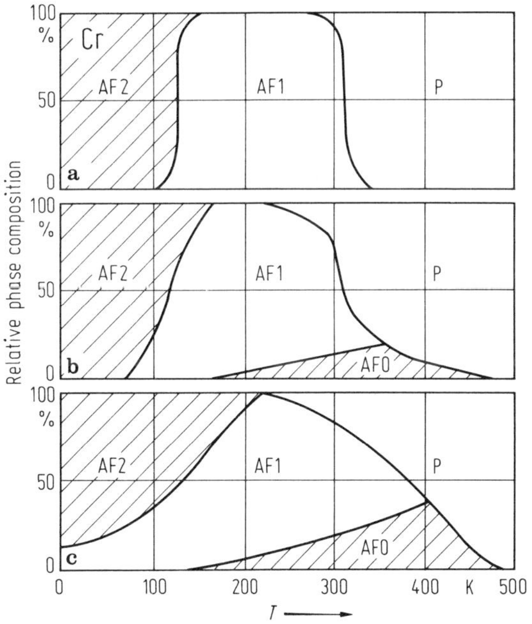

Fig. 4. Magnetic phase diagram in Cr for (a) annealed sample, (b) swaged sample, and (c) crushed powder sample [81 W 1]. P: paramagnetic, AF0: commensurate, $Q = 2\pi/a$, AF1: transverse incommensurate, AF2: longitudinal incommensurate.

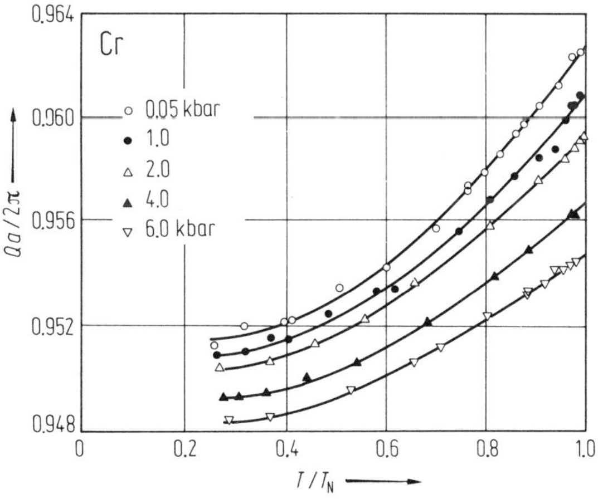

Fig. 5. Temperature dependence of the relative length $Qa/2\pi$ of the spin density wavevector for Cr as a function of the reduced temperature T/T_N, at various pressures [68 U 1].

Table 1. Data for the magnetic period of the spin density wave in Cr, based on the position of the (100) satellite lines in the neutron diffraction spectrum [64 K 1]. Q: spin density wavevector, $\delta^{-1} = (1 - Qa/2\pi)^{-1}$: length of antiferromagnetic modulation, see Fig. 3, divided by lattice constant a.

	T K	$Qa/2\pi$	δ^{-1}
Cr	197	0.9554	22.4(8)
	78	0.9519	20.8 *)
Cr – 0.45 at% V	197	0.9480	19.2(8)
	78	0.9431	17.6(8)

*) [62 S 1].

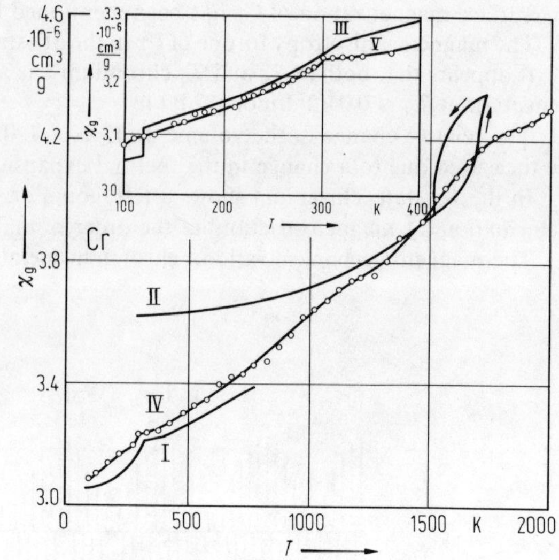

Fig. 6. Temperature dependence of the magnetic mass susceptibility χ_g for polycrystalline Cr. I: [58 L 1]; II: [52 M 1]; III: [64 M 2]; IV and V: [64 W 1].

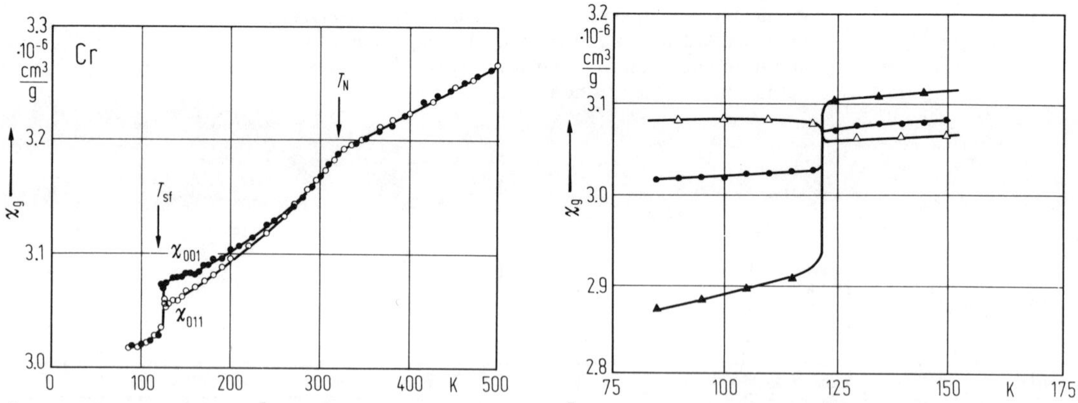

Fig. 7. Temperature dependence of the magnetic mass susceptibility χ_g of single-crystal Cr. (a) Cr cooled in zero applied field; χ_{001} and χ_{011} refer to measurements along cube edges and face diagonals, respectively.

(b) Cr cooled through T_N in an applied flux density of 5 T acting along [001]. Open triangles: measurements along [010]; solid triangles: measurements along [001]; solid circles: measurements along cube edge for Cr cooled in zero applied field, included for comparison [66 P 1].

1.1.4 Mn

The phase of manganese metal which is stable at room temperature, α-Mn, has probably the most complex crystallographic and magnetic structure of all elements. α-Mn has a cubic crystallographic symmetry, the cubic unit cell contains no less than 58 Mn atoms, distributed over 4 nonequivalent sites; the configuration of the surrounding of each site by the other Mn atoms is unique for each of the 4 different sites.

Below the Néel temperature $T_N = 95$ K, α-Mn is antiferromagnetic; since the magnitudes of the magnetic moments of the Mn atoms on different sites are highly different, an antiferromagnetic ordering can only be achieved if the atoms of each site order antiferromagnetically among themselves; the magnetization vectors of the 4 different sublattices are not collinear. This structure, which has been analyzed by Yamada in 1970 [70 Y 1–3], is depicted in Fig. 8.

From antiferromagnetic resonance measurements, Yamagata [72 Y 1] concluded that the site II atoms are divided into two subtypes (each 4 atoms/cubic cell) with a magnetic moment of 1.84 μ_B and 1.75 μ_B, respectively; also for sites III and IV the situation is probably more complicated than depicted above.

The magnetic susceptibility of α-Mn is only slightly temperature-dependent; it shows a broad maximum above the Néel point, but no anomaly near T_N, see Fig. 9. A good room temperature value is $\chi_g = 11.7 \cdot 10^{-8}\,m^3/kg$. A weak ferromagnetism, often observed below 45 K, is probably due to contamination of the sample with Mn_3O_4 [70 Y 3].

In antiferromagnetic-α-Mn the differential magnetic susceptibility increases by about 50% above its low-field value, if the applied field exceeds 11 T; this effect can possibly be interpreted as a change in the angle between the antiferromagnetic vectors of atoms on two different sites.

In paramagnetic α-Mn, the contributions of the different sites to the susceptibility has been analyzed by an interpretation of NMR measurements [81 M 1, 81 M 2]; the susceptibility of a site I atom is about twice that of a site II atom; the susceptibility of site III and site IV atoms is relatively small. Since there are four times as many site II as site I atoms, the major part of the total susceptibility is due to site II atoms.

In the temperature range $45\,K \leq T < T_N$ the principal axis of the magnetic symmetry is along a [100] direction [70 Y 3].

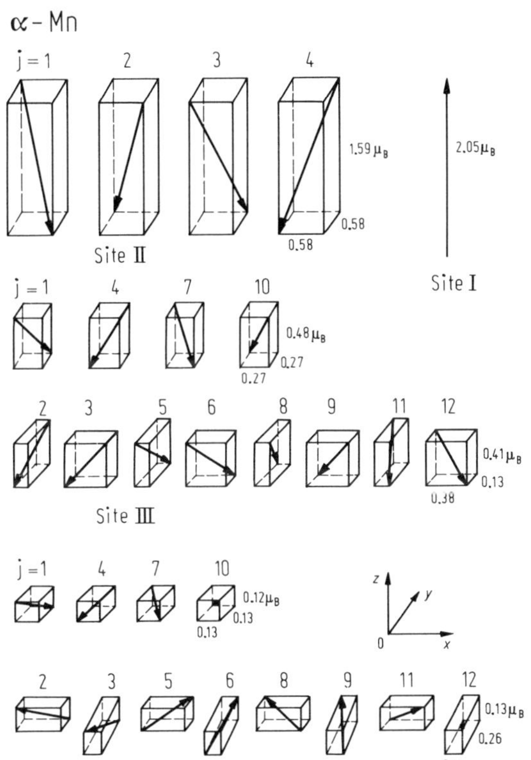

Fig. 8. Magnetic structure of α-Mn below 95 K as determined by Yamada. Vectors represent the magnetic moment for each of the 29 atoms in the primitive unit cell. The edges of the right prisms give, in units [μ_B], the components of the magnetic moment in, respectively, the x, y, z directions. Integers j number the atoms for each of the crystallographic sites I···IV [70 Y 1–3].

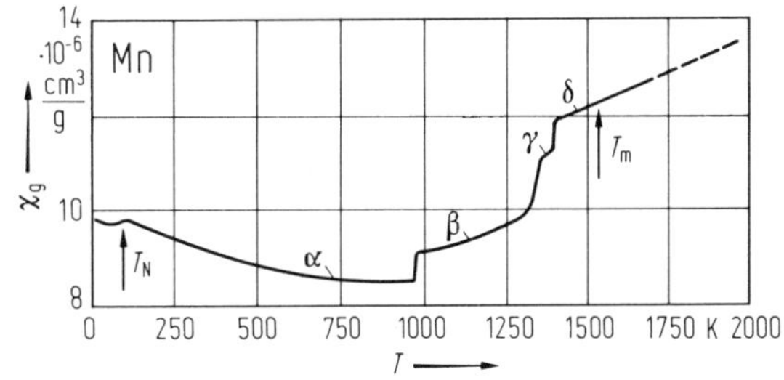

Fig. 9. Magnetic mass susceptibility χ_g vs. temperature for the various phases of Mn metal. T_N: Néel temperature; T_m: melting point [69 K 1].

Table 2. Low-temperature values of the magnetic moments p_{Mn} on the various atom sites in α-Mn [70 Y 2].

Site	Atoms/cubic cell	p_{Mn} [μ_B]
I	2	1.9
II	8	1.7
III	24	0.6
IV	24	0.25

1.1.5 References for 1.1

52 M 1 McGuire, T.R., Kriessman, C.J.: Phys. Rev. **85** (1952) 452.
53 K 1 Kriessman, C.J.: Rev. Mod. Phys. **25** (1953) 122.
55 R 1 Rostoker, W., Yamamoto, A.: Trans. Am. Soc. Met. **47** (1955) 1002.
56 M 1 McQuillan, A.D., McQuillan, M.K.: Titanium, London: Butterworth Scient. Publ. **1956**.
58 L 1 Lingelbach, R.: Z. Phys. Chem. N.F. **14** (1958) 1.
61 B 1 Burger, J.P., Taylor, M.A.: Phys. Rev. Lett. **6** (1961) 185.
62 O 1 Overhauser, A.W.: Phys. Rev. **128** (1962) 1437.
62 S 1 Shirane, G., Takei, W.J.: J. Phys. Soc. Jpn. **17**, BIII (1962) 35.
62 T 1 Taniguchi, S., Tebble, R.S., Williams, D.E.G.: Proc. R. Soc. London A **265** (1962) 502.
64 K 1 Komura, S., Kunitomi, N.: J. Phys. Soc. Jpn. **20** (1964) 103.
64 M 1 Montalvo, R.A., Marcus, J.A.: Phys. Lett. **8** (1964) 151.
64 M 2 Munday, B.C., Pepper, A.R., Street, R.: Brit. J. Appl. Phys. **15** (1964) 611.
64 W 1 Weiss, W.D., Kohlhaas, R.: Z. Naturforsch. A **19** (1964) 1631.
65 A 1 Arrot, A., Werner, S.A., Kendrick, H.: Phys. Rev. Lett. **14** (1965) 1022.
65 K 1 Kohlhaas, R., Weiss, W.D.: Z. Naturforsch. A **20** (1965) 1227.
66 P 1 Pepper, A.R., Street, R.: Proc. Phys. Soc. **87** (1966) 971.
67 E 1 Ebneter, A.E.: Thesis, Air Force Inst. of Techn., Wright-Patterson Air Force Base, Ohio USA **1967**.
68 U 1 Umebayashi, H., Shirane, G., Frazer, B.C., Daniels, W.B.: J. Phys. Soc. Jpn. **24** (1968) 368.
69 K 1 Kohlhaas, R., Weiss, W.D.: Z. Naturforsch. A **24** (1969) 287.
69 S 1 Steinitz, M.O., Schwartz, L.H., Marcus, J.A., Fawcett, E., Reed, W.A.: Phys. Rev. Lett. **23** (1969) 979.
70 Y 1 Yamada, T.: J. Phys. Soc. Jpn. **28** (1970) 596.
70 Y 2 Yamada, T., Kunitomi, N., Nakai, Y.: J. Phys. Soc. Jpn. **28** (1970) 615.
70 Y 3 Yamada, T., Tazawa, S.: J. Phys. Soc. Jpn. **28** (1970) 609.
71 C 1 Collings, E.W., Gehlen, P.C.: J. Phys. F **1** (1971) 908.
71 H 1 Huguenin, R., Pells, G.P., Baldock, D.N.: J. Phys. F **1** (1971) 281.
71 K 1 Kostina, T.I., Shafigullina, G.A., Kozlova, T.N., Kuznetsov, V.I.: Phys. Met. Metallogr. (USSR) **32** (1) (1971) 203.
72 Y 1 Yamagata, H., Asayama, K.: J. Phys. Soc. Jpn. **33** (1972) 400.
73 K 1 Kondorskii, E.I., Karstens, G.E., Kostina, T.I., Shafigullina, G.A., Ekonomova, L.N.: Proc. Int. Conf. Magnetism ICM-73 (Moscow) **I**(1) (1973) 310.
74 D 1 Degyareva, V.F., Kamirov, Yu.S., Rabin'kim, A.G.: Sov. Phys. Solid State **15** (1974) 2293.
77 A 1 Alikhanov, R.A., Zuy, V.N., Karstens, G.E., Smirnov, L.S.: Phys. Met. Metallogr. (USSR) **44**(3) (1977) 178.
80 W 1 Walker, M.B.: Phys. Rev. B **22** (1980) 1338.
80 W 2 Williams, I.S., Street, R.: J. Phys. F **10** (1980) 2551.
81 B 1 Barak, Z., Walker, M.B.: J. Phys. F **11** (1981) 947.
81 F 1 Fincher jr., C.R., Shirane, G., Werner, S.A.: Phys. Rev. B **24** (1981) 1312.
81 I 1 Iida, S., Tsunoda, Y., Nakai, Y., Kunimoto, N.: J. Phys. Soc. Jpn. **50** (1981) 2587.
81 M 1 Murayama, S., Nagasawa, H.: J. Phys. Soc. Jpn. **50** (1981) 1189.
81 M 2 Murayama, S., Nagasawa, H.: J. Phys. Soc. Jpn. **50** (1981) 1523.
81 W 1 Williams, I.S., Street, R.: Philos. Mag. B **43** (1981) 893.
82 B 1 Benediktsson, G., Åström, H.U.: Phys. Scr. (Sweden) **25** (1982) 671.
82 S 1 Siegmann, H.C.: J. Appl. Phys. **53** (1982) 2018.

1.2 Fe, Co, Ni

Survey

Section	Property	Fig.	Table
1.2.1	Phase diagrams, lattice constants and elastic moduli		1
1.2.2	Paramagnetic properties	1–5	2
1.2.3	Spontaneous magnetization, magnetic moments and high-field susceptibility	6, 7	3–5
1.2.4	Magnetocrystalline anisotropy constants	8, 9	6–8
1.2.5	Magnetostriction coefficients	10	9–11
1.2.6	Magnetic moments and densities	11–13	
1.2.7	Hyperfine fields		12
1.2.8	Spin wave properties		13
1.2.9	*g* factors		14
1.2.10	Fermi surfaces, band structures, exchange energies and electron spin polarizations		15–18
1.2.11	Kerr effect	14, 15	

1.2.1 Phase diagrams, lattice constants

Table 1. Lattice constants, interatomic distances, atomic volumes and thermal expansion coefficients of Fe, Co, and Ni [74 D 1].

	T °C	p bar	Phase	a Å	c Å	l Å	V_0 $Å^3$	α 10^{-6} K^{-1}
Fe	20	1	α, bcc	2.86638(190)	–	2.482(8)	11.78	11.7
	910	1	α, bcc	2.9044	–	2.515(8)	12.25	
	910	1	γ, fcc	3.6467	–	2.579(12)	12.12	
	1390	1	γ, fcc	3.6869	–	2.607(12)	12.53	
	1390	1	δ, bcc	2.9315	–	2.5388(8)	12.60	
	23	$130 \cdot 10^3$	α, bcc	2.805	–	2.429(8)	10.43	
	23	$130 \cdot 10^3$	ε, hcp	2.468	3.956	2.468(6)		
						2.408(6)	11.03	
Co	20	1	α, hcp	2.5070(3)	4.0698(9)	2.507(6)	11.08	
						2.497(6)		
	20	1	β, fcc	3.5445(4)	–	2.506(12)	11.13	
Ni	20	1	fcc	3.5241(7)	–	2.492(12)	10.94(1)	12.5
	20	1	hcp	2.495(12)	4.048(43)	2.495(6)		
						2.484(6)		
							10.91(16)	
	20	1	bcc	2.775(14)	–	2.403(8)	10.68(16)	

1.2.2 Paramagnetic properties

Table 2. Paramagnetic properties of Fe, Co, and Ni. ΔT is the temperature interval for which the parameters of the Curie-Weiss law are determined. p_{at} denotes the magnetic moment per atom in the ferromagnetic phase extrapolated to $T=0$ K. The ratio p/p_{at} gives an indication of the degree of localization or itineracy of the electrons forming the moment. A completely localized moment would have a value of one and a completely itinerant moment a large value >10; e.g. $Fe_{0.51}Cr_{0.49}$ has a ratio of 17.6 ($T_C=9$ K) and $Ni_{0.43}Pt_{0.57}$ a ratio of 17.2 ($T_C=23$ K) [78 W 1]. Thus the moments of Fe, Co, and Ni are seen to have a high degree of localized character.

	Fig.	Θ K	C_g 10^{-3} cm^3 K/g	ΔT K	p_{eff} μ_B	p/p_{at}	Ref.
α-Fe	1a, b, d	+1093(3)	22.0	1100···1180	3.13	1.01	60 A 1
β-Co		1403···1428	20.8	1430···1710	3.15	1.28	38 S 1
Ni [1])	3a	654.1	5.546	740···970	1.613	1.375	63 A 2
	3b, c		8.55	1528···1728			73 B 2
Liquid-Ni	3c		16.7	1728···1928			73 B 2

[1]) Ni does not obey a Curie-Weiss law; at temperatures above 970 K an additional temperature-dependent contribution is found.

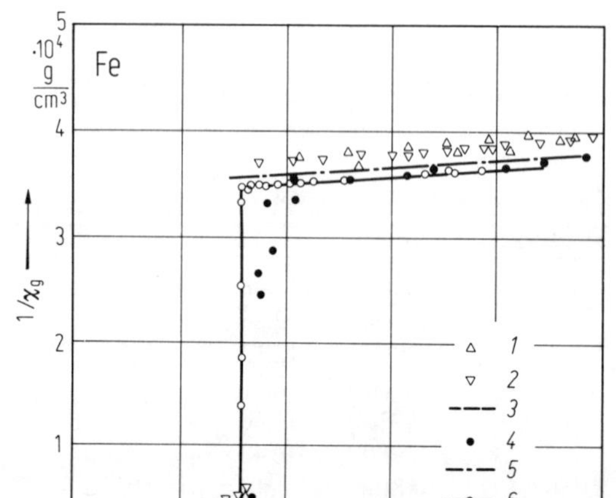

Fig. 1. Temperature dependence of the inverse paramagnetic mass susceptibility of Fe [60 A 1]. *1*: [11 W 1], *2*: [17 T 1], *3*: [34 P 1], *4*: [38 S 1], *5*: [56 N 1], *6*: [60 A 1].

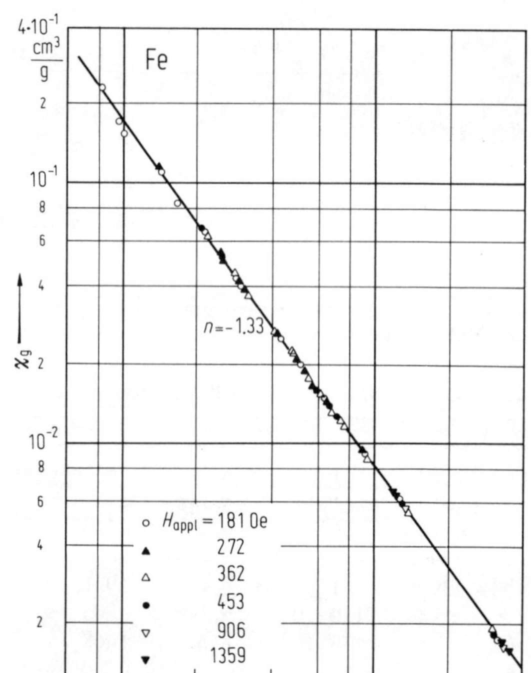

Fig. 2. Temperature dependence of the magnetic mass susceptibility of Fe above the ferromagnetic Curie temperature, $T_C=1044.1$ K. The straight line represents the relationship $\chi_g = K(T-T_C)^n$ with $n=-1.33$ [64 A 1]. Symbols indicate different applied magnetic fields.

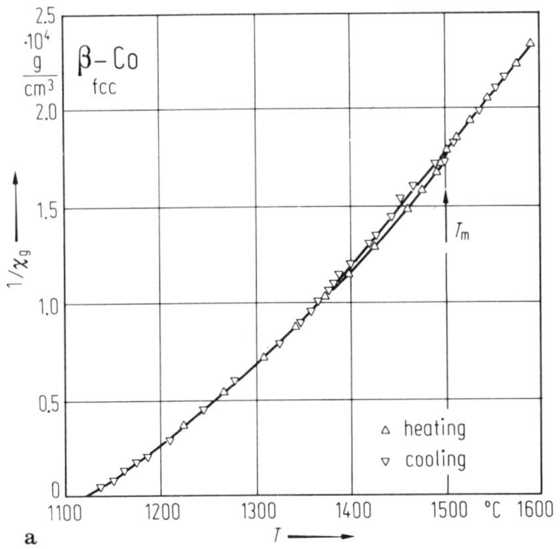

Fig. 3. Temperature dependence of the inverse paramagnetic mass susceptibility of fcc Co [56 N 1].

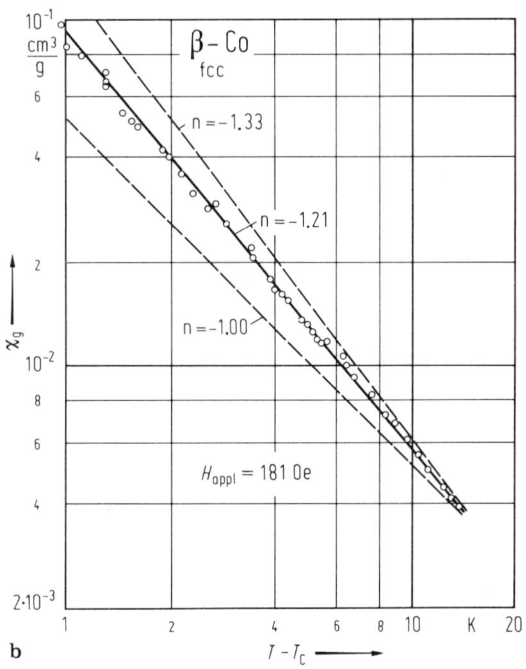

Fig. 4. Temperature dependence of the magnetic mass susceptibility of fcc Co above the ferromagnetic Curie temperature, $T_C = 1388.2$ K, in an applied magnetic field $H_{appl} = 181$ Oe. Straight lines represent fits to the data of the form $\chi_g = K(T - T_C)^n$ for various n [65 C 2].

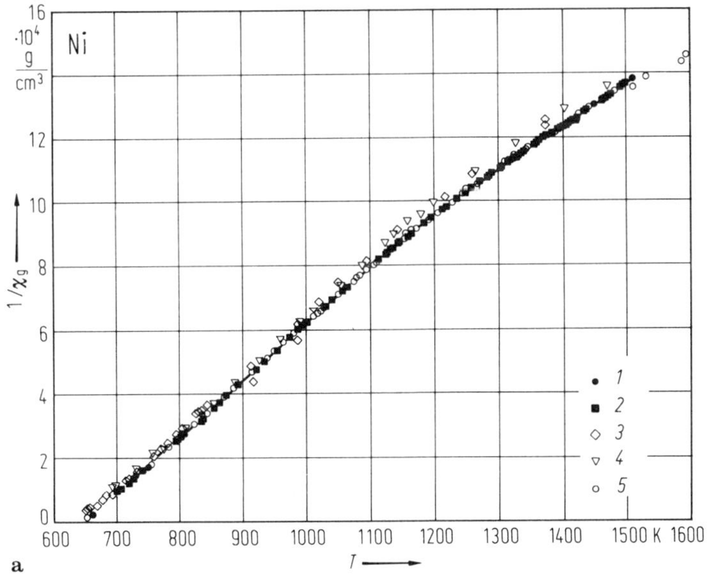

Fig. 5. Temperature dependence of the inverse paramagnetic mass susceptibility of Ni. Samples *1* and *2*: [63 A 2], *3*: [11 W 1], *4*: [38 S 2], *5*: [44 F 1].

1.2.3 Spontaneous magnetization, magnetic moments and high-field susceptibility

Table 3. The spontaneous magnetization σ_s and the high-field susceptibility χ_{HF}. Applied field direction is indicated. For Fe and Ni no anisotropy in σ_s or χ_{HF} was found [83 P 1], Co [83 P 2].

	T K	σ_s $G\,cm^3\,g^{-1}$	χ_{HF} $10^{-6}\,cm^3\,g^{-1}$
Fe [100]	4.21	222.671	3.60
	24.79	222.596	3.66
	51.06	222.367	3.70
	75.34	222.071	3.81
	100.51	221.825	3.84
	131.55	221.443	3.79
	165.81	220.937	3.76
	197.45	220.346	3.70
	226.34	219.736	3.81
	254.53	219.049	3.90
	286.41	218.210	3.95
Co [0001]	4.21	163.82	
	24.79	163.79	
	55.31	163.68	
	75.28	163.54	
	100.46	163.50	
	131.31	163.44	
	165.30	163.39	
	197.01	163.29	
	225.69	162.99	
	254.67	162.93	
	286.61	162.62	
Co [10$\bar{1}$0]	4.21	163.00	
	24.79	162.98	
	55.36	162.70	
	75.28	162.65	
	100.51	162.64	
	131.31	162.58	
	165.50	162.46	
	196.81	162.40	
	225.74	162.26	
	254.67	162.08	
	286.66	161.86	
Ni [111]	4.21	58.872	1.59
	25.00	58.810	1.64
	50.08	58.698	1.58
	75.15	58.550	1.52
	100.51	58.349	1.48
	131.08	58.063	1.37
	165.49	57.671	1.21
	196.71	57.223	1.11
	225.74	56.724	1.13
	254.82	56.121	1.18
	286.66	55.370	1.12

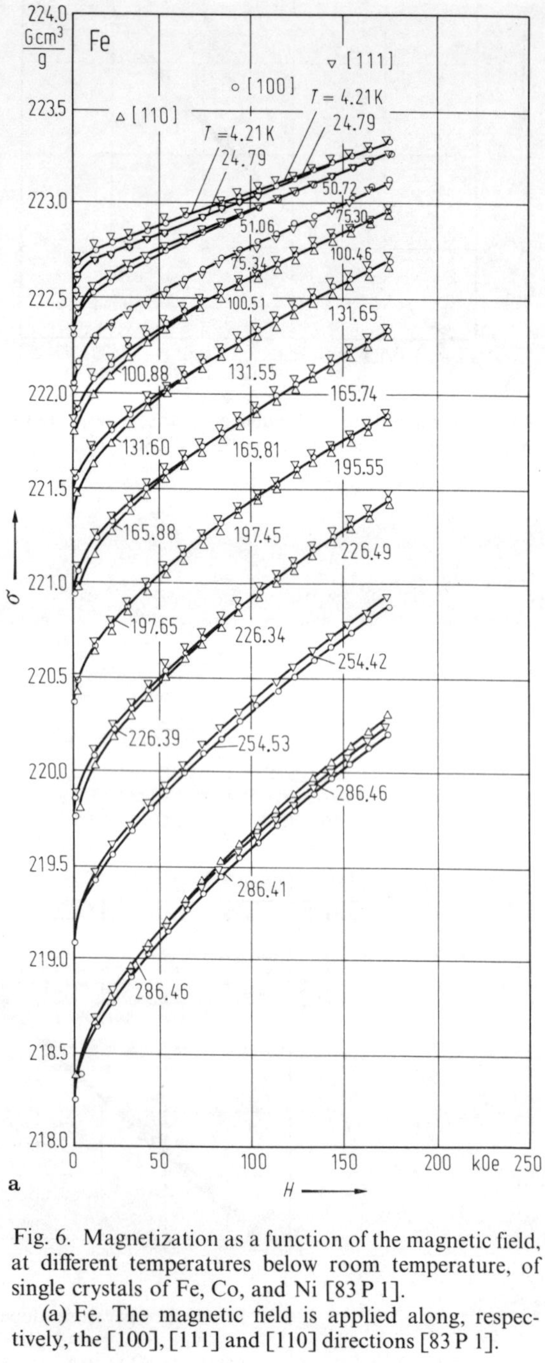

a

Fig. 6. Magnetization as a function of the magnetic field, at different temperatures below room temperature, of single crystals of Fe, Co, and Ni [83 P 1].

(a) Fe. The magnetic field is applied along, respectively, the [100], [111] and [110] directions [83 P 1].

For Figs. 6b and c, see next page.

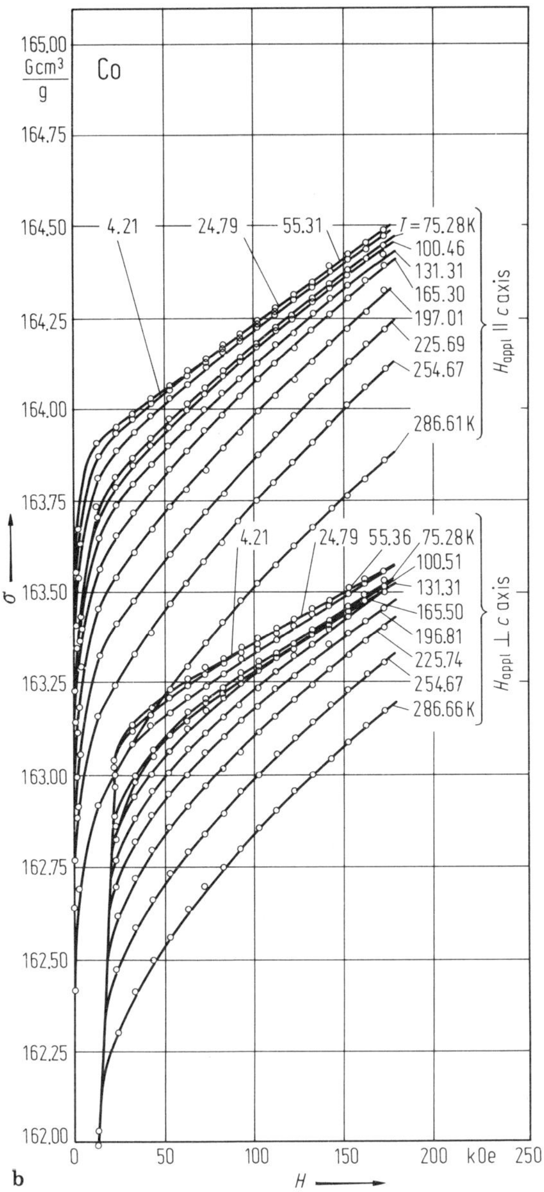

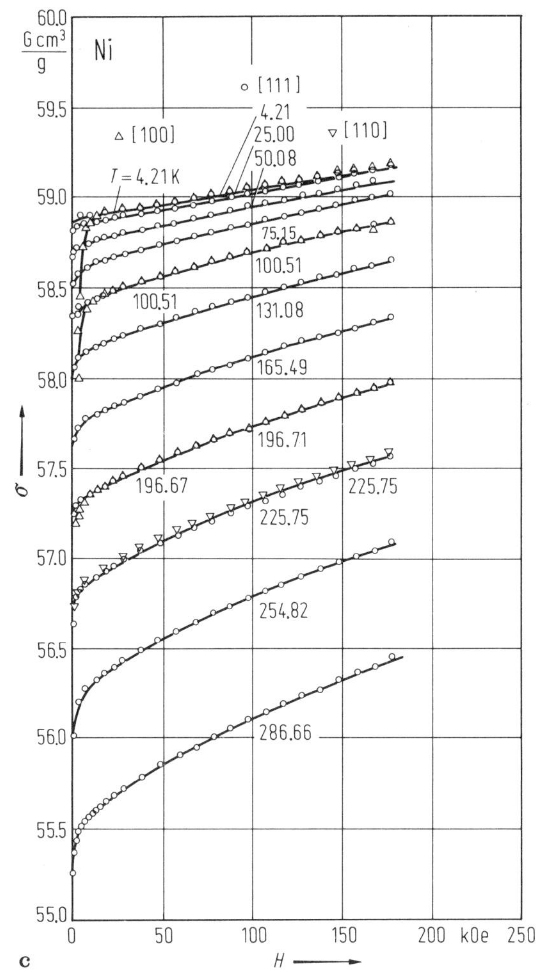

Fig. 6c. Ni. The magnetic field is applied along, respectively, the [111], [100] and [110] directions [83 P 1].

Fig. 6b. Co. The magnetic field is applied along, respectively, the c axis and the [100] direction in the basal plane [83 P 1].

Table 4. Survey of the spontaneous magnetization of Fe, Co, and Ni [82 P 1]. Direction of magnetization is indicated. See also Table 3.

	T K	σ_s $G\,cm^3\,g^{-1}$	ϱ $g\,cm^{-3}$	M_s G	$4\pi M_s$ G
Fe [100]	4.2	222.671	7.93	1766	22189
	286.41	218.210	7.87	1717	21580
Co [0001]	4.21	163.862	9.0	1475	18532
	286.61	162.624	8.9	1447	18188
Ni [111]	4.21	58.872	8.97	528	6636
	286.66	55.370	8.91	493	6200

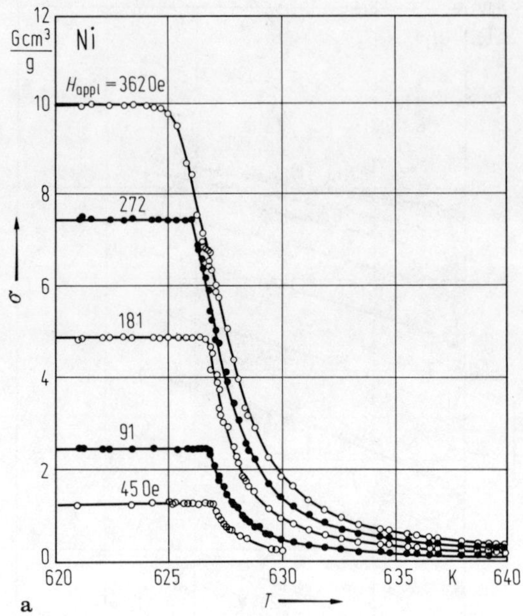

Fig. 7. Magnetization of Ni in various applied magnetic fields in the neighborhood of the ferromagnetic Curie temperature [65 A 1].

Table 5. Curie temperature and pressure dependence of the Curie temperature of Fe, Co, and Ni.

Metal	T_C K	$(dT_C/dp)_{p=0}$ K kbar^{-1}	Ref.
Fe, bcc	1044(2)		64 A 1
	1044	0.00(3)	72 L 1
	1045(3)		71 C 1
Fe, fcc	44 [1])	0.5 [2])	79 L 1
Co, fcc	1388(2)		65 C 1
	1398	0.00(5)	72 L 1
	1390.0		71 C 1
Ni, fcc	624.0(3)		61 M 1
	627	0.36(2)	72 L 1
	631(2)		71 C 1
	627(1)		65 A 1

[1]) T_N. [2]) $(dT_N/dp)_{p=0}$.

1.2.4 Magnetocrystalline anisotropy constants

Table 6. Magnetocrystalline anisotropy constants of Fe, as derived from magnetization ($\boldsymbol{M}_\perp$, $\boldsymbol{M}_\parallel$) and torque (TQ) measurements. $\boldsymbol{M}_\perp$: analysis, above technical saturation, of the magnetization perpendicular to the applied field vs. direction of $\boldsymbol{H}_{appl}$ in the plane indicated. $\boldsymbol{M}_\parallel$: analysis, below saturation, of the magnetization parallel to the applied field, direction as indicated, vs. H_{appl}.

T K	Method	K_1 10^5 erg cm^{-3}	K_2	K_3	Ref.
4.2	$\boldsymbol{M}_\perp$ (110)	4.70	1.90		82 T 1
	$\boldsymbol{M}_\parallel$ [110]	5.35		0.22	82 T 1
	$\boldsymbol{M}_\parallel$ [111]	5.64	1.94	−0.64	82 T 1
20	TQ	5.20	−0.158	0.202	68 G 1
77	$\boldsymbol{M}_\perp$ (110)	4.48	2.41		82 T 1
	$\boldsymbol{M}_\parallel$ [110]	5.31		0.25	82 T 1
	$\boldsymbol{M}_\parallel$ [111]	5.56	1.79	−0.62	82 T 1
	TQ	5.15	−0.154	0.711	68 K 1
	TQ [1])	5.15			66 K 1
	TQ [2])	5.02			
273	TQ	4.81	0.012	−0.012	73 E 1
	TQ	4.75	≈ −0.13	≈0.195	68 G 1
	TQ	4.50			66 K 1
	χ_{ac} [3])	4.77			81 H 1

[1]) K_1 vs. H at constant T, extrapolated to zero field.
[2]) T vs. H at constant K_1, extrapolated to zero field.
[3]) ac susceptibility measurements on whiskers.

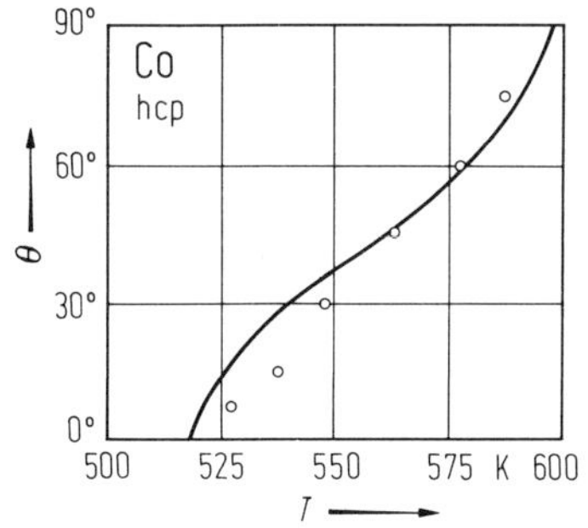

Fig. 8. Temperature dependence of the angle θ between the direction of spontaneous magnetization and the c axis of a single crystal of hcp Co [61 B 1]. Points: data. Curve: calculated from $\sin\theta = (-K_1/2K_2)^{1/2}$.

Table 7. Magnetocrystalline anisotropy constants for fcc Co [54 S 1].

T	K_1	K_2
°C	10^5 erg cm^{-3}	
500	−2.30	−4.31
550	−1.60	−3.80
600	−1.09	−3.21
650	−0.80	−2.50
700	−0.60	−1.91
750	−0.41	−1.35
800	−0.23	−1.03
850	−0.15	−0.76
900	−0.09	−0.57
950	−0.09	−0.36

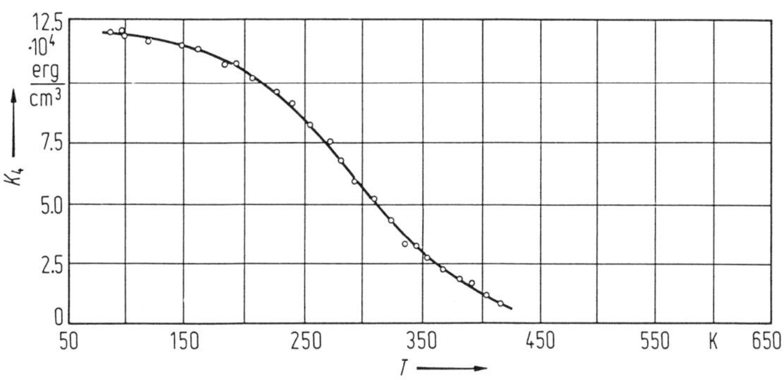

Fig. 9. Temperature dependence of magnetocrystalline anisotropy constant K_4 in the basal plane [84 P 1].

Table 8. Magnetocrystalline anisotropy constants for Ni as derived from magnetization ($M_\perp$, M) and torque (TQ) measurements. See caption to Table 6 for details on magnetization measurements.

T K	Method	K_1	K_2	K_3	K_4	Ref.
		10^5 erg cm^{-3}				
4.2	$\boldsymbol{M}_\perp$ (110)	−12.92(1)	4.79(7)	0.80(5)	0.32(7)	82 T 1
	$\boldsymbol{M}_\perp$ (100)	−12.82(6)		0.80(1)		82 T 1
	TQ	−12.97(2)	4.71(5)	1.32(8)	−1.0(1)	78 G 1
	$\boldsymbol{M}_\parallel$ [110]	−12.58	3.26	1.61		82 T 1
	$\boldsymbol{M}_\parallel$ [100]	−12.42	6.20	0.81		82 T 1
	TQ [2])	−12.64	2.48		2.95	76 A 1 [1])
77	$\boldsymbol{M}_\perp$ (110)	− 8.45(1)	0.96(8)	−1.43(6)	−0.66(8)	82 T 1
	$\boldsymbol{M}_\perp$ (100)	− 8.43(6)		−1.4(2)		82 T 1
	TQ	− 8.42	0.83	−1.64		68 F 1
	TQ [2])	− 8.36				76 A 1

[1]) The higher order anisotropy constants were shown to be strongly dependent on the truncation in the analysis procedure. It is concluded that it is never possible to describe the anisotropy of Ni with only K_1 and K_2.

[2]) $\boldsymbol{H}_{appl}$ in (110) plane.

1.2.5 Magnetostriction coefficients

Table 9. Linear saturation magnetostriction coefficients h_i and forced magnetostriction constants h'_i for a single crystal of Fe at various temperatures [61 G 1].

T K	h_1	h_2	h_3	h_4	h_5	h'_0 [1]	h'_1	h'_2
	$\cdot 10^{-6}$					$\cdot 10^{-10}\,\mathrm{Oe}^{-1}$		
4	35	−45	4	–	–	2.0	–	–
77	35.0	−44.8	3	–	–	1.6	0.6	−0.1
190	35.5	−39.9	4	–	–	1.7	1.0	0.0
293	36.2	−34.0	2	–	–	1.8	0.4	−0.2

[1]) $h'_0 = 1.5(1) \cdot 10^{-10}\,\mathrm{Oe}^{-1}$ and h'_1, h'_2 of the order of $\pm 1 \cdot 10^{-10}\,\mathrm{Oe}^{-1}$ was found by [65 S 1] over the whole temperature range.

Table 10. Room-temperature forced magnetostriction constants of Ni.

h'_0	h'_1	h'_2	Ref.
$10^{-10}\,\mathrm{Oe}^{-1}$			
0.26	−0.84	−0.33	64 L 1
0.40	−0.43	−0.18	71 L 1
0.2(1) [1])	0.0(1)	0.0(1)	65 S 1

[1]) At 1.5 K: $h'_0 = 0.4(1) \cdot 10^{-10}\,\mathrm{Oe}^{-1}$.

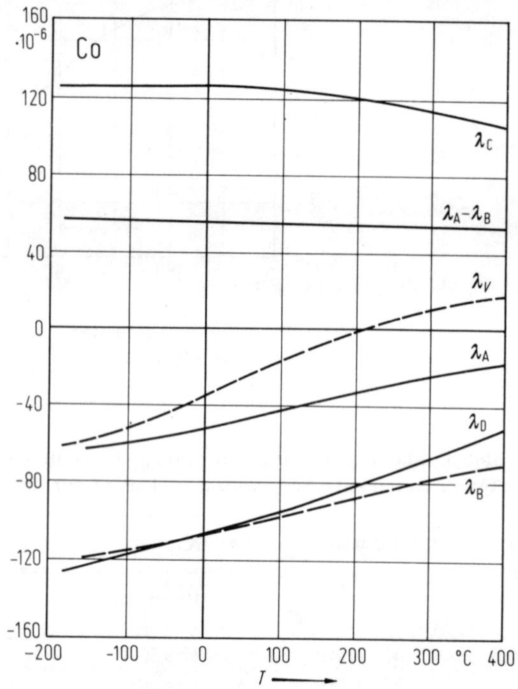

Fig. 10. Variation with temperature of the linear magnetostriction coefficients of single crystal Co. Also shown is the volume magnetostriction, $\lambda_V = \lambda_A + \lambda_B + \lambda_C$, and $\lambda_A - \lambda_B$ [69 H 1]. Room-temperature values: $\lambda_A = -50 \cdot 10^{-6}$, $\lambda_B = -107 \cdot 10^{-6}$, $\lambda_C = 126 \cdot 10^{-6}$, $\lambda_D = -105 \cdot 10^{-6}$, $\lambda_V = -31 \cdot 10^{-6}$, which agree with the earlier values of Bozorth [54 B 1].

Table 11. Room-temperature linear saturation magnetostriction coefficients for single crystal Ni.

h_1	h_2	h_3	h_4	h_5	Ref.
$\cdot 10^{-6}$					
−98 (3)	−41.5(10)	0.3(5)	6.3(10)	0.2(5)	70 F 1
−98.5(14)	−43.1(5)	0.1(9)	3.4(6)	0.2(9)	71 B 1
−94	−43	−0.5	0.2⋯1.4	1.4	71 L 1

1.2.6 Magnetic moments and densities

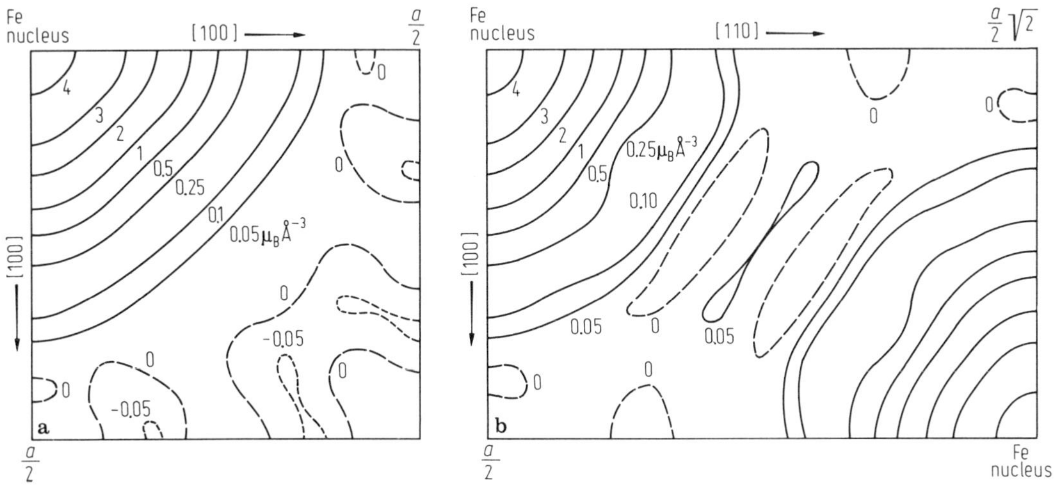

Fig. 11. Magnetic moment density distribution over (a) the (100) face and (b) the (110) diagonal plane of the Fe unit cell. The asymmetric contour lines show that the 3d electrons are asymmetrically distributed around the Fe nucleus. The values of the spin densities are in units of [$\mu_B/Å^3$], [62 S 1].

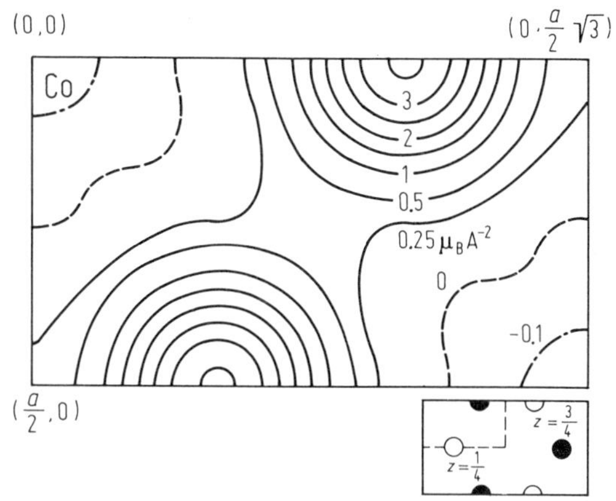

Fig. 12. Projection of magnetic moment density on basal plane of hcp Co. Lower right diagram shows projected position of atoms in orthorhombic unit cell. Dashed lines indicate portion of cell shown in density map [64 M 1].

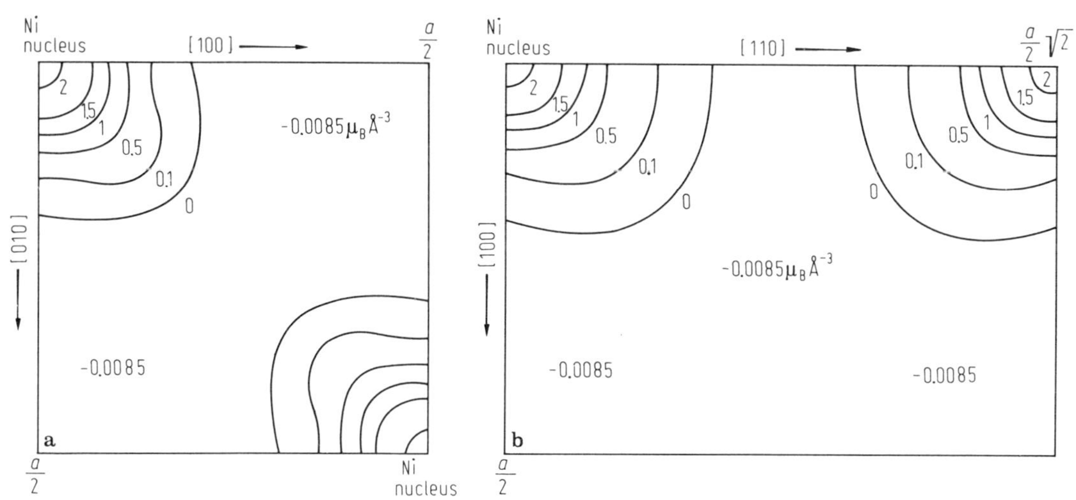

Fig. 13. Contour maps of the magnetic moment density in Ni obtained by Fourier inversion of the data for (a) the (100) plane and (b) the (110) plane. The numbers labelling the contours give the magnetization in [$\mu_B/Å^3$], [66 M 1].

1.2.7 Hyperfine fields

Table 12. NMR frequencies and magnetic hyperfine fields of Fe, Co, and Ni.

Metal	Nucleus position	T K	ν MHz	H_{hyp} kOe	Ref.
Fe, bcc		4.2	46.64	−339(1) [1]	61 B 2
				−339.0(3)	71 V 1
		RT	45.43	−330	61 B 1
				−330.4(3)	71 V 1
Co, hcp	wall center	4.2	228	−225.7 [2]	72 K 1
		RT	220.5	−218	72 K 1
	wall edge	4.2	219.9	−217.7	72 K 1
		RT	≈214.5	≈−212	72 K 1
Co, fcc		4.2	217.2 [3]	−215 [3]	60 P 1
		RT	213.1	−211	60 P 1
Ni		4.2	28.46	− 75(1)	63 S 1
		RT	26.04	− 69(1)	63 S 1

[1]) An upper limit for the anisotropy for a single crystal is 100 Oe [80 O 1].
[2]) $H_{hyp,\parallel c} - H_{hyp,\perp c} = +8.0(1)$ kOe [72 K 1].
[3]) Extrapolated from high-temperature values.

1.2.8 Spin wave properties

Table 13. Spin wave stiffness constants derived from the temperature dependence of the spontaneous magnetization (MAG) and from the magnetic hyperfine field (HYP).

Metal	$a_{3/2}$ $\cdot 10^{-6}\,K^{-3/2}$	$a_{5/2}$ $\cdot 10^{-9}\,K^{-5/2}$	$a_{3/2}/a_{5/2}$ K	D_0 meV Å^2	D_1 $\cdot 10^{-3}$ meV Å^2 K^{-2}	Meas. method	Ref.
Fe	3.42(30)	2.2(10)	1550	280		MAG	82 P 1, 83 P 1
				311(10)	0.60(8)	MAG	72 A 1
	3.4(2)	1(1)				MAG	63 A 1
	3.01(15)			308(10)		HYP	73 R 1
			4540			Theory	56 D 1
hcp Co	≈1.5 [1]	≈ 0 [1]		580		MAG	82 P 1, 83 P 1
Ni	6.64(60)	18.5(30)	360	422		MAG	82 P 1, 83 P 1
				362	+1.05	MAG	75 A 1
	7.5(2)	15(2)				MAG	63 A 1
	7.38(11)			393(6)		HYP	77 R 1
			2600			theory	56 D 1

[1]) For $T > 100$ K only.

1.2.9 g factors

Table 14. Spectroscopic splitting factor g and ferromagnetic resonance damping parameter λ for Fe, Co, and Ni.

Metal	Sample	T K	ν GHz	g	λ $\cdot 10^8 s^{-1}$	Fig.	Ref.
Fe	whisker	RT···950	9.6; 23; 31	2.09	1.3	1	72 B 1
	whisker	4···300	70		0.7	2	74 B 1
Co hcp	crystal [1])	RT	71.52	2.18(1)			64 F 1
	whisker	4···700	60	2.18	1	4 [2])	74 B 1
		350	37	2.18			74 B 1
		300	135	2.18			74 B 1
Ni	crystal	RT···650	23; 32	2.21	2.3		69 B 1
	crystal	4···300	22	2.2	2.3	5 [3])	74 B 1

[1]) Magnetcrystalline anisotropy constants: $K_1 = 5.22 \cdot 10^6$ erg cm^{-3} and $K_2 = 0.91 \cdot 10^6$ erg cm^{-3}.
[2]) Rapid increase <100 K and slow increase from 100···700 K.
[3]) Rapid increase <100 K which saturates at ≈ 35 K.

1.2.10 Fermi surfaces, band structures, exchange energies and electron spin polarizations

Table 15. Measured dHvA frequencies f, Fermi wavevectors k_F, number of itinerant d electrons per atom assuming free electron behavior, $n(d_i)$, polarization and the paramagnetic Fermi wavevector k_F^{eff} for the sphere-like Fermi surfaces of Fe [73 B 1]. Spin up: majority spin, spin down: minority spin.

Property	Spin	Crystal plane (100)	(111)	(110)
f [MG] [1])	up	436	370	349
	down	71	52	58
k_f [$2\pi/a$]	up	0.51	0.495	0.43
	down	0.24	0.18	0.19
$n(d_i)$	up	0.28	0.25	0.17
	down	0.030	0.012	0.014
Polarization [%]		80	90	85
k_F^{eff} [$2\pi/a$]		0.42	0.40	0.35

[1]) Accuracy $\pm 1\%$.

Table 16. Measured excited-states exchange energies per spin, J_m^{dd}, and calculated ground state exchange energies J_c^{dd} for the same regions of k-space as the quoted measured values for Fe, Co, and Ni. KS: Kohn-Sham potential, vBH: von Barth-Hedin potential. J in [eV].

	Fe (Γ'_{25})	Ref.	Co (Γ)	Ref.	Ni (L_3)	Ref.
J_m^{dd} [1])	1.00(5)		0.8(2)		0.63(5)	
J_c^{dd}	0.85 (vBH)	77 C 1			0.92 (vBH)	77 W 1
	1.1 (KS)	77 C 1	0.88 (KS)	75 B 1	1.27 (KS)	77 W 1

[1]) Obtained from the measured exchange splitting Δ_m divided by the total spin, $J_m^{dd} = \Delta_m / p_{at}$ in μ_B.

Table 17. Spin polarizations, in [%], of Fe, Co, and Ni as derived from various experiments. $\hbar\omega$: photon energy, Φ: work function.

Method	Fe	Co	Ni	$\hbar\omega - \Phi$ eV	Ref.
Photo-emission	34(1)	17(2)	3(2)	2	73 A 1
			5(3)	5	76 E 1
			8(3)	16	79 B 1
[1])	26.6	17.5	5.6	> 10	
[2])	30	20	6	≪ 10	
Tunneling between poly-crystalline film and supercon-ductor	0.44(2)	0.34(4)	0.11		73 T 1
	0.45 [3])	–	0.10 [3])		77 S 1

[1]) Calculated for the case that all the valence band electrons are excited, i.e. p_{spin}/n_{3d+4s}, p_{spin} in μ_B.
[2]) Calculated for the case that mainly d-band electrons are excited, i.e. p_{spin}/n_{3d}, p_{spin} in μ_B.
[3]) Calculated.

Table 18. Measured room-temperature values Δ_m and calculated values Δ_c [77 C 1] of the exchange splitting for Fe.

Δ_m eV	Δ_c eV	Symmetry point	Crystal surface	Ref.
1.5(2)	1.3	P_4	(111)	80 E 1
2.08(10)	1.8	Γ'_{25}	(110)	82 T 1
2		Γ'_{25}	(100)	83 F 1

1.2.11 Kerr effect

For Fig. 14, see next page.

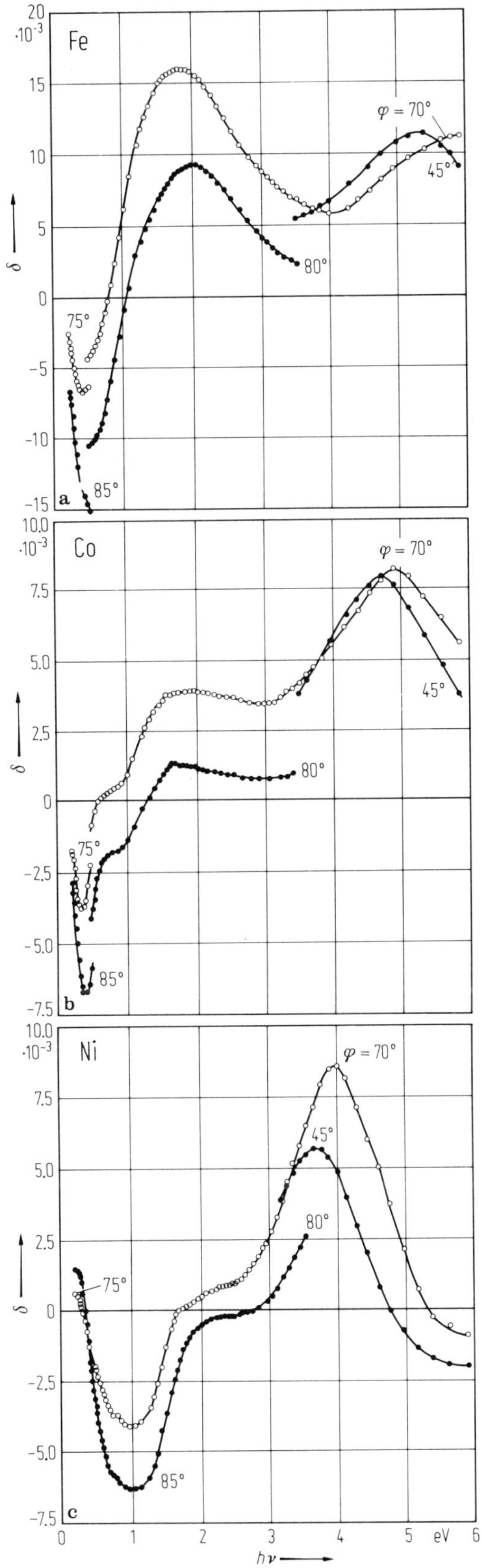

Fig. 15. Equatorial Kerr effect of (a) Fe, (b) Co, and (c) Ni for different angles φ of incident light [68 K 1].

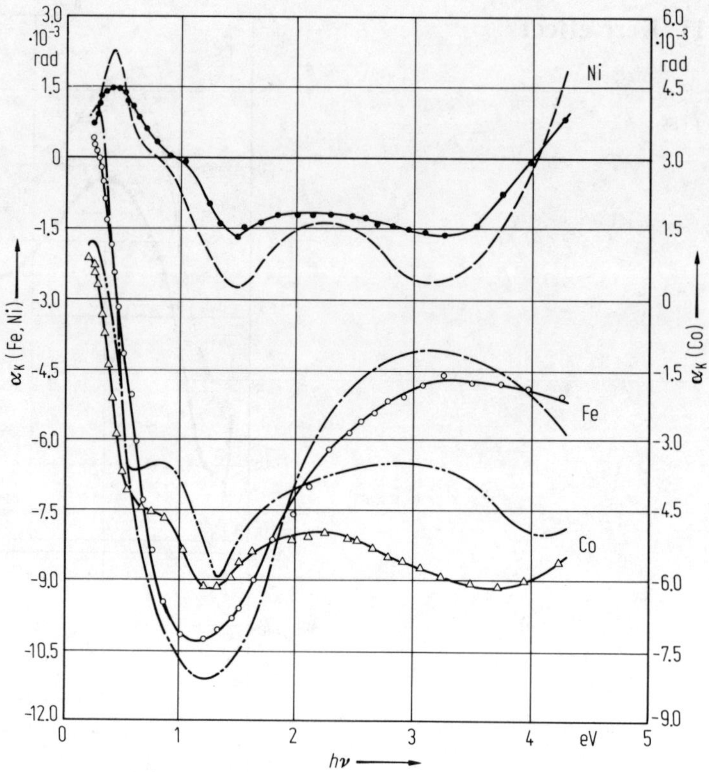

Fig. 14. Polar Kerr effect curves α_K derived from the arithmetic mean of the s and p-wave data for Fe, Co, and Ni. The samples were 99.99% pure plates subjected to mechanical polishing, annealing and electrolytic polishing before measurements. The dashed curves were calculated from values of ε'_{xy} and ε''_{xy} determined from equatorial Kerr effect measurements [68 K 1].

1.2.12 References for 1.2

11 W 1 Weiss, P., Foex, G.: Arch. Sci. Natl. **31** (1911) 89.
17 T 1 Terry, E.M.: Phys. Rev. **9** (1917) 394.
34 P 1 Potter, H.H.: Proc. R. Soc. London Ser. A **146** (1934) 362.
38 S 1 Stoner, E.C.: Proc. R. Soc. London Ser. A **165** (1938) 372.
38 S 2 Sucksmith, W., Pearce, R.R.: Proc. R. Soc. London Ser. A **167** (1938) 189.
44 F 1 Fallot, M.: J. Phys. Radium **5** (1944) 153.
54 B 1 Bozorth, R.M.: Phys. Rev. **96** (1954) 311.
54 S 1 Sucksmith, W., Thompson, J.E.: Proc. R. Soc. London Ser. A **225** (1954) 362.
56 D 1 Dyson, F.J.: Phys. Rev. **102** (1956) 1217.
56 N 1 Nakagawa, Y.: J. Phys. Soc. Jpn. **11** (1956) 855.
60 A 1 Arajs, S., Miller, D.S.: J. Appl. Phys. **31** (1960) 986.
60 P 1 Portis, A.M., Gossard, A.C.: J. Appl. Phys. **31** (1960) 205S.
61 B 1 Barnier, Y., Pauthenet, R., Rimet, G.: C.R. Acad. Sci. Ser. B **253** (1961) 400.
61 B 2 Benedek, G.B., Armstrong, J.: J. Appl. Phys. **32** (1961) 106 S.
61 G 1 Gersdorf, R.: Thesis, University of Amsterdam, The Netherlands **1961**.
61 M 1 Meyer, A.J.P., Asch, G.: J. Appl. Phys. **32** (1961) 330.
62 S 1 Shull, C.G., Yamada, Y.: J. Phys. Soc. Jpn. **17**, Supp. B-III, (1962) 1.
63 A 1 Argyle, B.E., Charap, S.H., Pugh, E.W.: Phys. Rev. **132** (1963) 2051.
63 A 2 Arajs, S., Colvin, R.V.: J. Phys. Chem. Solids **24** (1963) 1233.

63 S 1 Streever, R.L., Bennett, L.H.: Phys. Rev. **131** (1963) 2000.
64 A 1 Arajs, S., Colvin, R.V.: J. Appl. Phys. **35** (1964) 2424.
64 F 1 Frait, Z.: Brit. J. Appl. Phys. **15** (1964) 993.
64 L 1 Lourens, J.H.J., Alberts, L.: Solid State Commun. **2** (1964) 141.
64 M 1 Moon, R.M.: Phys. Rev. **136** (1964) A 195.
65 A 1 Arajs, A.: J. Appl. Phys. **36** (1965) 1136.
65 C 1 Cowan, D.L., Anderson, L.W.: Phys. Rev. A **139** (1965) 424.
65 C 2 Colvin, R.V., Arajs, S.: J. Phys. Chem. Solids **26** (1965) 435.
65 S 1 Stoelinga, J.H.M., Gersdorf, R., DeVries, G.: Physica **31** (1965) 349.
66 K 1 Klein, H.-P., Kneller, E.: Phys. Rev. **144** (1966) 372.
66 M 1 Mook, H.A.: Phys. Rev. **148** (1966) 495.
68 F 1 Franse, J.J.M., deVries, G.: Physica **39** (1968) 477.
68 G 1 Gengnagel, H., Hofmann, U.: Phys. Status Solidi **29** (1968) 91.
68 K 1 Krinchik, G.S., Artem'es, V.A.: Sov. Phys. JETP **26** (1968) 1080.
69 B 1 Bhagat, S.M., Chicklis, E.P.: Phys. Rev. **178** (1969) 828.
69 H 1 Hubert, A., Unger, W., Kranz, J.: Z. Physik **224** (1969) 148.
70 F 1 Franse, J.J.M., Stolp, M.: Phys. Lett. **32** A (1970) 316.
71 B 1 Bower, D.I.: Proc. R. Soc. London Ser. A **326** (1971) 87.
71 C 1 Crangle, J., Goodman, G.M.: Proc. R. Soc. London Ser. A **321** (1971) 477.
71 L 1 Lee, E.W., Asgar, M.A.: Proc. R. Soc. London Ser. A **326** (1971) 73.
71 V 1 Violet, C.E., Pipkorn, D.N.: J. Appl. Phys. **42** (1971) 4339.
72 A 1 Aldred, A.T., Froehle, P.H.: Int. J. Magn. **2** (1972) 195.
72 B 1 Bhagat, S.M., Rothstein, M.S.: Solid State Commun. **11** (1972) 1535.
72 K 1 Kawakami, M., Hihara, T., Koi, Y., Wakiyama, T.: J. Phys. Soc. Jpn. **33** (1972) 1591.
72 L 1 Leger, J.M., Loriers-Susse, C., Vodar, B.: Phys. Rev. B **6** (1972) 4250.
73 A 1 Alder, H., Campagna, M., Siegmann, H.C.: Phys. Rev. B **8** (1973) 2075.
73 B 1 Baraff, D.R.: Phys. Rev. B **8** (1973) 3439.
73 B 2 Briane, M.: C.R. Acad. Sci. Paris **276** (1973) 3789.
73 E 1 Escudier, P.: Thesis, Grenoble, quoted in [80 W].
73 R 1 Riedi, P.C.: Phys. Rev. B **8** (1973) 5243.
73 T 1 Tedrow, P.M., Meservey, R.: Phys. Rev. B **7** (1973) 318.
74 B 1 Bhagat, S.M., Lubitz, P.: Phys. Rev. B **10** (1974) 179.
74 D 1 Donohue, J.: The Structure of the Elements, New York: J. Wiley & Sons Ltd. **1974**.
75 A 1 Aldred, A.T.: Phys. Rev. B **11** (1975) 2597.
75 B 1 Batallan, F., Rosenman, I., Sommers, C.B.: Phys. Rev. B **11** (1975) 545.
76 A 1 Aubert, G., Ayant, Y., Belorizky, E., Casalegno, R.: Phys. Rev. B **14** (1976) 5314.
76 E 1 Eib, W., Alvarado, S.F.: Phys. Rev. Lett. **37** (1976) 444.
77 C 1 Callaway, J., Wang, C.S.: Phys. Rev. B **16** (1977) 2095.
77 R 1 Riedi, P.C.: Phys. Rev. B **15** (1977) 5197.
77 S 1 Stearns, M.B.: J. Mag. Magn. Mater. **5** (1977) 167.
77 W 1 Wang, C.S., Callaway, J.: Phys. Rev. B **15** (1977) 298.
78 G 1 Gersdorf, R.: Phys. Rev. Lett. **40** (1978) 344.
78 W 1 Wohlfarth, E.P.: J. Mag. Magn. Mater. **7** (1978) 113.
79 B 1 Bringer, A., Campagna, M., Feder, R., Gudat, W., Kisker, E., Kuhlmann, E.: Phys. Rev. Lett. **42** (1979) 1705.
79 L 1 Liu, C.M., Ingalls, R.: J. Appl. Phys. **50(3)** (1979) 1751.
80 E 1 Eastman, D.E., Himpsel, F.J., Knapp, J.A.: Phys. Rev. Lett. **44** (1980) 95.
80 O 1 Oppelt, A., Kaplan, N., Fekete, D.: J. Mag. Magn. Mater. **15–18** (1980) 660.
81 H 1 Hanham, S.D., Arrott, A.S., Heinrich, B.: J. Appl. Phys. **52** (1981) 1941.
82 P 1 Pauthenet, R.: J. Appl. Phys. **53** (1982) 2029 and 8187: C. R. Acad. Sci. Ser. B **295** (1982) 331, 1067.
82 T 1 Tung, C.J., Said, I., Everett, G.E.: J. Appl. Phys. **53** (1982) 2044.
83 F 1 Feder, R., Gudat, W., Kisker, E., Rodriguez, A., Schroder, K.: Solid State Commun. **46** (1983) 619.
83 P 1 Pauthenet, R.: Conf. on High Field Magn. (Date, M., ed.), Amsterdam: North-Holland Publ. Co. **1983**, p. 77.
83 P 2 Pauthenet, R.: C. R. Acad. Sci. Paris V **297(II)** (1983) 13.
84 P 1 Paige, D.M., Szpunar, B., Tanner, B.K.: J. Mag. Magn. Mater **44** (1984) 239.

2 Alloys between 3d elements

2.1 Alloys between Fe, Co or Ni

In this chapter the magnetic properties of binary and ternary alloys between the elements Fe, Co or Ni are given, as well as the influence of small amounts of other elements on the properties of the alloys. Secondary magnetic properties like permeability, coercive force, hysteresis losses, etc., which depend to a large degree on the preparation technique, crystal size, and on the various treatments of a polycrystalline sample, can be found in subvolume i of volume III/19 of Landolt-Barnstein (New Series) dealing with the properties of technically applied magnetic materials.

Survey

Section	Properties	$Fe_{1-x}Co_x$ Fig.	Table	$Fe_{1-x}Ni_x$ Fig.	Table	$Co_{1-x}Ni_x$ Fig.	Table
2.1.1	Phase diagrams, lattice parameters	1–3		4, 6		5	
2.1.2	Paramagnetic properties	7, 8		10, 11		9	
2.1.3	Hyperfine magnetic field, isomer shifts	12–14	1	12, 15, 16	1		
2.1.4	Spin waves		2		3–6	17	6, 7
2.1.5	Atomic magnetic moment, g and g' factor	18–21	11	20, 23, 24	8, 9	20, 22	19, 12
2.1.6	Spontaneous magnetization, Curie temperature	25, 26, 28		25–27, 29–31		25	
2.1.7	High-field susceptibility	33		32			
2.1.8	Magnetocrystalline anisotropy	34, 36, 37		34, 35, 38, 39		34, 40, 41	
2.1.9	Magnetostriction	42, 45	13, 14	43, 44	15, 16	46, 47	

2.1.1 Phase diagrams, lattice parameters

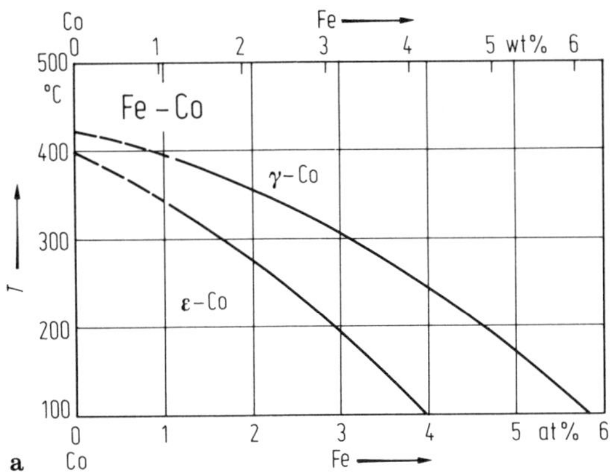

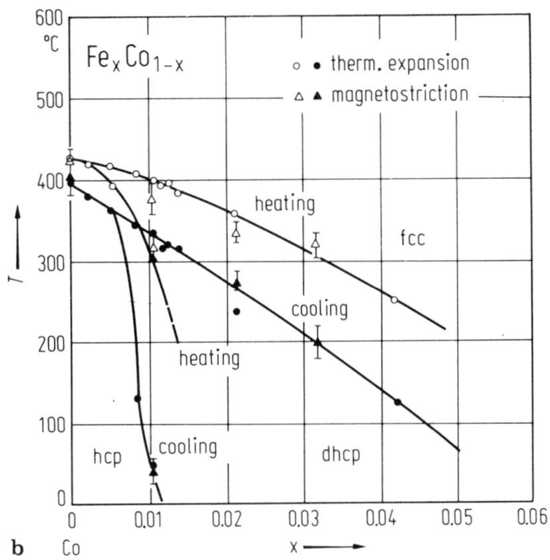

Fig. 1(a). Temperature hysteresis of the phase boundary between the Co-rich fcc phases (γ-Co) and hcp phases (ε-Co) [82 K 1], (b) Phase diagram in the dilute Fe_xCo_{1-x} alloy system as derived from dilatation curves [84 I 1].

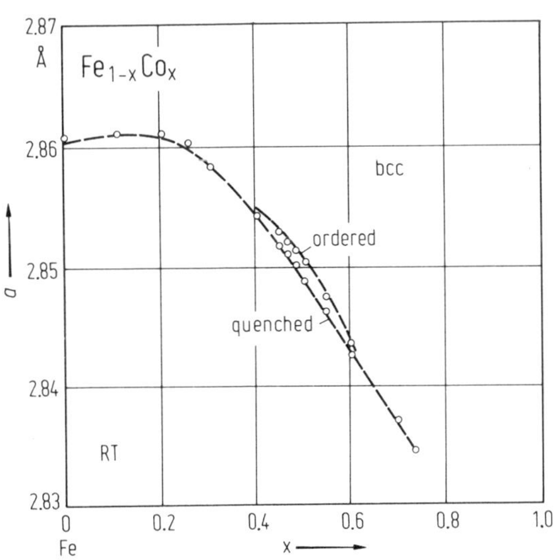

Fig. 2. Room-temperature values of the lattice parameter a of bcc $Fe_{1-x}Co_x$ alloys [72 S 1], data from [41 E 1].

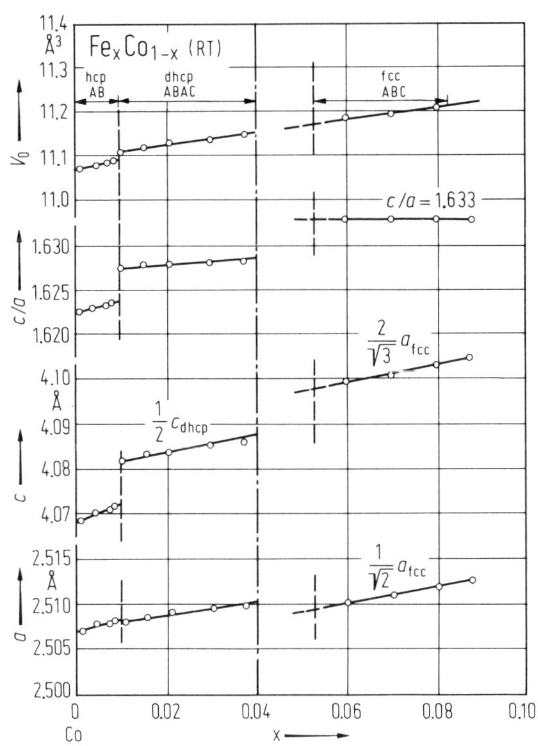

Fig. 3. Lattice constants, c/a and volume per atom, V_0, of Fe_xCo_{1-x} alloys at room temperature for less than 10 at% Fe [73 W 1, 74 O 1].

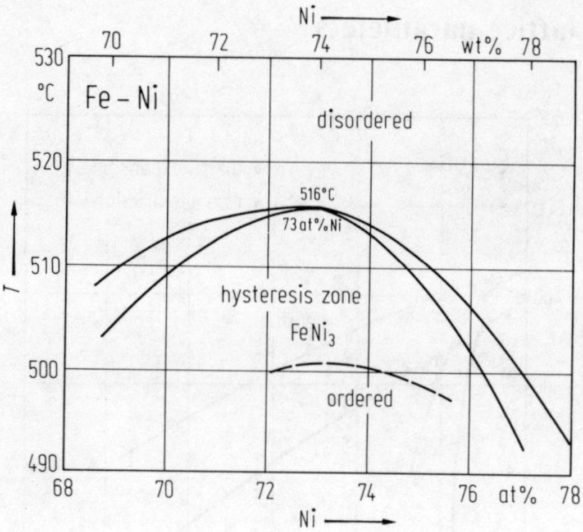

Fig. 4. Occurrence of superstructure in the $FeNi_3$ region, as deduced from Mössbauer effect spectroscopy [82 K 1]. See also the original work [77 D 1, 77 D 2, 79 D 1].

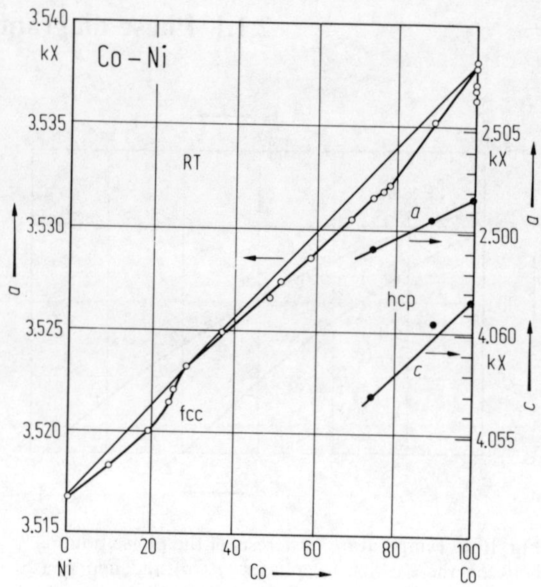

Fig. 5. Lattice parameters of Co–Ni alloys annealed at 900 °C and slowly cooled to room temperature in 14 days [50 T 1]. 1 kX ≙ 1.002 Å.

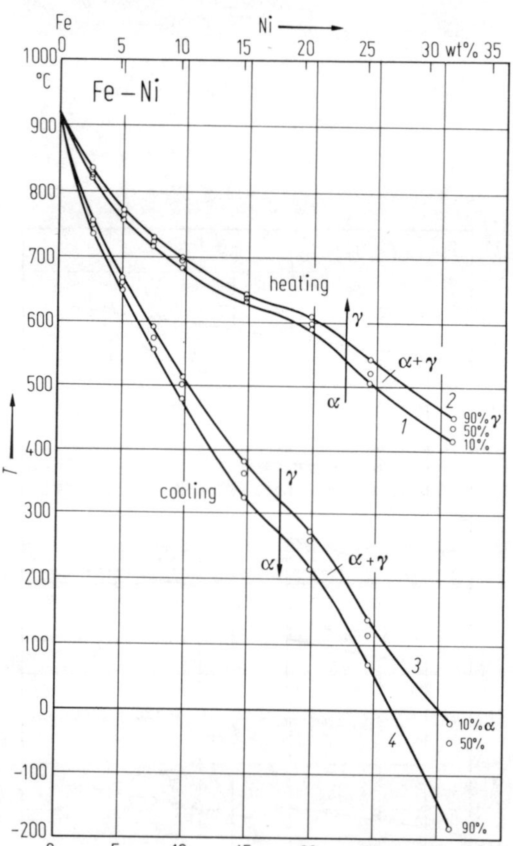

Fig. 6. Temperature hysteresis of the α–γ transition of Fe-rich Fe–Ni alloys. The field between the curves *1* and *3* are the points with more than 90% α- or γ-phase when the alloy is heated or cooled through the corresponding temperature range, respectively, as is indicated by the arrows [58 H 1].

2.1.2 Paramagnetic properties

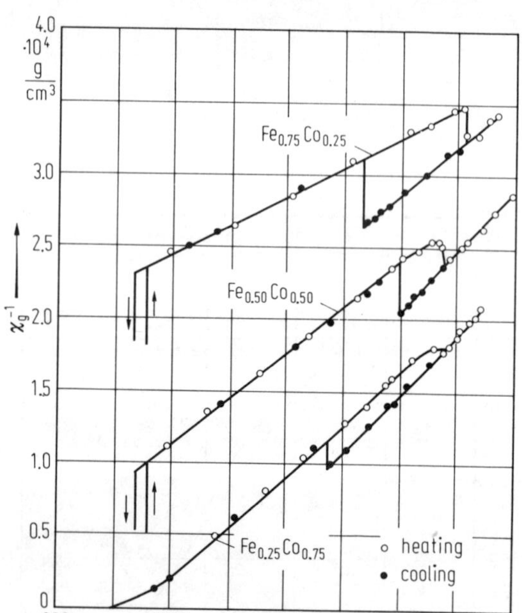

Fig. 7. Temperature dependence of the inverse paramagnetic mass susceptibility χ_g^{-1} for different Fe–Co alloys, measured during cooling and heating the specimens [56 N 1].

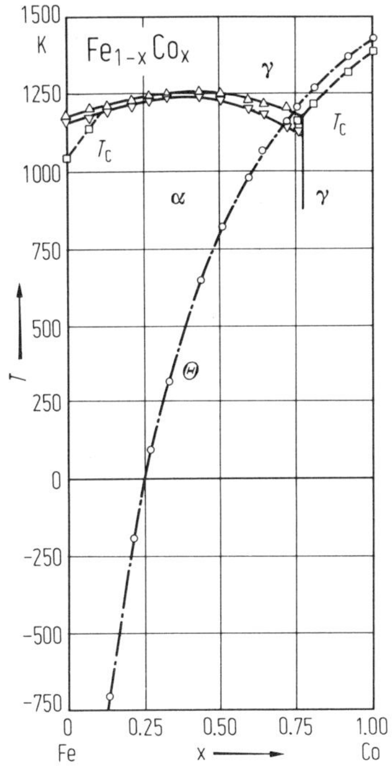

Fig. 8. Phase diagram of $Fe_{1-x}Co_x$ alloys, with indication of the extrapolated paramagnetic Curie temperature Θ of the high-temperature phase [43 F 1]. Triangles: phase transformation temperature, squares: T_C, circles: Θ.

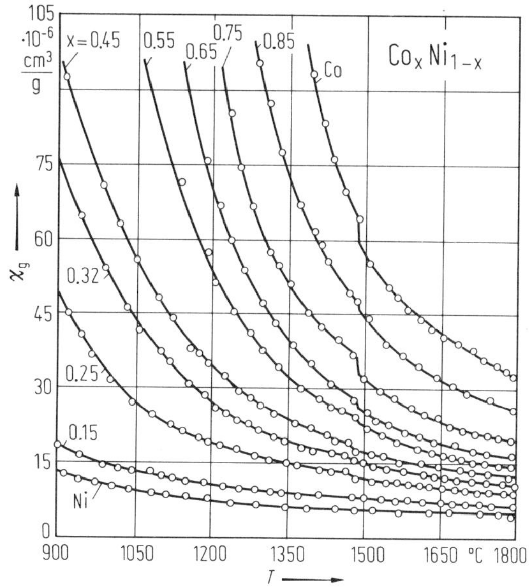

Fig. 9. Temperature dependence of the paramagnetic mass susceptibility χ_g for the alloys Co_xNi_{1-x} [77 S 1].

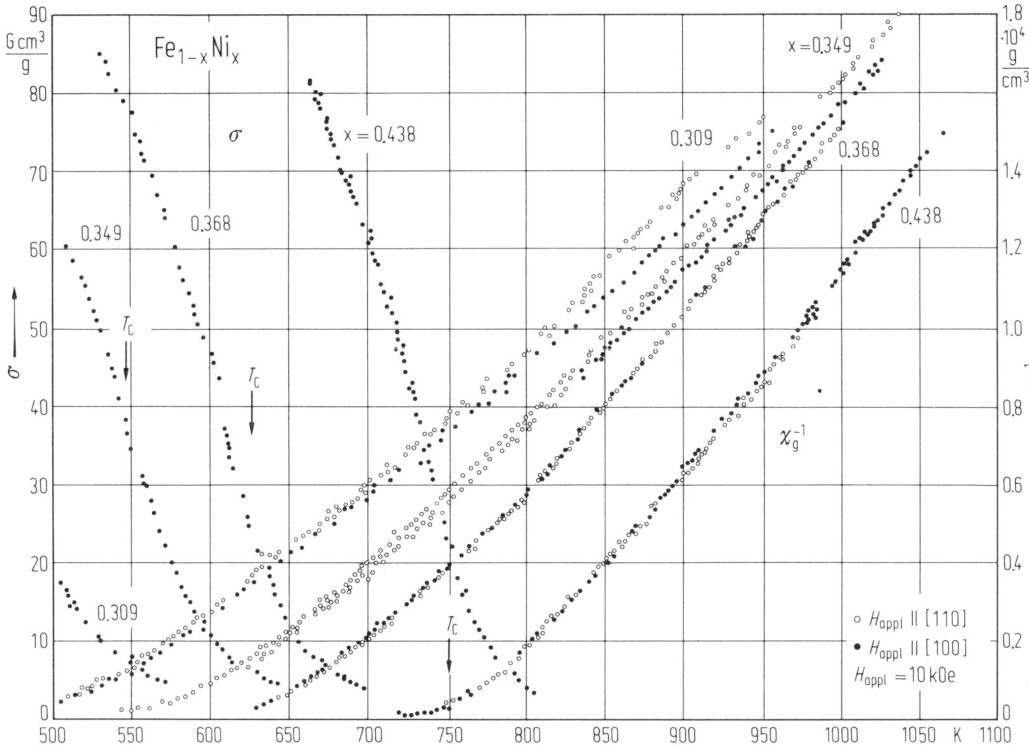

Fig. 10. Temperature dependence of the inverse mass susceptibility χ_g^{-1} and the magnetic moments per unit mass σ of single crystals of $Fe_{1-x}Ni_x$ invar alloys. T_C is the ferromagnetic Curie temperature [79 C 1].

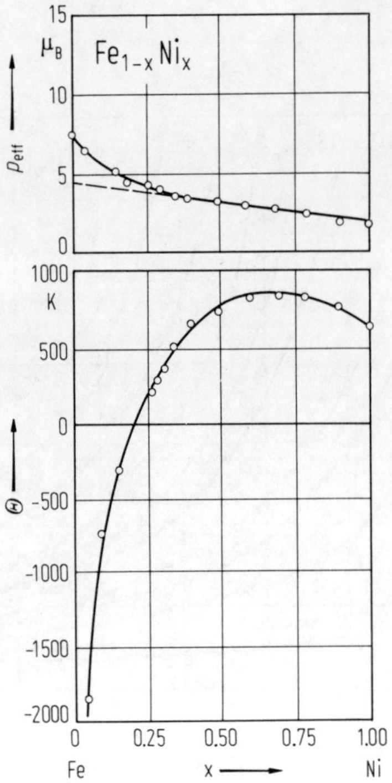

Fig. 11. Paramagnetic Curie temperature Θ and the effective paramagnetic moment per atom p_{eff} for the γ-phase alloys $Fe_{1-x}Ni_x$ [62 C 1].

2.1.3 Hyperfine magnetic fields, isomer shifts

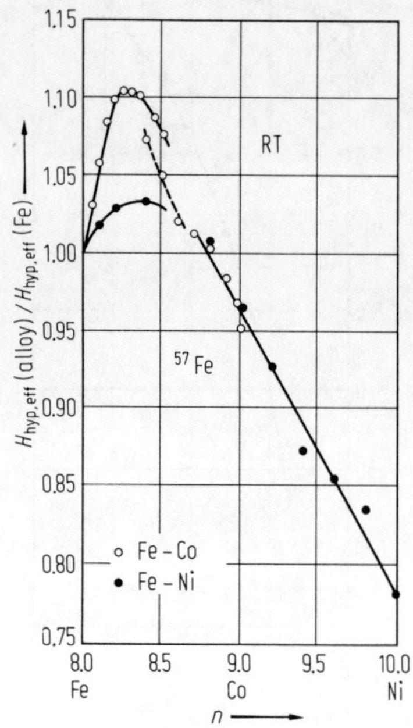

Fig. 12. Effective magnetic hyperfine field $H_{hyp,eff}$ at ^{57}Fe nuclei in Fe–Co and Fe–Ni alloys at room temperature relative to the field in metallic iron, plotted as a function of the number n of 4s and 3d electrons per atom. The data in the range $Fe_{0.8}Ni_{0.2}$ to $Fe_{0.5}Ni_{0.5}$ ($n = 8.4 \cdots 9$) where the Curie points are low have been corrected to take account of incomplete saturation at room temperature [61 J 1].

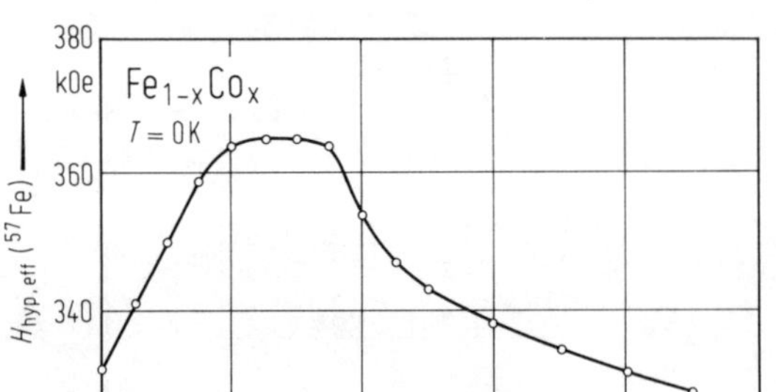

Fig. 13. Variation of the effective magnetic hyperfine field $H_{hyp,eff}$ at ^{57}Fe nuclei vs. Co concentration of $Fe_{1-x}Co_x$ alloys. Values obtained by extrapolating room-temperature data to 0 K according to the increase of the spontaneous magnetization [63 J 1], see also [70 M 1].

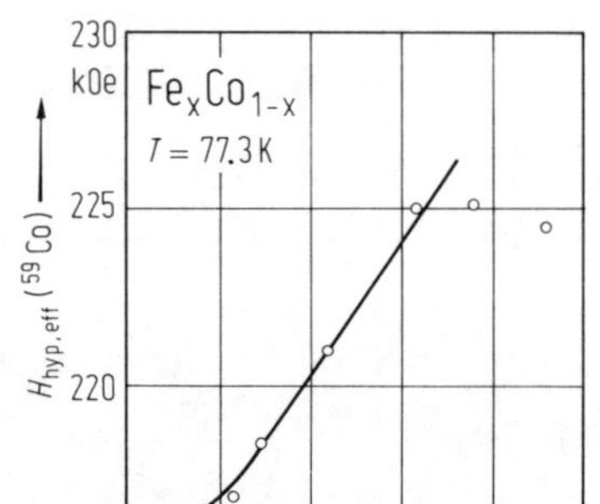

Fig. 14. Effective magnetic hyperfine field $H_{hyp,eff}$ at ^{59}Co nuclei in fcc Fe_xCo_{1-x} single crystals plotted as a function of the composition [77 B 1]. $T = 77.3$ K. The solid line represents calculated values.

Table 1. Isomer shifts for ^{57}Fe in Fe–Co and Fe–Ni alloys.

Alloy	Composition	IS (^{57}Fe) $mm\,s^{-1}$	Ref.
$Fe_{1-x}Co_x$	$x \leqq 0.9$	0.05 [1])	63 J 1, 70 M 1
$Fe_{1-x}Ni_x$	$0.18 < x < 0.32$	ferromagnetic part: +0.29(5) [2])	74 C 1
		antiferromagnetic part: +0.18(5)	74 C 1, 68 A 1
	$\alpha \rightarrow \gamma$ phase boundary	ferromagnetic part: −0.12 [3])	63 J 1
	$x = 0.34$	±0.2 [4])	64 N 1

[1]) Positive with respect to pure Fe.
[2]) Relative to ^{57}Fe in Cr.
[3]) Shifts from phase to phase.
[4]) As compared to stainless steel.

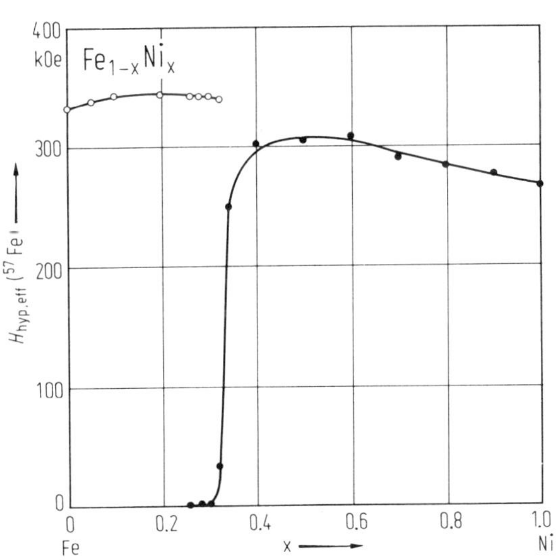

Fig. 15. Room-temperature values of the ^{57}Fe hyperfine field $H_{hyp,eff}$ in $Fe_{1-x}Ni_x$ alloys obtained from arc-casted and spectroscopically standardized materials [63 J 1]. See also [73 M 1]; for an explanation on the basis of two electronic configurations of the Fe atoms, see [63 W 1]. For discussions on the asymmetry of the Mössbauer spectra of γ–$Fe_{1-x}Ni_x$ alloys, see [74 W 1, 75 B 1, 77 H 1].

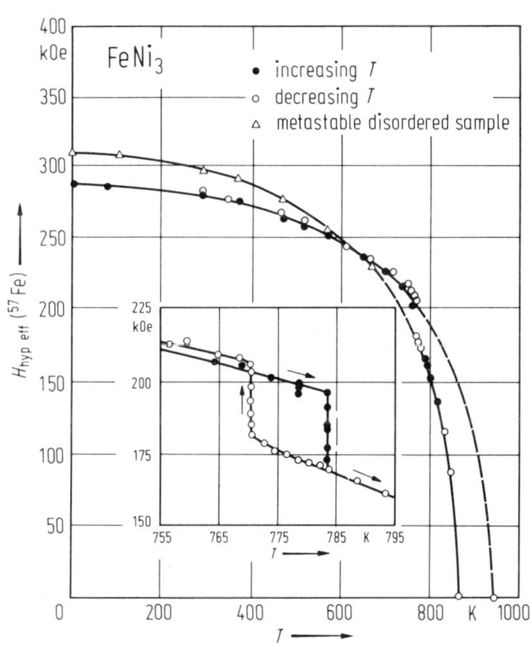

Fig. 16. Effective hyperfine field $H_{hyp,eff}$ vs. temperature T at ^{57}Fe nuclei in $FeNi_3$. In the insert the transition region for the ordering is given on a larger scale [77 D 2], see also [79 D 1].

2.1.4 Spin waves

Table 2. The second-moment exchange integrals J, in [K], between atom pairs in Fe–Co alloys 1 K ≙ 0.0862 meV.

	2nd-moment exchange integral [K]	
	bcc [1])	fcc [2])
$J^{(2)}_{Fe-Fe}$	≈500	<0
$J^{(2)}_{Fe-Co}$	≈800	>0
$J^{(2)}_{Co-Co}$	≈400	>0

[1]) As derived from Curie and ordering temperatures [80 L 1].
[2]) [78 B 1].

Table 3. Exchange integrals J between atom pairs in $Fe_{1-x}Ni_x$ alloys as derived from inelastic small-angle scattering of neutrons by spin waves.

	Exchange integral [meV]		
	bcc [1])	fcc [1])	fcc [2])
J_{Fe-Fe}	+18.2(5)	− 9.0(26)	− 8(1)
J_{Fe-Ni}	−41(13)	+39(5)	+38(4)
J_{Ni-Ni}	–	+52(3)	+52(5)

[1]) Second moments [64 H 1].
[2]) [75 M 1].

Table 4. Spin wave stiffness constant D of Fe_xNi_{1-x} alloys.

[75 H 1]: x	Ni	0.05	0.4	0.5	0.6	0.68
D_{293K} [meV Å²]	460(15)	365(10)	235(10)	200(5)	140(5)	70(5)
$D_{4.2K}$ [meV Å²]	525(15)	400(10)	250(10)	200(5)	160(5)	
293 K/T_C	0.47	0.43	0.34	0.38	0.46	0.77
$D_{293K}/D_{4.2K}$	0.88	0.91	0.94	1	0.88	

[83 P 1]: x	0.658	0.653	0.646	0.630	0.614	0.598	0.550
T_N [K]	21(1)	21.0(15)	19.5(20)	17(1)	14.0(15)	12(2)	5(1)
D [meV Å²]	77.0	86.8	87.8	102.0	123.7	131.7	170.4

Table 5. Spin wave stiffness constant D_0 at 0 K, quadratic temperature coefficient D_2 of the spin wave constant and the coefficient β of the quadratic term in the spin wave dispersion relation as derived from low-temperature magnetization curves measured on single crystals of Fe_xNi_{1-x} alloys [83 N 1].

x	ϱ g cm⁻³	σ_s G cm³ g⁻¹	$\bar{p}_{at}$ μ_B	χ_{HF} 10^{-6} cm³ g⁻¹	D_0 meV Å²	D_2 10^{-7} K$^{-5/2}$	β Å²
0.496	8.256(5)	162.11	1.663	4.71	245(20)	0.3(1)	5.0(10)
0.406	8.362(5)	145.70	1.501	3.67	300(20)	0.25(10)	5.0(10)
0.302	8.510(5)	125.88	1.304	2.88	370(20)	0.8(2)	3.5(10)
0.198	8.640(5)	101.87	1.060	2.22	390(20)	0.6(2)	4.0(10)
0.102	8.772(5)	80.345	0.8403	2.16	450(20)	0.8(2)	3.5(10)
Ni	8.917(5)	58.549	0.6154	2.06	530(20)	1.1(1)	2.5(5)

Table 6. Spin wave stiffness constant D of Co–Ni alloys and $Fe_{0.91}Ni_{0.81}$.

	D meV Å^2	T K	Measuring method	Ref.
$Fe_{0.19}Ni_{0.81}$	125(12)	RT	photoacoustically	83 D 1
	139(12)	77	photoacoustically	83 D 1
Ni	160(50)	RT	spin wave resonance	77 C 1
$Co_{0.1}Ni_{0.9}$	68(20)	RT	spin wave resonance	77 C 1
$Co_{0.2}Ni_{0.8}$	83(17)	RT	spin wave resonance	77 C 1
$Co_{0.3}Ni_{0.7}$	94(20)	RT	spin wave resonance	77 C 1
$Co_{0.4}Ni_{0.6}$	90(24)	RT	spin wave resonance	77 C 1
	116(10)	RT	photoacoustically	83 D 1

Table 7. The second-moment exchange integrals J between atom pairs in Co–Ni alloys.

	2nd-moment exchange integral [meV]		
	fcc [1])	fcc [2])	hcp [2])
$J^{(2)}_{Co-Co}$	21(2)	1(3)	10.0(5)
$J^{(2)}_{Co-Ni}$	37(4)	6(4)	− 45(25)
$J^{(2)}_{Ni-Ni}$	51(5)	52(2)	738(500)

[1]) Magnetization measurements [76 M 1].
[2]) Spin waves [72 H 1].

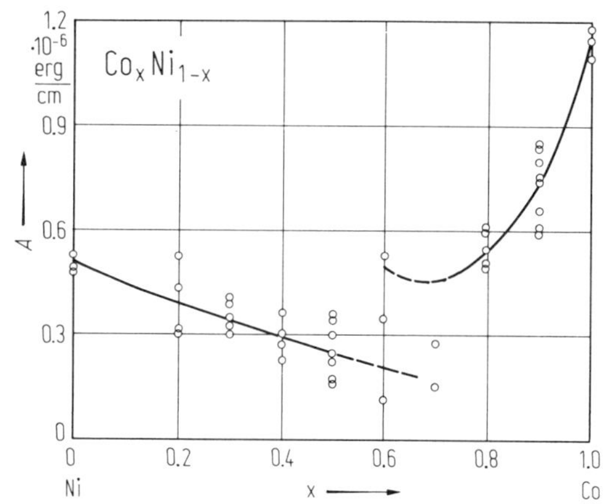

Fig. 17. Exchange stiffness constant A vs. composition for Co_xNi_{1-x} alloys [72 H 2]. The relation between A and the spin wave stiffness constant D is given by $D=2A\hbar\gamma/M_s$, where γ is the gyromagnetic ratio and M_s is the spontaneous magnetic moment per unit of volume. Single crystal measurement: x = 0.85, $A=0.8\cdot10^{-6}$ erg cm^{-1} [75 W 1].

2.1.5 Atomic magnetic moment, g and g' factor

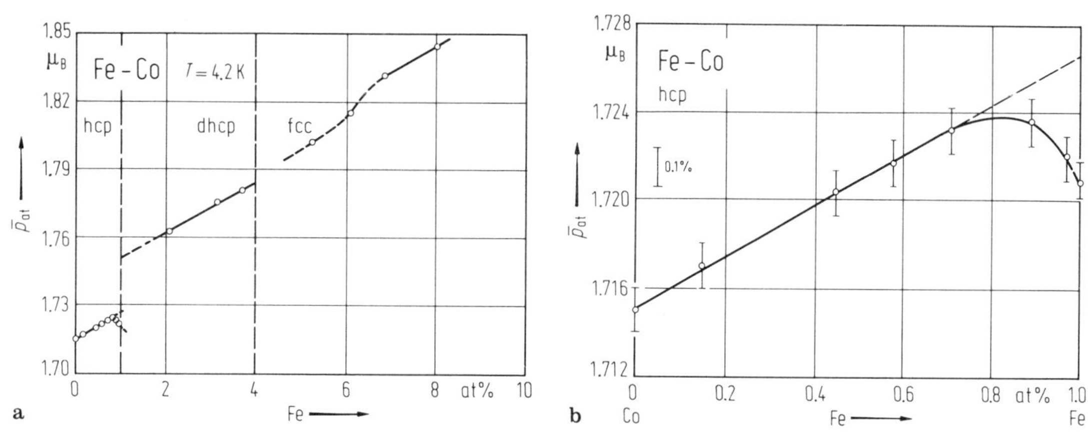

Fig. 18. (a) Mean magnetic moment $\bar{p}_{at}$ in hcp, dhcp and fcc Fe–Co alloys as derived from the saturation magnetization of polycrystalline samples at 4.2 K. (b) Enlarged graph for the hcp region [73 W 2].

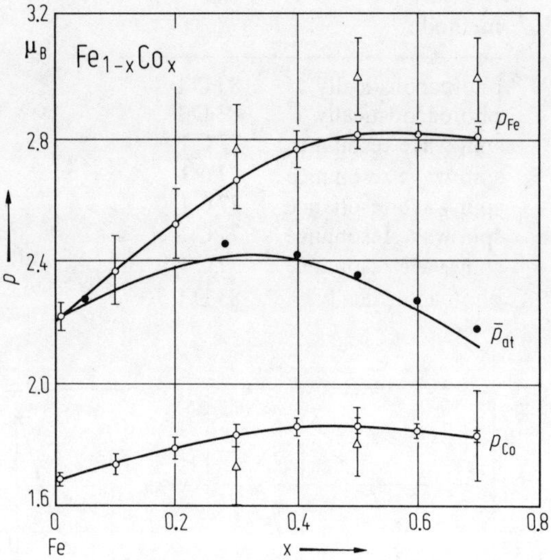

Fig. 19. Atomic magnetic moments p_{Fe} and p_{Co} in bcc $Fe_{1-x}Co_x$ alloys. Triangles: neutron diffuse scattering data [63 C 1, 64 G 1], open circles: calculation results based on the tight-binding model of 3d electrons with allowance for local-environment effects [79 H 2], solid circles: mean atomic magnetic moment for the alloy [69 B 1].

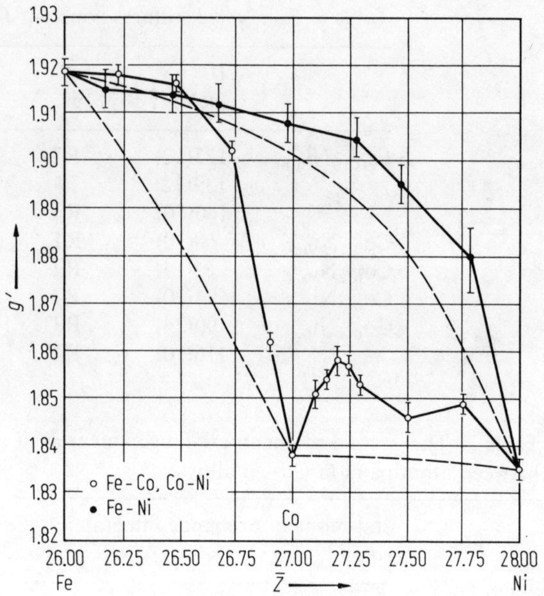

Fig. 20. Magnetomechanical factor g' for Fe, Co, Ni, and their binary alloys as dependent on the mean atomic number $\bar{Z}$. The dashed lines are computed from the properties of the constituent elements [69 S 1].

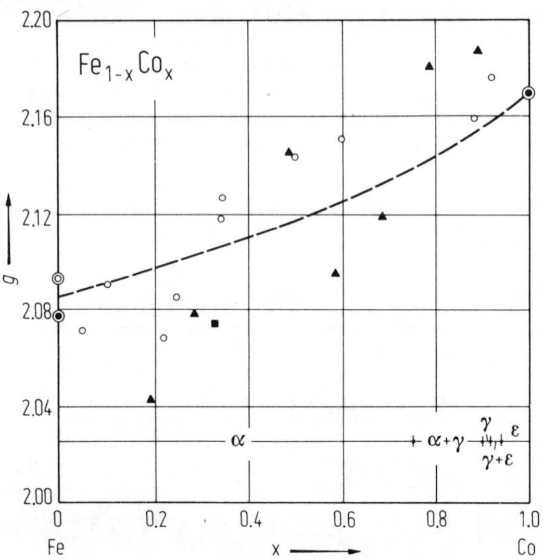

Fig. 21. Spectroscopic splitting factor g for $Fe_{1-x}Co_x$ alloys. Solid circles are the g values measured by ferromagnetic resonance [61 M 1]. Solid symbols indicate g values derived from the magneto-mechanical factor g' through the relation $1/g + 1/g' = 1$. Squares: [44 B 1], triangles: [52 B 1]. Dashed line: calculated from the properties of the constituent elements. For a calculation of the electronic spin and orbital magnetization, see [69 R 1].

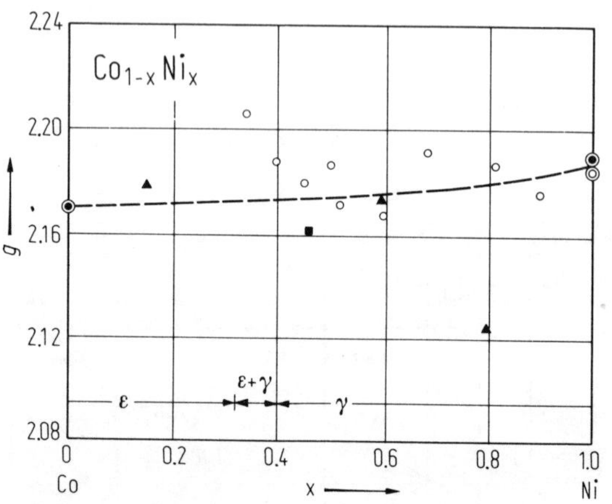

Fig. 22. Spectroscopic splitting factor g of $Co_{1-x}Ni_x$ alloys vs. composition. Open circles are the g values measured by ferromagnetic resonance [61 M 1]. Solid symbols indicate g values derived from the magnetomechanical factor g' through the relation $1/g + 1/g' = 1$. Squares: [44 B 1], triangle: [52 B 1]. Dashed line: calculated from the properties of constituent elements.

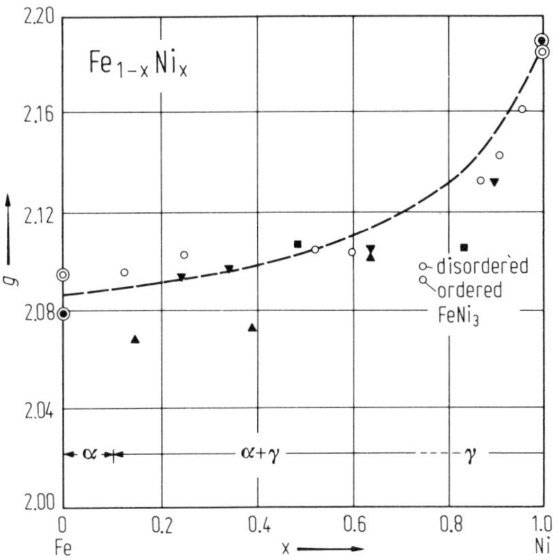

Fig. 23. Spectroscopic splitting factor g of $Fe_{1-x}Ni_x$ alloys. Open circles are the g values measured by ferromagnetic resonance [61 M 1]. Solid symbols indicate g values derived from the magnetomechanical factor g' through the relation $1/g+1/g'=1$. Squares: [44 B 1], triangles (upward): [52 B 1], triangles (downward): [56 S 1]. Dashed line: calculated from the properties of the constituent elements.

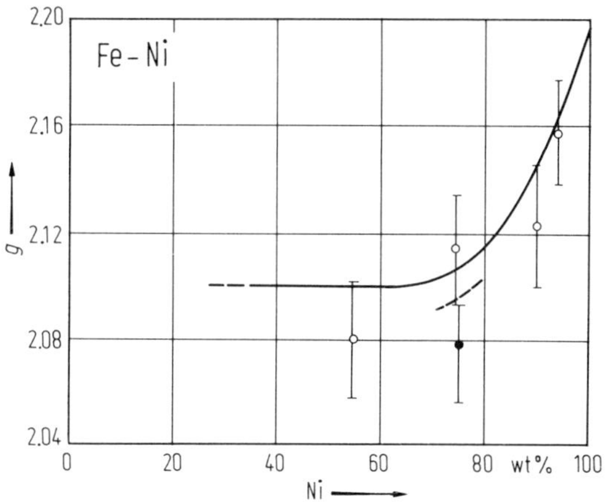

Fig. 24. Spectroscopic splitting factor g of Fe–Ni alloys. Circles [76 B 1], disordered single crystals. The solid curve designates mean values obtained by [73 P 1, 73 M 1, 61 O 1]. For an $FeNi_3$ annealed ordered sample: dashed line. The lower point at 75 wt% Ni applies to an irradiated sample, considered to result in a higher degree of ordering [76 B 1].

Table 8. Magnetic moments p_{Fe} and p_{Ni} attributable to Fe and Ni atoms in $Fe_{1-x}Ni_x$ alloys, as derived from magnetic structure factors obtained from neutron scattering with either polarized (pol) or unpolarized (unpol) neutrons at room temperature. For theoretical calculations based on various theories, see [71 H 1, 72 H 3, 75 R 1, 79 I 1, 81 K 1] for coherent potential approximation; [74 M 1, 78 H 1, 79 H 1, 79 H 2, 80 H 1, 83 K 1, 83 K 2] for similar calculations taking into account a chemical local environment effect; [79 M 1] for Hartree-Fock approximation in the itinerant electron theory. The samples are polycrystalline, unless otherwise stated.

$Fe_{1-x}Ni_x$ x	Sample	p_{Fe} μ_B	p_{Ni} μ_B	$p_{Fe}-p_{Ni}$ μ_B	Measuring method	Ref.
0.10	disordered	2.41(8)	0.94(20)		pol	63 C 1
0.10	disordered	2.41(4)	0.93(10)		unpol	62 C 2
0.32	crystal, invar	1.10	0.64	0.46	pol	82 C 1
0.34	invar	1.80(15)	0.87(20)		pol	79 I 2
0.35	crystal, invar	2.3(25)			unpol	65 C 1
0.40	disordered	2.44(8)	0.83(6)		unpol	62 C 2
0.40	disordered	2.42(17)	0.82(13)		pol	63 C 1
0.499	disordered	2.60	0.67	1.93(20)	unpol	55 S 1
0.50	crystal, disordered	2.54(16)	0.78(4)		unpol	73 C 1
0.601	disordered	2.65	0.65	2.06(21)	unpol	55 S 1
0.70	annealed	3.02(12)	0.63 ± 0.09		unpol	62 C 2
0.70	disordered	2.66(9)	0.63 ± 0.08		unpol	62 C 2
0.743	disordered	2.91	0.60	2.31(22)	unpol	55 S 1
0.75	crystal, disordered	3.13(15)	0.63(5)		pol	73 C 1
	crystal, ordered	3.10(1)	0.68(5)		pol	73 C 1
	polycrystalline	2.97(15)	0.62(5)	2.35(20)	unpol	55 S 1
0.80	disordered	2.99(1)			unpol	74 N 1
0.85	disordered	2.56(33)			unpol	74 N 1
0.90	disordered	2.58(14)	0.64(5)		unpol	62 C 2

Table 9. Analyses of the magnetic moments of $Fe_{1-x}Ni_x$ alloys. $\bar{p}_{loc}$ is the mean localized atomic moment, p_{Fe} and p_{Ni} are the localized moments of the Fe and Ni atom, respectively. p_{ce} is the conduction electron polarization per atom and $\bar{p}_{at}$ the mean atomic moment derived from the magnitude of the volume magnetization. γ is the fraction of 3d electrons in E_g orbitals.

x	Sample	T K	p_{Fe} μ_B	p_{Ni} μ_B	$\bar{p}_{loc}$ μ_B	p_{ce} μ_B	$\bar{p}_{at}$ μ_B	γ	Ref.
0.34	single crystal, invar [1])	300	1.80(15)	0.87(20)	1.48(2)	−0.18(3)	–	0.446	79 I 2
		77	2.41(15)	0.82(5)	1.87(1)	−0.22(5)	–	0.479	79 I 2
0.50	single crystal, disordered	300	2.54(16)	0.78(4)	1.66(8)	−0.10	1.56	0.456(12)	73 C 1
0.505(10)	single crystal, 3° mosaic spread	300	–	–	1.78(2)	−0.17(3)	1.61(3)	0.49(3)	74 M 2
0.75	single crystal, disordered	300	3.13(15)	0.63(5)	1.25	–	1.15 [2])	0.375(17)	73 C 1
	ordered	300	3.10(1)	0.680(5)	1.29	−0.07	1.22 [2])	0.462(13) [3])	73 C 1

[1]) In the disordered invar alloys, x = 0.40 and x = 0.37, a mean antiferromagnetic component is found at 4.2 K of $\bar{p}_{at} = 0.5(1)\,\mu_B$ and $0.63(1)\,\mu_B$, respectively, where the bar refers to the average related to the various surroundings of the atoms in the disordered alloy [80 D 1]. A Néel point $T_N = 15$ K was derived from the temperature dependence of the peak intensity of the antiferromagnetic neutron scattering spectrum [73 D 1].

[2]) See [53 W 1].

[3]) Applies to Fe atoms only.

Table 10. Mean magnetic moment contributions per atom, in [μ_B], for fcc Co–Ni alloys at room temperature [71 A 1, 70 A 1], Ni [66 M 1].

	$Co_{0.91}Fe_{0.09}$	$Co_{0.65}Ni_{0.35}$	$Co_{0.50}Ni_{0.50}$	$Co_{0.25}Ni_{0.75}$	$Co_{0.10}Ni_{0.90}$	Ni
$\bar{p}_{at}$	1.820(10)	1.342(6)	1.173(23)	0.876(5)	0.712(5)	0.579(5)
p_{orb}	0.135(8)	0.105(6)	0.097(5)	0.072(4)	0.062(4)	0.053(3)
p_{spin}	1.993(21)	1.421(9)	1.217(8)	0.927(6)	0.737(6)	0.621(5)
p_{Co}	−0.308(25)	−0.184(12)	−0.141(25)	−0.123(9)	−0.087(9)	−0.095(7)

Table 11. Magnetic moment distribution in $Fe_{1-x}Co_x$ alloys [70 D 1, 72 M 1, 62 S 1].

g: gyromagnetic ratio derived from magnetic form factors as obtained from polarized neutron scattering

$\bar{p}_{at}$: mean atomic moment

$\bar{p}_{ce}$: average conduction electron polarization per atom

γ: fraction of 3d electrons in an E_g state

x	g	γ	$\bar{p}_{at}$ μ_B	$\bar{p}_{ce}$ μ_B	Ref.
0.9175	2.17	0.472(1)	2.057(40)	−0.39(3)	70 D 1
0.8865	2.14	0.474(2)	2.15(6)	−0.43(7)	70 D 1
0.883	2.16	0.476(1)	2.15(4)	−0.46(3)	70 D 1
0.858	2.16	0.478(1)	2.06(3)	−0.33(2)	70 D 1
0.91		0.471(11)	1.99(2)	−0.31(3)	72 M 1
0.91 [1])		0.463(13)	1.75(2)	−0.28(7)	72 M 1
1.00			1.86(7)	−0.28(7)	62 S 1

[1]) At 600 °C.

Table 12. Atomic magnetic moments for $Co_{0.25}Ni_{0.75}$ and $Co_{0.5}Ni_{0.5}$ [65 C 2].

Alloy	$\bar{p}_{at}$	$\lvert p_{Co} - p_{Ni}\rvert$	p_{Co}		p_{Ni}
	μ_B				
$Co_{0.25}Ni_{0.75}$	0.89(2)	1.26(4)	1.84(5) −0.05(5)	or	0.58(2) 1.21(2)
$Co_{0.5}-Ni_{0.5}$	1.14(2)	1.12(4)	1.70(3) 0.58(3)	or	0.58(3) 1.70(3)

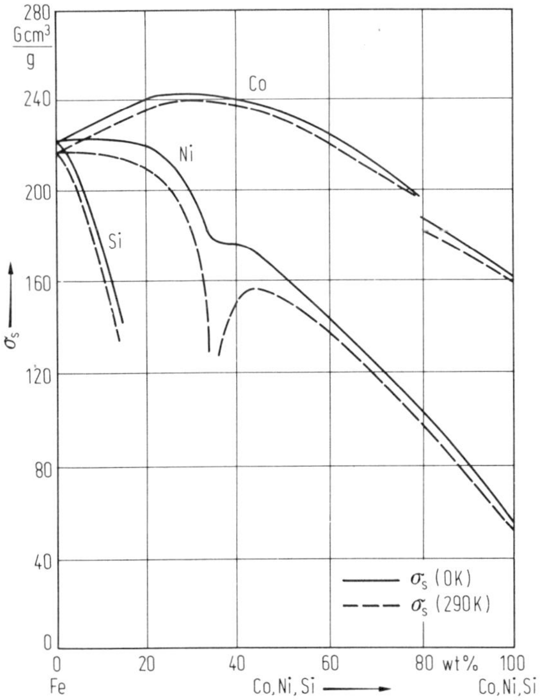

2.1.6 Spontaneous magnetization, Curie temperature

Fig. 25. Spontaneous magnetic moment σ_s (0 K) and σ_s (290 K) for Fe–Co [29 W 1], Fe–Ni [25 P 1, 2] and Fe–Si [36 F 1] alloys, as well as the density ϱ of these alloys [62 K 1].

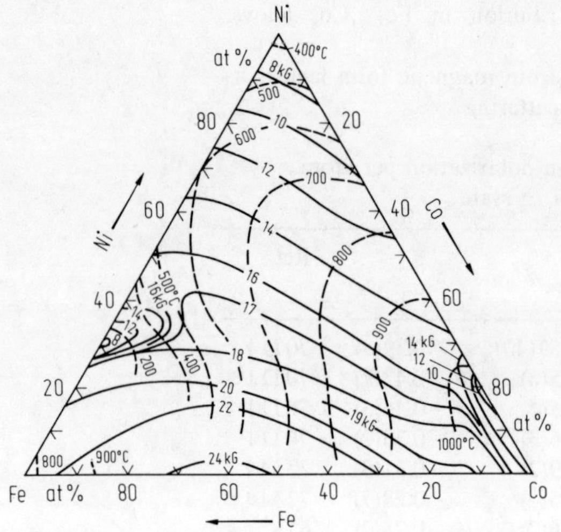

Fig. 26. Room-temperature volume magnetization, $4\pi M$, in a magnetic field of 1.5 kOe for Fe–Co–Ni alloys [62 K 1], originally from [27 K 1, 28 E 1, 29 E 1, 29 M 1]. Broken curves are the Curie temperatures T_C.

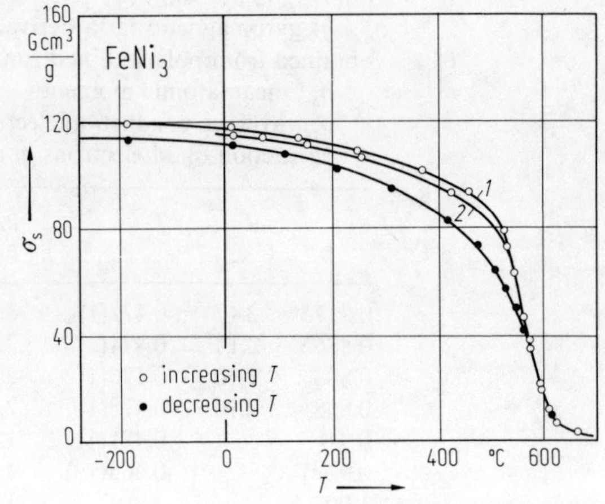

Fig. 27. Spontaneous magnetic moment σ_s for $FeNi_3$. Open circles: measured with increasing temperature; solid circles: measured with decreasing temperature, the cooling rate being too rapid for ordering. Curve *1*: sample cooled at 0.23 °C/h, *2*: sample cooled at 1.37 °C/h, both rates small enough to obtain a relatively high degree of ordering [53 W 1].

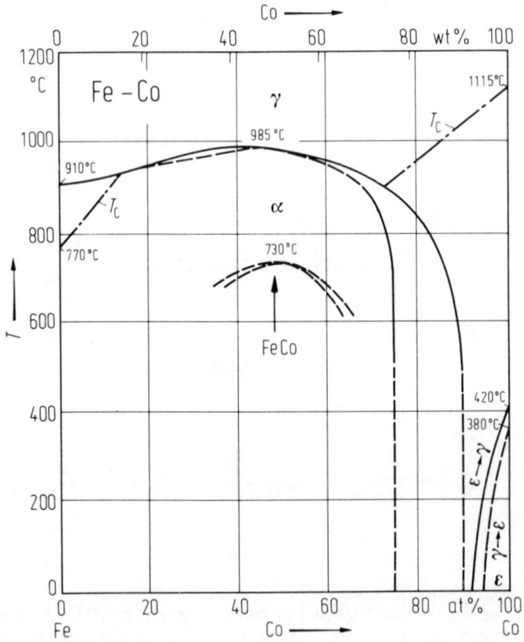

Fig. 28. Curie temperatures in the binary alloy system Fe–Co [58 H 1].

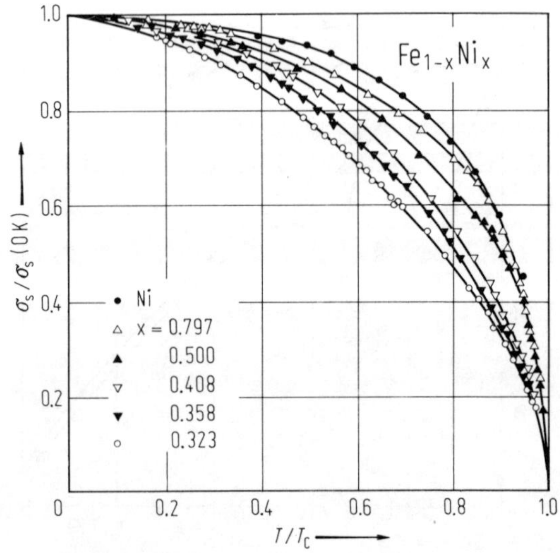

Fig. 29. Reduced spontaneous magnetic moment $\sigma_s/\sigma_s(0\,K)$ as a function of reduced temperature T/T_C for fcc alloys $Fe_{1-x}Ni_x$ [63 C 2].

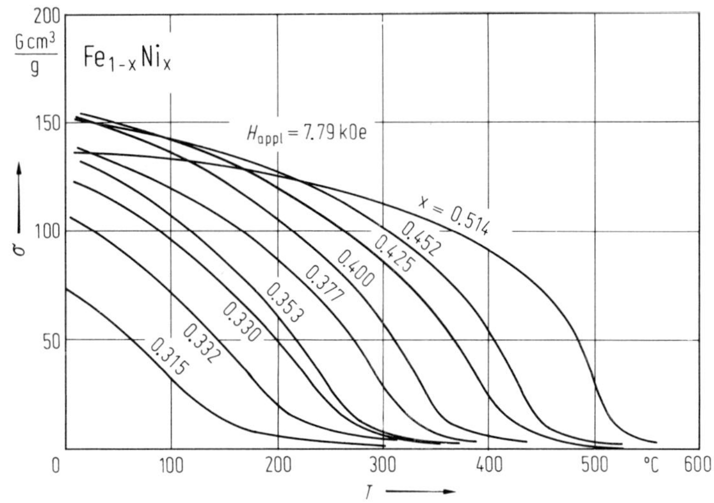

Fig. 30. Magnetic moment per gram σ in a magnetic field of 7.79 kOe measured on vacuum-annealed single crystals of $Fe_{1-x}Ni_x$ invar alloys [73 H 1].

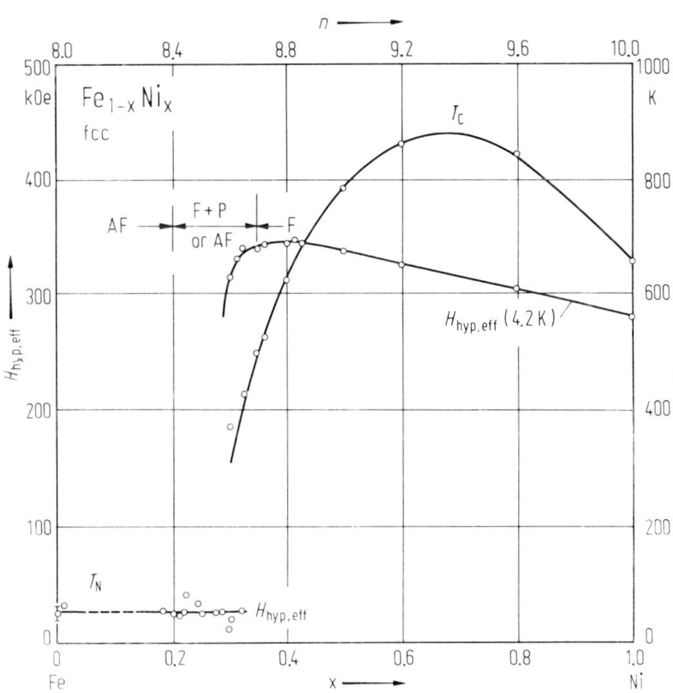

Fig. 31. Fcc magnetic phase diagram of the $Fe_{1-x}Ni_x$ alloys, showing the Curie temperature and the Néel temperature (T_N for $x \leq 0.22$), as well as the effective hyperfine fields $H_{hyp,eff}$ for ^{57}Fe at 4.2 K. n indicates the number of 4s and 3d electrons per atom [79 G 1].

2.1.7 High-field susceptibility

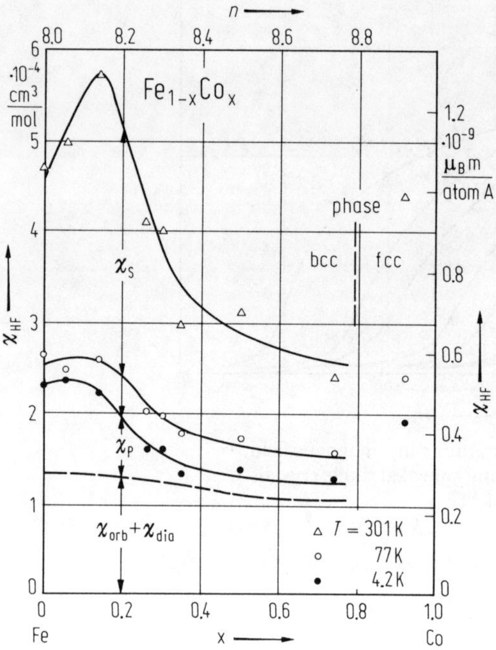

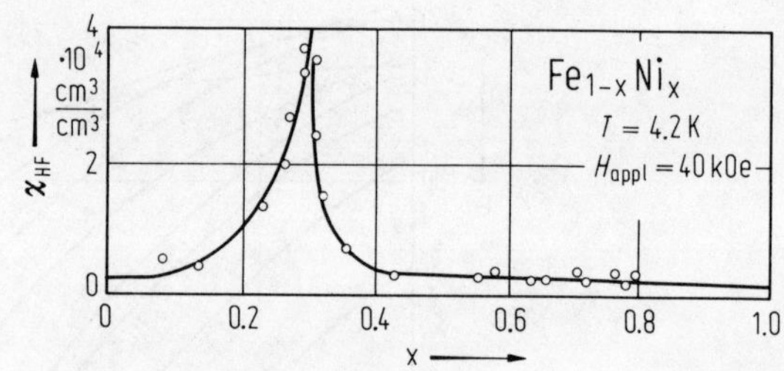

Fig. 32. High-field susceptibility χ_{HF} of $Fe_{1-x}Ni_x$ alloys, measured in a magnetic field of 40 kOe at liquid helium temperature [79 R 1, 72 R 1, 74 R 1].

Fig. 33. High-field susceptibility χ_{HF} for $Fe_{1-x}Co_x$ alloys [66 S 1, 69 S 2]. *n*: number of 4s and 3d electrons per atom, χ_s: spin wave contribution, χ_p: Pauli-paramagnetism contribution, χ_{orb}: orbital contribution, χ_{dia}: total diamagnetic contribution.

2.1.8 Magnetocrystalline anisotropy

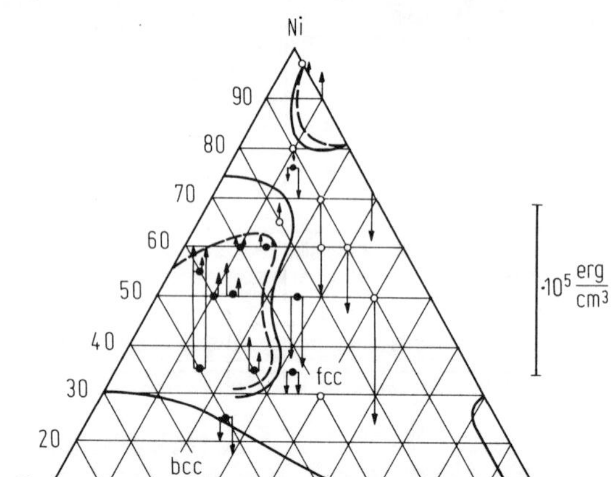

Fig. 34. Survey of the room-temperature magnetocrystalline anisotropy energy of fcc crystals of Fe–Co–Ni alloys. Arrows indicate the anisotropy constant K_1 of the various alloys. Arrow up denotes positive anisotropy. Solid circles: [63 P 1], open circles: [37 M 1]. The lefthand and righthand arrows apply to, respectively, quenched and annealed samples of the same composition. Solid and broken lines indicate the boundaries separating positive and negative anisotropy fields for quenched and annealed samples, respectively [63 P 1]. For earlier results, see [37 M 1], where a similar survey for room-temperature values and the values at 200 °C are given.

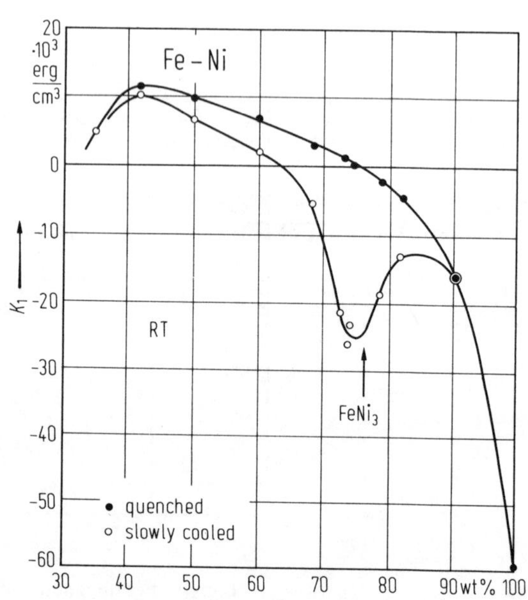

Fig. 35. Magnetocrystalline anisotropy constant K_1 vs. composition for fcc Fe–Ni alloys at room temperature. Cooling rate between 600 and 300 °C either 10^5 °C/h (quenched), or 2.5 °C/h (slowly cooled) [53 B 1].

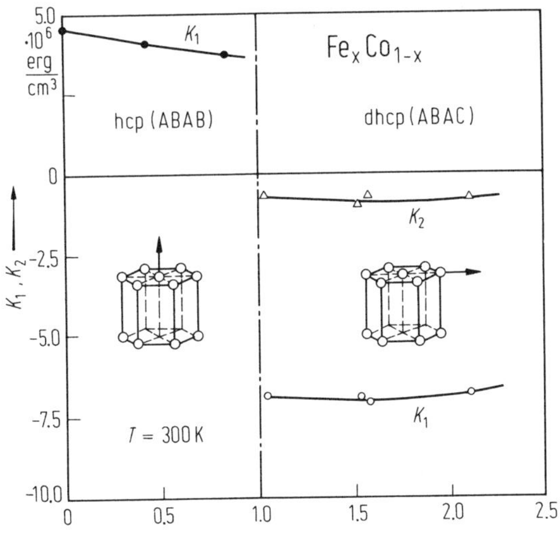

a

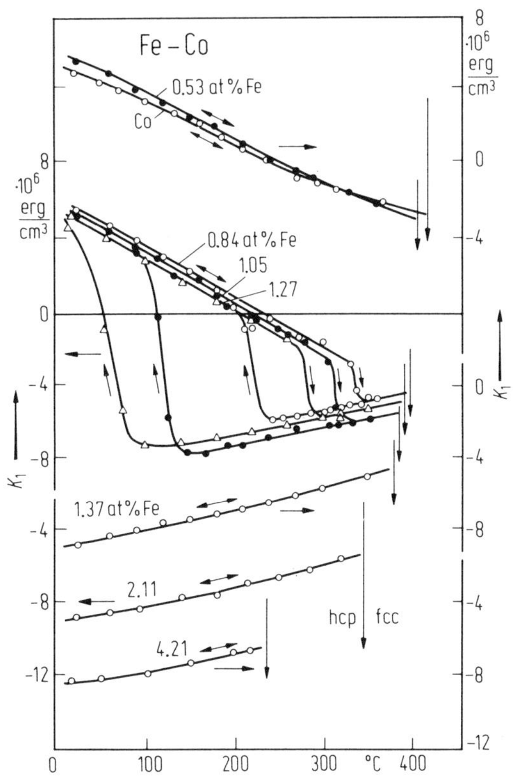

b

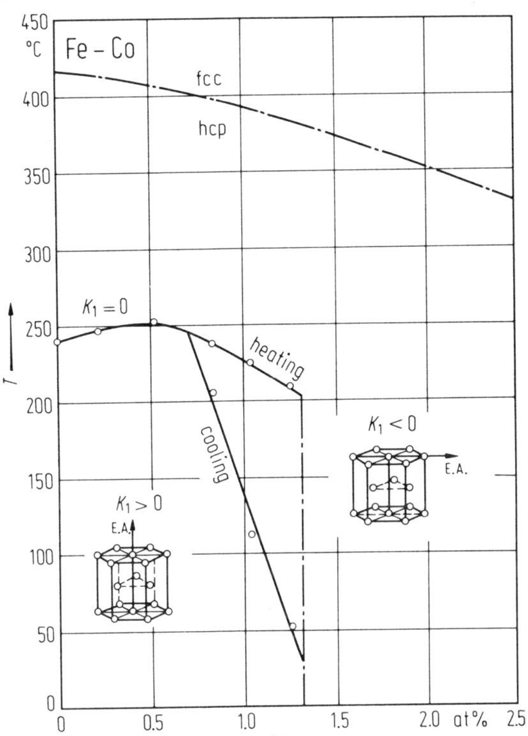

c

Fig. 36a. Room-temperature values of the magnetocrystalline anisotropy constants K_1 and K_2 for hexagonal Fe_xCo_{1-x} alloys in hcp (ABAB) and dhcp (ABAC) configuration [64 C 1, 73 W 1]. From the calculation of the anisotropy constants from the energy bands of the electrons in the dhcp configuration it is found that $K_1 = -8.3 \cdot 10^6\ \mathrm{erg\,cm^{-3}}$ and $K_2 = -1.1 \cdot 10^6\ \mathrm{erg\,cm^{-3}}$ [74 M 3]. See also [83 M 1].

Fig. 36b. Magnetocrystalline anisotropy constant K_1 vs. temperature T for hcp Fe–Co alloys. The inversibility with respect to the change of temperature is indicated by the arrows along the curves. Phase transitions are indicated by vertical arrows [80 T 1].

Fig. 36c. Diagram showing the direction of the easy axis (E.A.) of magnetization for hcp Fe–Co alloys [80 T 1].

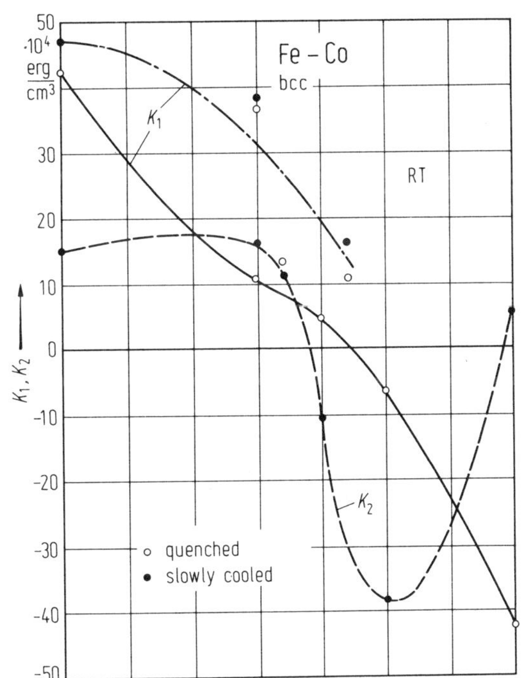

Fig. 37. Magnetocrystalline anisotropy constants K_1 and K_2 of bcc Fe–Co alloys at room temperature [62 L 1, 36 S 1, 37 M 1]. Symbols: results obtained by [59 H 1].

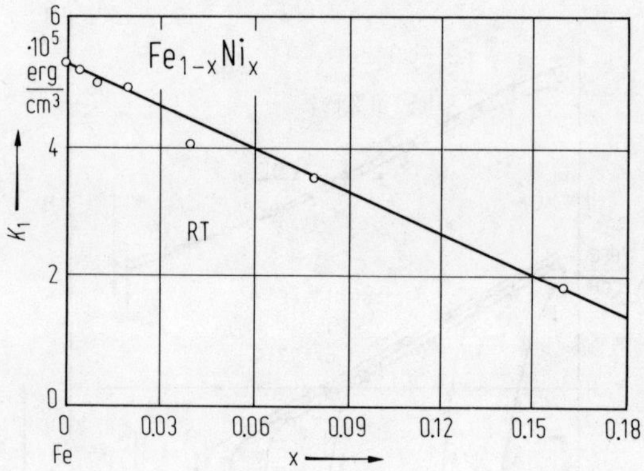

Fig. 38. Magnetocrystalline anisotropy constant K_1 vs. composition for bcc $Fe_{1-x}Ni_x$ alloys at room temperature [39 T 1].

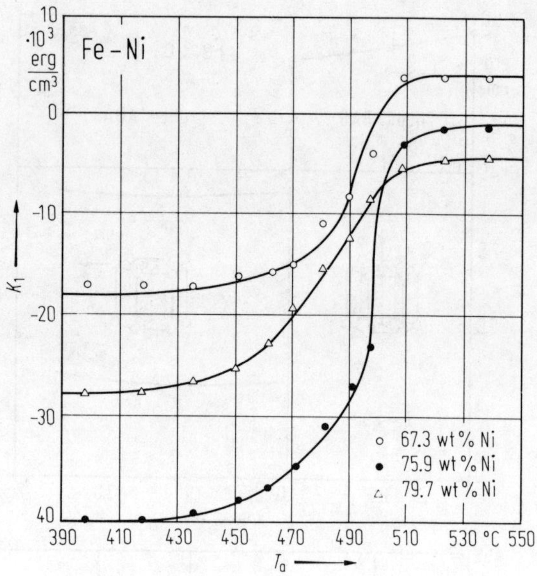

Fig. 39. Room-temperature first order magnetocrystalline anisotropy constant K_1 vs. annealing temperature T_a for various $FeNi_3$-type alloys. The samples were in the perfectly ordered state before annealing for 1 h in sequence at each T_a. For $T_a = 510\,°C$ the samples become disordered [83 H 1].

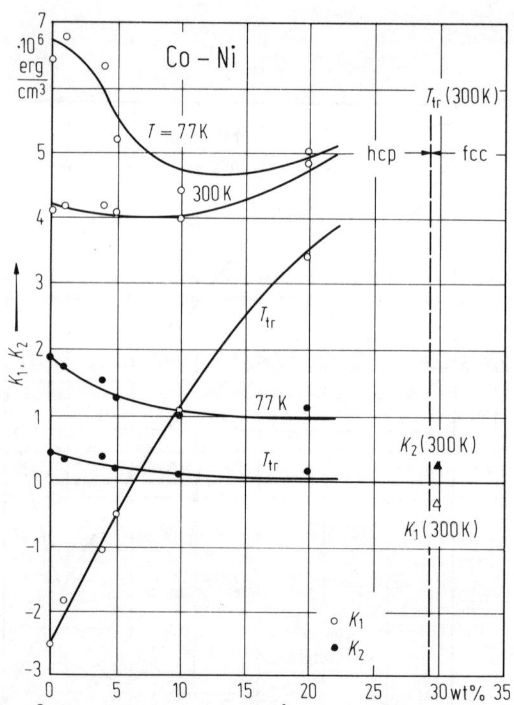

Fig. 40. First and second order uniaxial magnetocrystalline anisotropy constants K_1 and K_2 vs. Ni concentration in hcp Co–Ni alloys at 77 K, 300 K and at the temperature T_{tr}, just below the temperature of the ε(hcp) → γ(fcc) phase transformation [78 T 1].

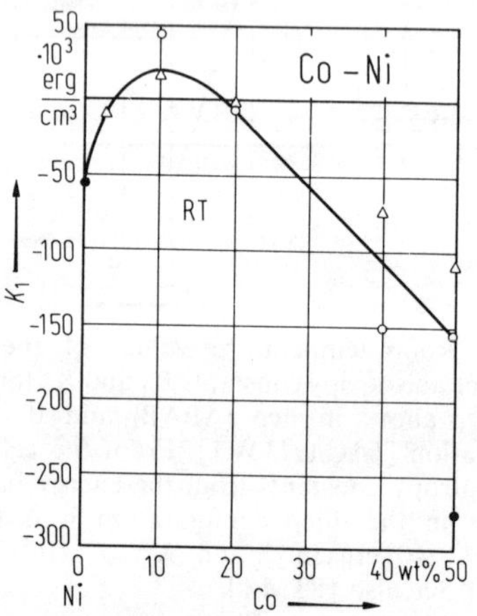

Fig. 41. Magnetocrystalline anisotropy constant K_1 of fcc Co–Ni alloys at room temperature. Solid circles: [59 H 1], open circles: [36 S 1], triangles: [37 M 1].

2.1.9 Magnetostriction

Table 13. Linear forced magnetostriction constants h'_0 of polycrystalline $Fe_{1-x}Co_x$ alloys [65 S 1].

T [K]	h'_0 [10^{-10} Oe^{-1}]								
	x: 0	0.054	0.164	0.262	0.303	0.35	0.503	0.745	0.925 [1])
293	1.5(1)	1.7(1)	1.9(1)	1.4(1)	1.1(1)	0.8(1)	0.8(1)	0.9(1)	1.7(1)
77	1.5(1)	1.7(1)	1.8(1)	1.2(1)	0.8(1)	0.7(1)	0.7(1)	0.8(1)	1.7(1)
1.5	1.5(1)	1.6(1)	1.8(1)	1.1(1)	0.6(1)	0.6(1)	0.8(1)	0.8(1)	1.6(1)

[1]) X-ray analysis for the γ-phase.

Table 14. Room-temprature values of the linear magnetostriction constants for hcp and dhcp Fe–Co alloys, defined by the equation:

$$\lambda = \lambda_1^{\alpha 2}(\beta_1^2+\beta_2^2)(\alpha_3^2-\tfrac{1}{3}) + \lambda_2^{\alpha 2}\beta_3^2(\alpha_3^2-\tfrac{1}{3}) + \lambda^{\gamma 2}\{\tfrac{1}{2}(\beta_1^2-\beta_2^2)(\alpha_1^2-\alpha_2^2)+2\beta_1\beta_2\alpha_1\alpha_2\} + 2\lambda^{\varepsilon 2}(\beta_1\alpha_1+\beta_2\alpha_2)\alpha_3\beta_3,$$

where α_i and β_i are the direction cosines of the magnetization and of the measured change in length, respectively, see [65 C 3, 84 I 1] and also [63 C 3]. The constants are only weakly dependent on composition.

	$10^6 \cdot \lambda_1^{\alpha 2}$	$10^6 \cdot \lambda_2^{\alpha 2}$	$10^6 \cdot \lambda^{\gamma 2}$	$10^6 \cdot \lambda^{\varepsilon 2}$
hcp	78	−134	*)	−234
dhcp	28	− 85	*)	− 51

*) $\lambda^{\gamma 2}$ gradually decreases with increasing Fe content, without an appreciable change at the phase transition (see Fig. 42).

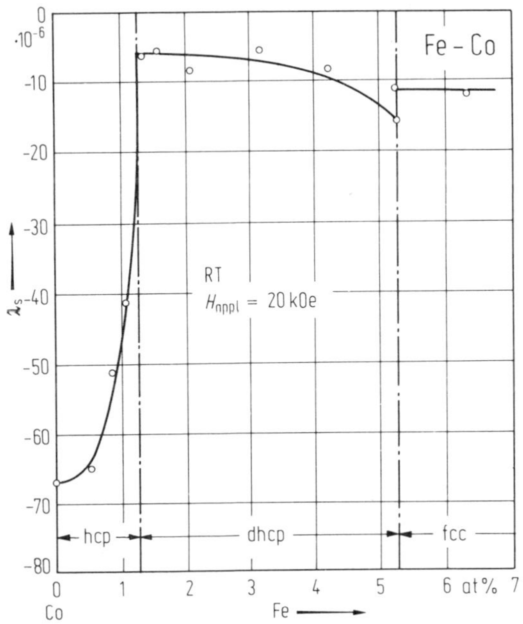

Fig. 42. Dependence of the saturation magnetostriction $\lambda_s = \frac{2}{3}(\lambda_\parallel - \lambda_\perp)$ on Fe concentration in polycrystalline Fe–Co alloys at room temperature, measured in an applied magnetic field of 20 kOe, $\lambda_\parallel$ and $\lambda_\perp$ are the magnetostrictions parallel and perpendicular to the field direction, respectively [84 I 1].

Table 15. Linear forced magnetostriction constants h'_0 as in Table 13, but now for polycrystalline $Fe_{1-x}Ni_x$ alloys [65 S 1].

T [K]	h'_0 [10^{-10} Oe^{-1}]			
	x: 0	0.094	0.192	0.287
293				56(5) [1])
293	1.5(1)	3.3(1)	3.3(1)	8.0(1) [2])
234				38(2) [1])
173				20(2) [1])
77	1.5(1)	3.1(1)	3.6(1)	8.0(2) [2])
1.5	1.5(1)	3.1(1)	3.6(1)	8.0(2) [2])

[1]) bcc + fcc. [2]) bcc.

Table 16. Linear saturation magnetostriction constant λ_s of polycrystalline Fe–Ni alloys [05 H 1].

wt% Ni	$\lambda_s \cdot 10^6$	
	T = 20 °C	−186 °C
36	20.3	30.5
46	25.4	30.7
50	24.3	26.8
70	11.6	12.6

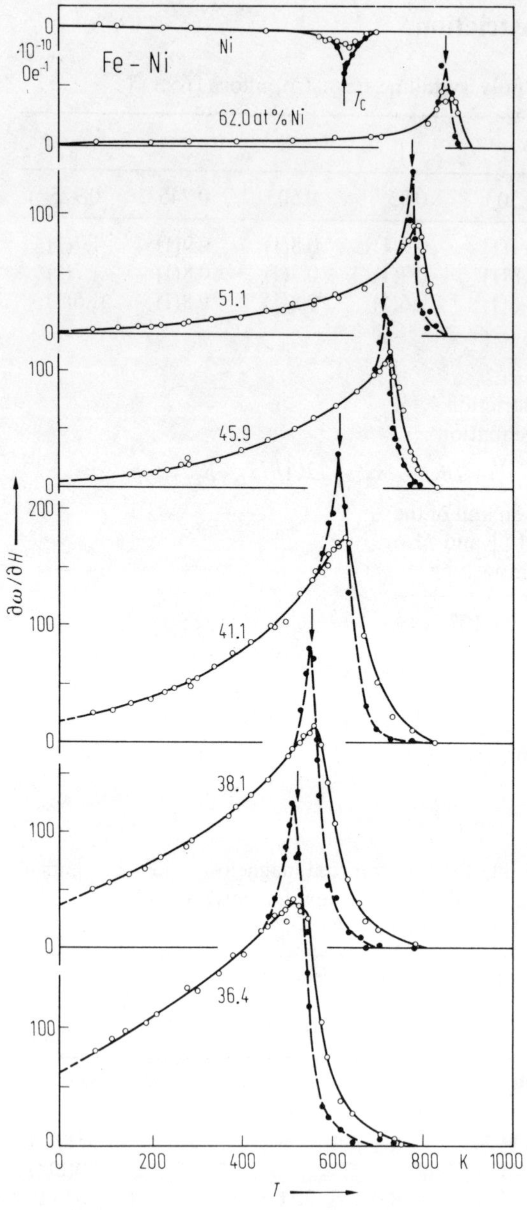

Fig. 43. Forced volume magnetostriction $\partial\omega/\partial H$ vs. temperature T for Fe–Ni alloys. Solid circles: derived from low-field measurements, $0 < H < 6$ kOe, open circles: derived from higher-field measurements, $10 < H < 20$ kOe. The arrows indicate the Curie temperatures derived from Arrott plots [85 I 1], see also [84 Y 1].

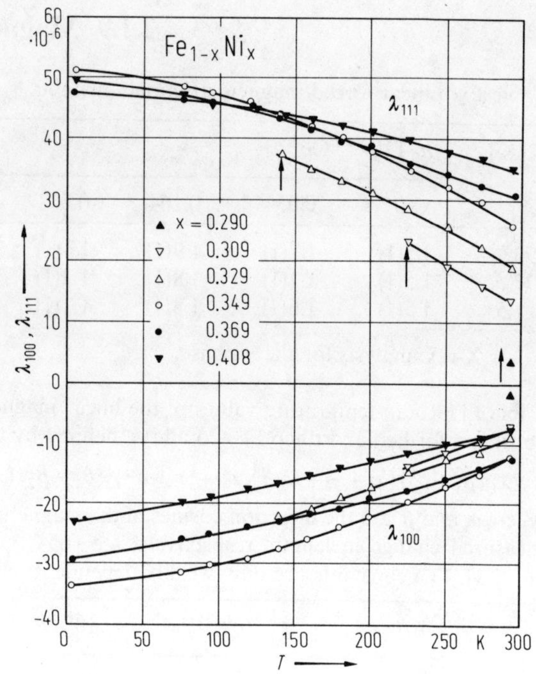

Fig. 44. Temperature dependence of the linear magnetostriction coefficients, λ_{100} and λ_{111}, as a function of the composition of single crystals of the Fe–Ni invar alloys. The arrows indicate the temperature of martensitic transformation fcc to bcc [78 K 1].

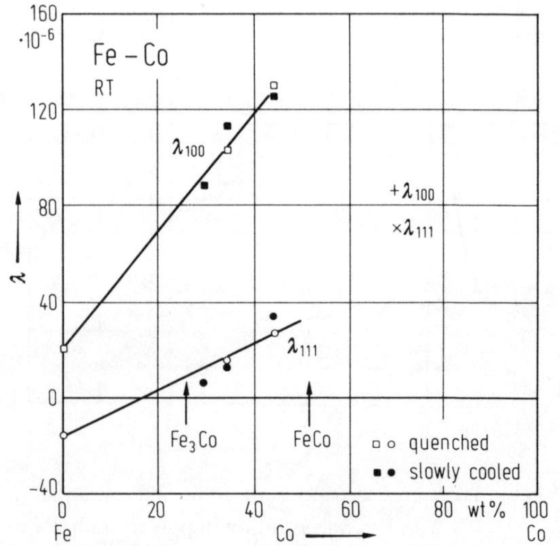

Fig. 45. Magnetostriction constants λ_{100} and λ_{111} of fcc Fe–Co alloys at room temperature [62 K 1]. Open symbols: [59 H 1], solid symbols: [52 U 1], crosses: [62 K 1].

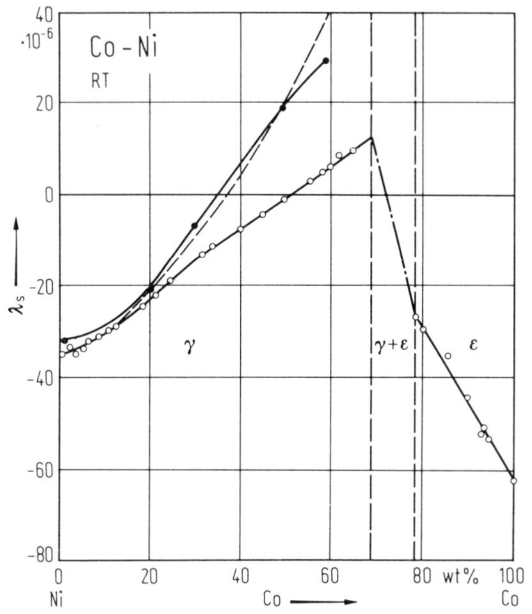

Fig. 46. Linear saturation magnetostriction constant of polycrystalline Co–Ni alloys at room temperature [62 L 1]. Open circles: [53 Y 1], solid circles: [51 W 1]. Dashed curve: calculated from data depicted in Fig. 47, $\lambda_s = (2\lambda_{100} + 3\lambda_{111})/5$.

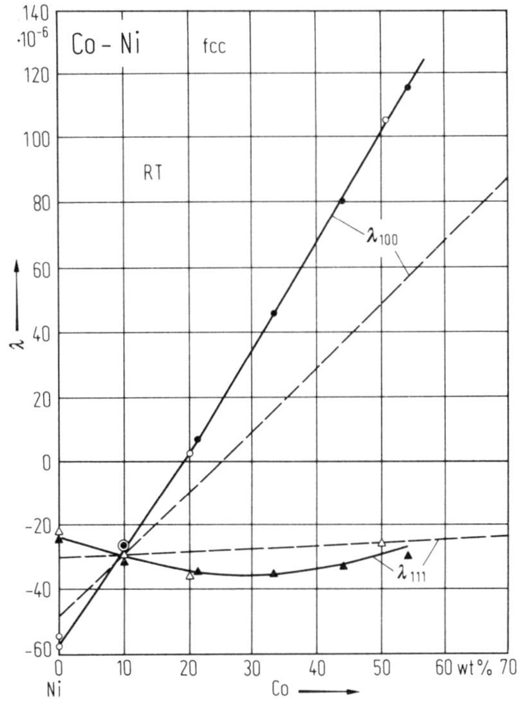

Fig. 47. Magnetostriction constants λ_{100} and λ_{111} of fcc Co–Ni alloys at room temperature [62 K 1]. Crosses: [58 Y 1], circles and triangles: [59 H 1], dashed line: [53 Y 1].

2.1.10 References for 2.1

05 H 1 Honda, K., Shimizu, S.: Philos. Mag. **10** (1905) 548.
25 P 1 Peschard, M.: Rev. Met. Paris **8** (1925) 490.
27 K 1 Kasé, T.: Sci. Rept. Tohoku Univ. **16** (1927) 491.
28 E 1 Elmen, G.W.: J. Franklin Inst. **206** (1928) 317.
29 E 1 Elmen, G.W.: J. Franklin Inst. **207** (1929) 583.
29 M 1 Masumoto, H.: Sci. Rept. Tohoku Univ. **18** (1929) 195.
29 W 1 Weiss, P., Forrer, R.: Ann. Phys. Paris **12** (1929) 279.
36 F 1 Fallot, M.: Ann. Phys. Paris **6** (1936) 305.
36 S 1 Shih, J.W.: Phys. Rev. **50** (1936) 376.
37 M 1 McKeehan, L.W.: Phys. Rev. **51** (1937) 136.
39 T 1 Tarasov, L.P.: Phys. Rev. **56** (1939) 1245.
41 E 1 Ellis, W.C., Greiner, E.S.: Trans. Am. Soc. Met. **29** (1941) 415.
43 F 1 Fallot, M.: Metaux Corrosion-Ind. **18** (1943) 214.
44 B 1 Barnett, S.J.: Phys. Rev. **66** (1944) 224.
50 T 1 Taylor, A.: J. Inst. Metals **77** (1950) 585.
51 W 1 Went, J.J.: Physica **17** (1951) 98.
52 B 1 Barnett, S.J., Kenny, G.S.: Phys. Rev. **87** (1952) 723.
52 U 1 Urquhart, H.M.A., Goldman, J.E.: Phys. Rev. **87** (1952) 210.
53 B 1 Bozorth, R.M., Walker, J.G.: Phys. Rev. **89** (1953) 624.
53 W 1 Wakelin, R.J., Yates, E.L.: Proc. Phys. Soc. (London) Sect. B **66** (1953) 221.
53 Y 1 Yamamoto, M., Misyasawa, R.: Sci. Rept. Tohoku Univ. A **5** (1953) 113.
55 S 1 Shull, C.G., Wilkinson, M.K.: Phys. Rev. **97** (1955) 304.

56 N 1 Nakagawa, Y.: J. Phys. Soc. Jpn. **11** (1956) 855.
56 S 1 Scott, G.G.: Phys. Rev. **103** (1956) 561.
58 H 1 Hansen, M.: Constitution of binary alloys, New York: McGraw-Hill (1958) 319, 486, 677.
58 Y 1 Yamamoto, M., Nakamichi, T.: J. Phys. Soc. Jpn. **13** (1958) 228.
59 H 1 Hall, R.C.: J. Appl. Phys. **30** (1959) 816.
60 W 1 Wertheim, G.K.: Phys. Rev. Lett. **4** (1960) 403.
61 J 1 Johnson, C.E., Ridout, M.S., Cronshaw, T.E., Madson, P.E.: Phys. Rev. Lett. **6** (1961) 450.
61 M 1 Meyer, A.J.P., Asch, G.: J. Appl. Phys. **32** (1961) 330 S.
61 O 1 Ondris, M., Frait, Z.: Czech. J. Phys. B **11** (1961) 88.
62 C 1 Chechernikov, V.I.: Sov. Phys. JETP **15** (1962) 659.
62 C 2 Collins, M.F., Jones, R.V., Lowde, R.D.: J. Phys. Soc. Jpn. **17** (1962) 19.
62 K 1 Kneller, E.: Ferromagnetism, Berlin, Göttingen, Heidelberg: Springer **1962**, p. 148.
62 L 1 Landolt-Börnstein, 6th ed.: Magnetische Eigenschaften I (Hellwege, K.-H., Hellwege, A.M., eds.), Berlin, Göttingen, Heidelberg: Springer, vol. II/9 **1962**.
62 S 1 Shull, C.G., Yamada, Y.: J. Phys. Soc. Jpn. **17** (1962) 1.
63 C 1 Collins, M.F., Forsyth, J.B.: Philos. Mag. **8** (1963) 401.
63 C 2 Crangle, J., Hallam, G.C.: Proc. R. Soc. London Ser A **272** (1963) 119.
63 C 3 Callen, E.R., Callen, H.B.: Phys. Rev. **129** (1963) 578.
63 J 1 Johnson, C.E., Ridout, M.S., Cranshaw, T.E.: Proc. Phys. Soc. (London) **81** (1963) 1079.
63 P 1 Puzey, I.M.: Phys. Met. Metallogr. (USSR) **16** (1963) 19.
63 W 1 Weiss, R.J.: Proc. Phys. Soc. (London) **82** (1963) 281.
64 C 1 Chikazumi, S., Wakiyama, K., Yosida, K.: Proc. Int. Conf. Magn. **1964**, p. 756.
64 G 1 Goman'kow, V.I., Loshmanov, A.A.: Bull. Acad. Sci. USSR, Phys. Series **28** (1964) 357.
64 H 1 Hatherly, M., Hirakawa, K., Lowde, R.D., Mallett, J.F., Stringfellow, M.W., Torrie, B.H.: Proc Phys. Soc. **84** (1964) 55.
64 N 1 Nakamura, Y., Shiga, M., Shikazono, N.: J. Phys. Soc. Jpn. **19** (1964) 1177.
65 C 1 Collins, M.F.: Proc. Phys.Soc. **86** (1965) 973.
65 C 2 Cable, J.W., Wollan, E.O., Koehler, W.C.: Phys. Rev. **138** (1965) A 755.
65 C 3 Clark, A.E., DeSavage, B.F., Bozorth, R.: Phys. Rev. **138** (1965) A 216.
65 S 1 Stoelinga, J.H.M., Gersdorf, R., De Vries, G.: Physica **31** (1965) 349.
66 M 1 Mook, H.A.: Phys. Rev. 2. Ser. **148** (1966) 495.
66 S 1 Stoelinga, J.H., Gersdorf, R.: Phys. Lett. **19** (1966) 640.
68 A 1 Asano, H.: J. Phys. Soc. Jpn. **25** (1968) 286.
69 B 1 Bardos, D.I.: J. Appl. Phys. **40** (1969) 1371.
69 R 1 Reck, R.A., Fry, D.L.: Phys. Rev. **184** (1969) 492.
69 S 1 Scott, G.G., Sturner, H.W.: Phys. Rev. **184** (1969) 490.
69 S 2 Stoelinga, J.H.M., Gersdorf, R., Vries de, G.: Physica **41** (1969) 457.
70 A 1 Antonini, B., Lucari, F., Menzinger, F.: Solid State Commun. **8** (1970) 1.
70 D 1 Dobrzynski, L., Maniawski, F., Modrzejewski, A., Sikorska, D.: Phys. Status Solidi **38** (1970) 103.
70 M 1 Mayo de, B., Forester, D.W., Spooner, S.: J. Appl. Phys. **41** (1970) 1319.
71 A 1 Antonini, B., Menzinger, F., Paoletti, A., Sacchetti, F.: Int. J. Magn. **1** (1971) 183.
71 H 1 Hasegawa, H., Kanamori, J.: J. Phys. Soc. Jpn. **31** (1971) 382.
72 H 3 Hasegawa, H., Kanamori, J.: J. Phys. Soc. Jpn. **33** (1972) 1599.
72 H 1 Hausch, G., Warlimont, H.: Z. Metallkd. **63** (1972) 547.
72 H 2 Hinoul, M., Witters, J.: Solid State Commun. **10** (1972) 749.
72 M 1 Menzinger, F., Paoletti, A.: Il nuovo Cimento **10** B (1972) 565.
72 R 1 Rode, V.E., Krynetskaya, I.B.: Sov. Phys. Solid State **14** (1972) 533.
72 S 1 Shiga, M.: IEEE Trans. Magn. (1972) 666.
73 C 1 Cable, J.W., Wollan, E.O.: Phys. Rev. B **7** (1973) 2005.
73 D 1 Dubinin, S.F., Sidorov, S.K., Teploukhov, S.G., Arkhipov, V.E.: JETP Lett. **18** (1973) 324.
73 H 1 Hausch, G., Warlimont, H.: Acta Metall. **21** (1973) 401.
73 M 1 Maedo, T., Yamauchi, H., Watanabe, H.: J. Phys. Soc. Jpn. **35** (1973) 1635.
73 P 1 Pokatilov, V.S., Puzei, I.M.: Sov. Phys. JETP **36** (1973) 108.
73 W 1 Wakiyama, T.: AIP Conf. Proc. Mag. Magn. Mater. **2** (1973) 921.
73 W 2 Window, B.: J. Appl. Phys. **44** (1973) 2853.
74 C 1 Crowell, J.M., Walker, J.C.: J. Mag. Magn. Mater. **2** (1974) 427.
74 M 1 Miwa, H.: Progress of Theor. Phys. **52** (1974) 1.

74 M 2 Menzinger, F., Sacchetti, F., Leoni, F.: Il Nuovo Cimento **20** B (1974) 1.
74 M 3 Mori, N., Ukai, T., Kono, S.: J. Phys. Soc. Jpn. **37** (1974) 1278.
74 N 1 Nishi, M., Nakai, Y., Kunitomi, N.: J. Phys. Soc. Jpn. **37** (1974) 570.
74 O 1 Onozuka, T., Yamaguchi, S., Hirabayashi, M., Wakiyama, T.: J. Phys. Soc. Jpn. **37** (1974) 687.
74 R 1 Rode, U.Ye., Krynetskaya, I.B.: Phys. Met. Metallogr. (USSR) **38** (1974) 183.
74 W 1 Window, B.: J. Phys. F **4** (1974) 329.
75 B 1 Billard, L., Chamberod, A.: Solid State Commun. **17** (1975) 113.
75 H 1 Hennion, M., Hennion, B., Castets, A., Tochetti, D.: Solid State Commun. **17** (1975) 899.
75 M 1 Menshikov, A.Z., Kazantsev, V.A., Kuzmin, N.N., Sidorov, S.K.: J. Mag. Magn. Mater. **1** (1975) 91.
75 R 1 Riedinger, R., Nauciel-Bloch, M.: J. Phys. F **5** (1975) 732.
75 W 1 Wu, C.Y., Quach, H.T., Yelon, A.: AIP Conf. Proc. Mag. Magn. Mater. **29** (1975) 681.
76 B 1 Bastian, D., Biller, E.: Phys. Status Solidi (a) **35** (1976) 465.
76 M 1 Maeda, I., Yamauchi, H., Watanabe, H.: J. Phys. Soc. Jpn. **40** (1976) 1559.
77 B 1 Bessmertnyi, A.M., Mushailov, E.S., Pyn'ko, V.G., Suvorov, A.V.: Sov. Phys. Solid State **19** (1977) 1473.
77 C 1 Cullis, I.G., Heath, M.: Solid State Commun. **23** (1977) 891.
77 D 1 Drijver, J.W., Woude van der, F., Radelaar, S.: Phys. Rev. B **16** (1977) 985.
77 D 2 Drijver, J.W., Woude van der, F., Radelaar, S.: Phys. Rev. B **16** (1977) 995.
77 H 1 Hesse, J., Müller, J.B.: Solid State Commun. **22** (1977) 637.
77 S 1 Singer, V.V., Radovskiy, I.Z.: Russ. Metall. **1** (1977) 65.
78 B 1 Bansal, C.: Phys. Status Solidi (a) **48** (1978) K 119.
78 H 1 Hamada, N., Miwa, H.: Progress of Theor. Phys. **59** (1978) 1045.
78 K 1 Kim, C.-D., Matsui, M., Chikazumi, S.: J. Phys. Soc. Jpn. **44** (1978) 1152.
78 S 1 Shirakawa, Y., Tanji, Y.: Phys. and Appl. of Invar Alloys. Honda Mem. Series Mat. Science **3** (1978) 137.
78 T 1 Takahashi, M., Kadowaki, S., Wakiyama, T., Anayama, T., Takahashi, M.: J. Phys. Soc. Jpn. **44** (1978) 825.
79 C 1 Chikazumi, S.: J. Mag. Magn. Mater. **10** (1979) 113.
79 D 1 Deen van, J.K., Woude van der, F.: Phys. Rev. B **20** (1979) 296.
79 E 1 Endoh, Y.: J. Mag. Magn. Mater. **10** (1979) 177.
79 G 1 Gonser, Y., Nasu, S., Kappes, W.: J. Mag. Magn. Mater. **10** (1979) 244.
79 H 1 Hamada, N.: J. Phys. Soc. Jpn. **47** (1979) 797.
79 H 2 Hamada, N.: J. Phys. Soc. Jpn. **46** (1979) 1759.
79 I 1 Inone, J., Yamada, H., Shimizu, M.: J. Phys. Soc. Jpn. **46** (1979) 1496.
79 I 2 Ito, Y., Akimitsu, J., Matsui, M., Chikazumi, S.: J. Mag. Magn. Mater. **10** (1979) 194.
79 M 1 Miwa, H.: J. Mag. Magn. Mater. **10** (1979) 223.
79 R 1 Rode, V.E.: Phys. Status Solidi (a) **56** (1979) 407.
80 D 1 Dubinin, S.F., Teplouchov, S.G., Sidorov, S.K., Izyumov, Yu.A. Syromyatnikov, V.N.: Phys. Status Solidi (a) **61** (1980) 159.
80 H 1 Hennion, B., Hennion, M.: J. Phys. F **10** (1980) 2289.
80 L 1 Morán-López, J.L., Falicov, L.M.: J. Phys. C: Solid State Phys. **13** (1980) 1715.
80 T 1 Takahashi, M., Kadowaki, S.: J. Phys. Soc. Jpn. **48** (1980) 1391.
81 H 1 Harada, S., Sohmura, T., Fujita, F.E.: J. Phys. Soc. Jpn. **50** (1981) 2909.
81 K 1 Kakehashi, Y.: J. Phys. Soc. Jpn. **50** (1981) 2236.
82 C 1 Cable, J.W., Brundage, W.E.: J. Appl. Phys. **53** (1982) 8085.
82 K 1 Kakehashi, Y.: J. Phys. Soc. Jpn. **51** (1982) 3183.
83 D 1 Davies, M., Heath, M.: J. Mag. Magn. Mater. **31–34** (1983) 661.
83 H 1 Hatafuku, H., Takashashi, S., Sasaki, T., Ichinohe, H.: J. Mag. Magn. Mater. **31–34** (1983) 847.
83 K 1 Kakahashi, Y.: J. Mag. Magn. Mater. **31–34** (1983) 53.
83 K 2 Kakehashi, Y.: J. Mag. Magn. Mater. **37** 1983) 189.
83 M 1 Mori, N., Ukai, T., Oktsuka, S.: J. Mag. Magn. Mater. **31** (1983) 43.
83 N 1 Nakai, I.: J. Phys. Soc. Jpn. **52** (1983) 1781.
83 P 1 Pauthenet, R., Maruyama, H.: J. Mag. Magn. Mater. **31–34** (1983) 835.
84 I 1 Ishio, S., Takahashi, M.: J. Mag. Magn. Mater. **46** (1984) 142.
84 Y 1 Yamada, O., Du Tremolet De Lacheisserie, E.: J. Phys. Soc. Jpn. **53** (1984) 729.
85 I 1 Ishio, S., Takahashi, M.: J. Mag. Magn. Mater. **50** (1985) 271.

2.2 Alloys between Ti, V, Cr or Mn

2.2.1 V–Ti

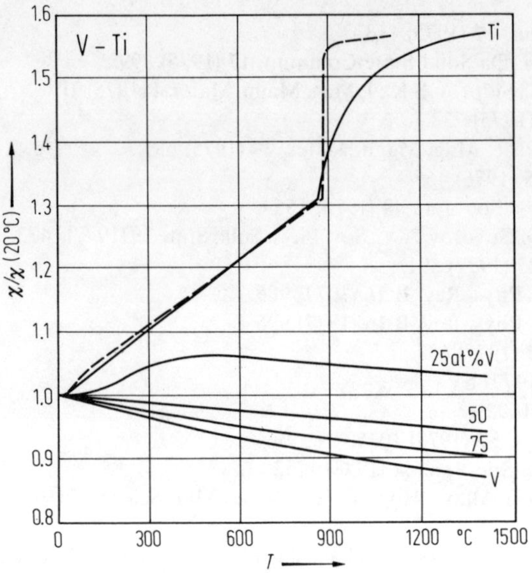

Fig. 1. Temperature dependence of the relative magnetic susceptibility for V–Ti alloys [62 T 1]. The broken line represents data of McQuillan and Evans (1960) for Ti.

2.2.2 Cr–Ti

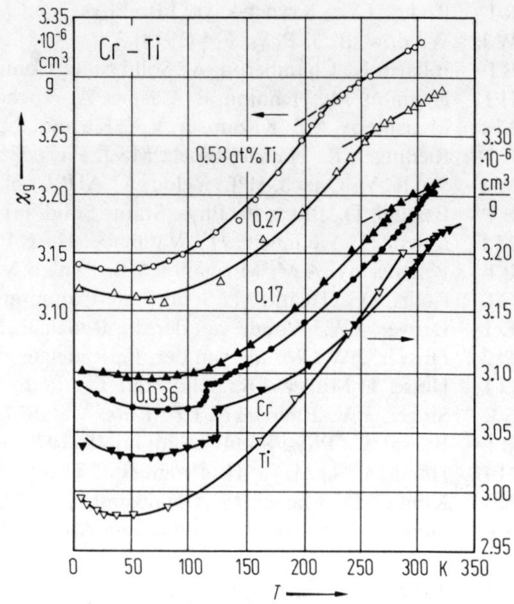

Fig. 2. Magnetic mass susceptibility χ_g vs. temperature for Cr–Ti alloys [71 C 1].

2.2.3 Cr–V

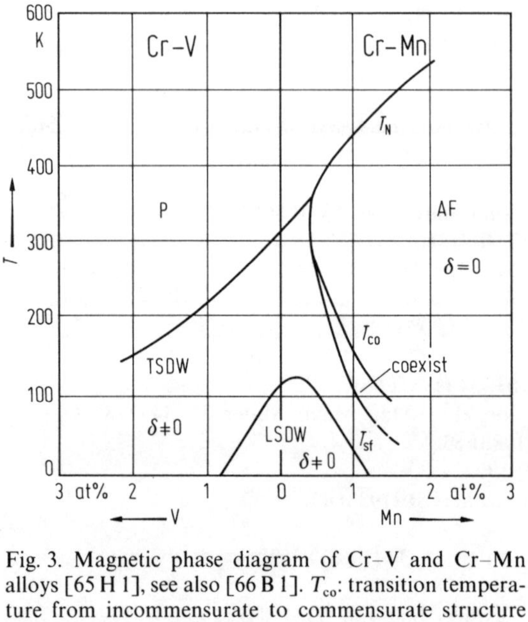

Fig. 3. Magnetic phase diagram of Cr–V and Cr–Mn alloys [65 H 1], see also [66 B 1]. T_{co}: transition temperature from incommensurate to commensurate structure (see Fig. 4), T_{sf}: spin flip transition temperature from longitudinal (L) to transverse (T) spin density wave (SDW) state, $\delta=0$: commensurate phase, $\delta \neq 0$: incommensurate phase, P: paramagnetic, AF: antiferromagnetic.

Table 1. Néel temperature T_N, average magnetic moment per atom, $\bar{p}$, and wavevector Q of the spin density waves in Cr–V and Cr–Mn alloys at 77 K [67 K 1].

	V	Mn	T_N	$\bar{p}$	$Qa/2\pi$
	at%		K	μ_B	
Cr [1])	–	–	310	0.40(2)	0.9518
Cr–V	0.45	–	268(5)	0.36(3)	0.9431(25)
	1.00	–	220(5)	0.28(3)	0.9300(25)
Cr–Mn	–	0.70	440(5)		
	–	1.85	545(5)		

[1]) [62 W 1].

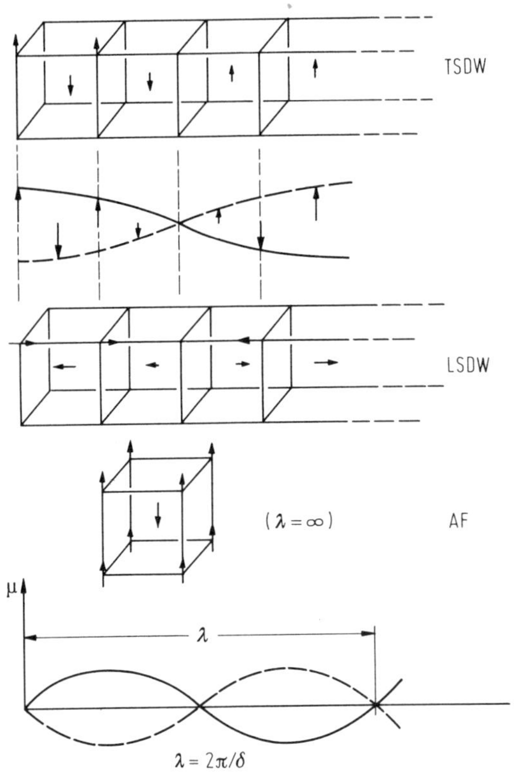

Fig. 4. Spin structure of Cr and Cr-based alloys: TSDW: transverse spin density wave (incommensurate), LSDW: longitudinal spin density wave (incommensurate), AF: ordinary antiferromagnetic structure (commensurate), λ: wavelength of SDW ($\lambda = 2\pi/\delta$), cf. caption to Fig. 5.

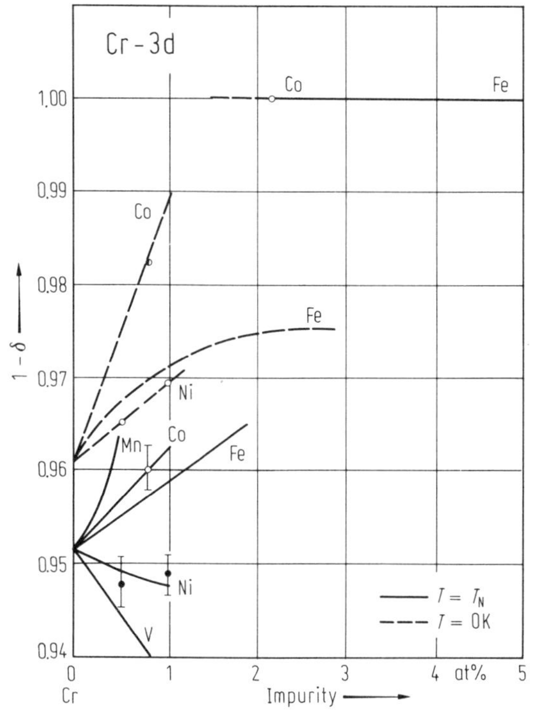

Fig. 5. Variation of $1-\delta$ for Cr alloys containing V, Mn, Fe, Co and Ni impurities at T_N (dashed curve) and at 0 K (solid curve) [68 E 1]. The satellite of the magnetic reflection appears at $\frac{2\pi}{a}(1 \pm \delta, 0, 0)$, etc. The wavelength of the SDW is $\lambda = 2\pi/\delta$, see Fig. 4.

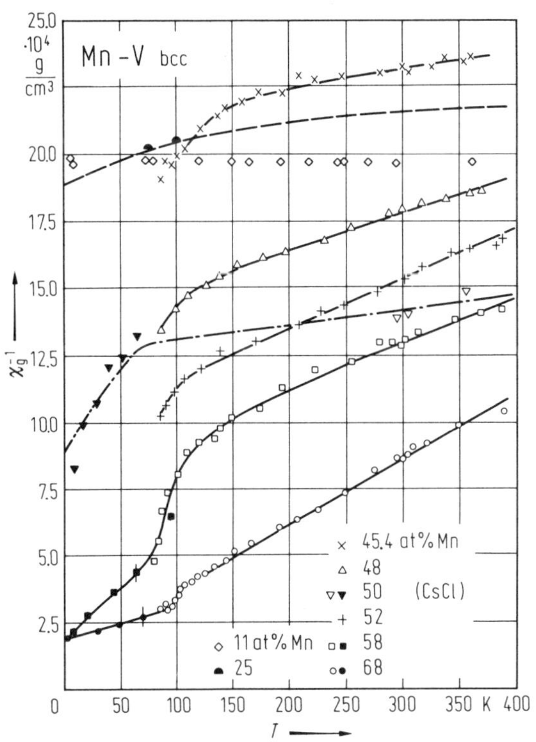

2.2.4 Mn–V

Fig. 6. Inverse magnetic mass susceptibility χ_g^{-1} vs. temperature for bcc Mn–V alloys. Solid symbols: high-field extrapolation values, open triangles (downward): $V_{0.5}Mn_{0.5}$, ordered CsCl-type [71 V 1].

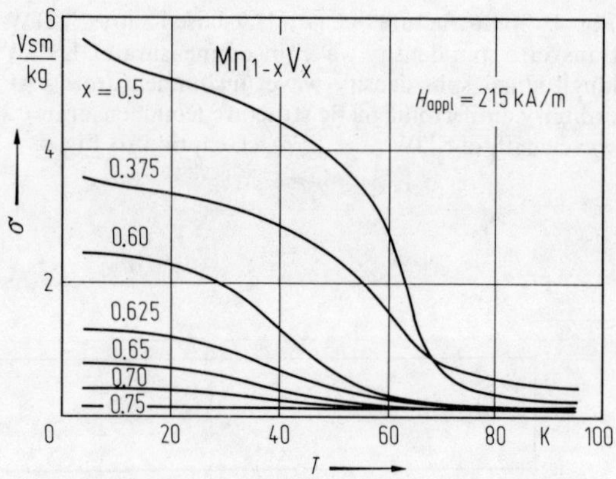

Fig. 7. Magnetic moment σ for $Mn_{1-x}V_x$ alloys in an applied magnetic field of 215 kA/m as a function of temperature [80 M 1].

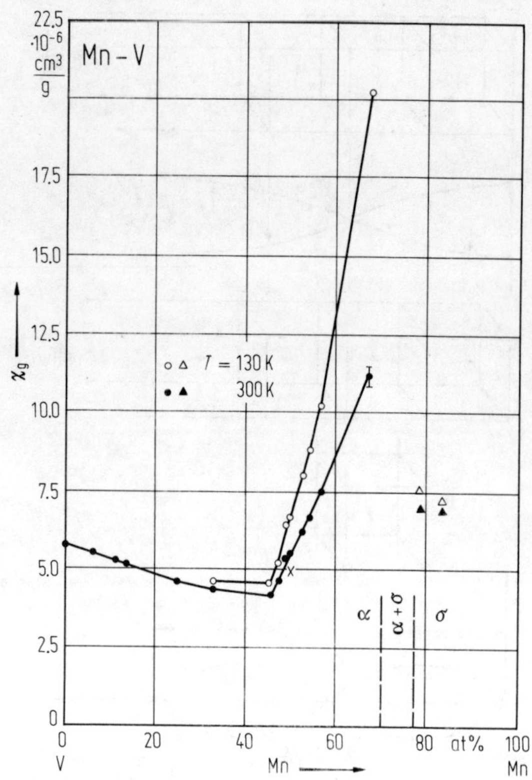

Fig. 8. Magnetic mass susceptibility χ_g for Mn–V alloys at 130 K (open symbols) and 300 K (solid symbols). Circles: bcc phase, triangles: σ-phase, cross: ordered CsCl-type [71 V 1].

2.2.5 Mn–Cr

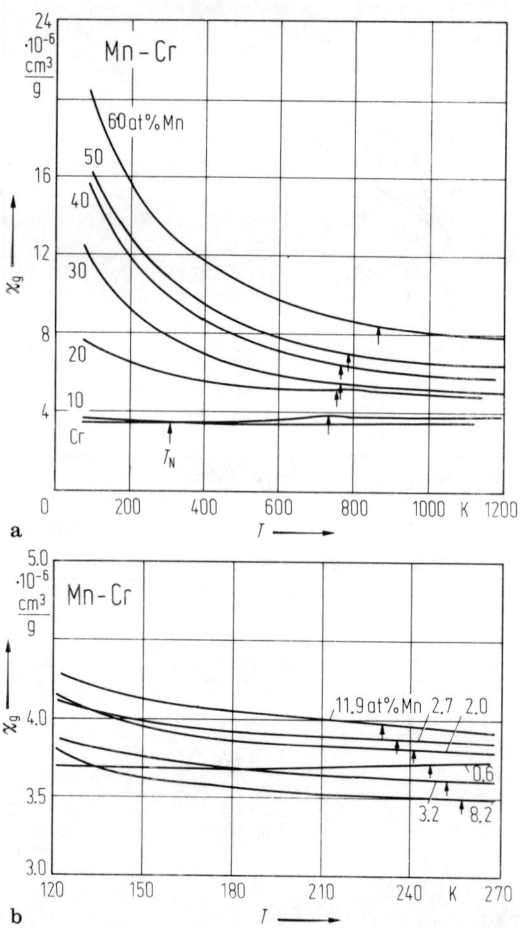

Fig. 9. Magnetic mass susceptibility χ_g vs. temperature for Mn–Cr alloys. (a) [79 M 1], see also [62 T 1], (b) [66 S 1]. The arrows indicate Néel temperatures determined by electrical resistivity measurements.

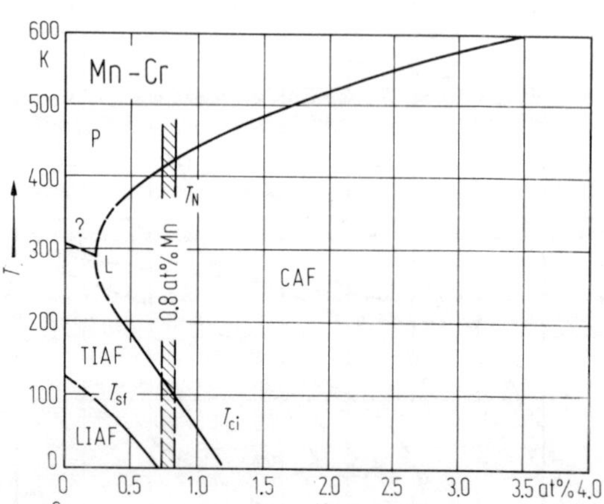

Fig. 10. Magnetic phase diagram of dilute Mn–Cr system [81 P 1]. Solid lines: obtained by neutron diffraction [66 K 1], dotted lines: guess based on [80 H 1], shaded area: [81 P 1]. P: paramagnetic phase, CAF: commensurate antiferromagnetic phase, TIAF: transverse incommensurate antiferromagnetic phase, LIAF: longitudinal incommensurate antiferromagnetic phase, L: Lifshitz point.

2.3 Alloys of Fe, Co or Ni and Ti, V, Cr or Mn

2.3.1 Fe–Ti

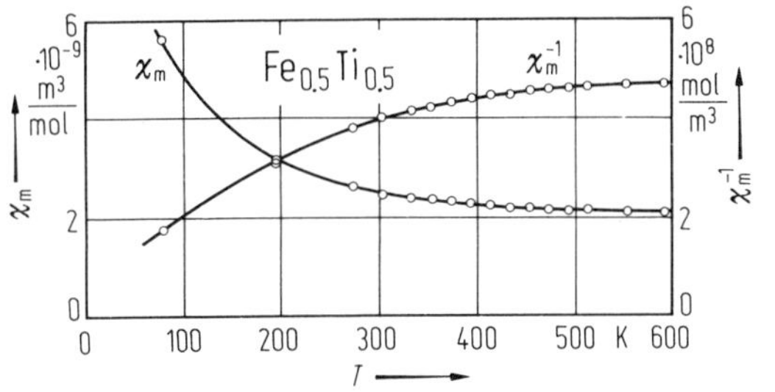

Fig. 11. Magnetic molar susceptibility χ_m and its inverse value, χ_m^{-1}, as a function of temperature for $Fe_{0.5}Ti_{0.5}$ [77 H 1].

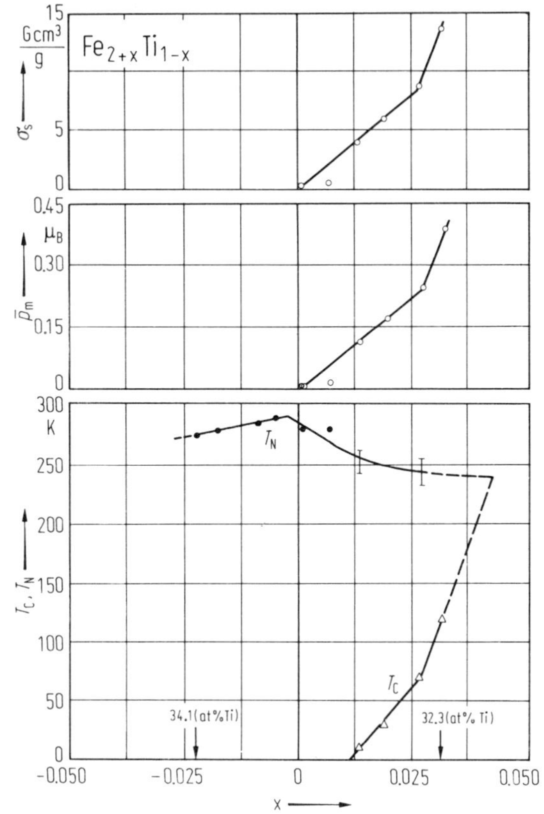

Fig. 12. Néel temperature T_N, Curie temperature T_C and the spontaneous magnetic moment σ_s ($\bar{p}_m$: average magnetic moment per formula unit) at 0 K for $Fe_{2+x}Ti_{1-x}$ alloys [71 O 1]. Solid circles: T_N defined as the temperature for the maximum in the $\chi(T)$ curve, bars: T_N derived from Mössbauer experiments.

Table 2. Transition from ferromagnetism in Fe-rich to antiferromagnetism in Ti-rich Fe–Ti alloys [67 W 1]. Isomer shift *IS*, relative to Fe at room temperature. Quadrupole shift *dQ*, Mössbauer linewidth δ and effective hyperfine field H_{eff} for ^{57}Fe, T_N: Néel temperature.

	T_N K	T K	IS(RT) [1] mm s^{-1}	dQ(RT) mm s^{-1}	δ mm s^{-1}	$H_{hyp,eff}(^{57}Fe)$ kOe
Ti-rich:						
$Fe_{0.66}Ti_{0.34}$	275	RT	−0.286(5)	0.404(5)	0.29	
		20.4	−0.17(1) [1]	0.40(1) [2]	0.27···0.29	97 [3]
Fe-rich:						
$Fe_{0.69}Ti_{0.31}$		298				247 [4]
		20.4	≈0.25 [4] [5]			

[1]) For Fe on 6*h* and 2*a* sites.
[2]) For Fe on 2*a* site.
[3]) For Fe on 6*h* site.
[4]) For excess Fe on Ti sites.
[5]) Relative to α-Fe at 20 K.

2.3.2 Fe–V

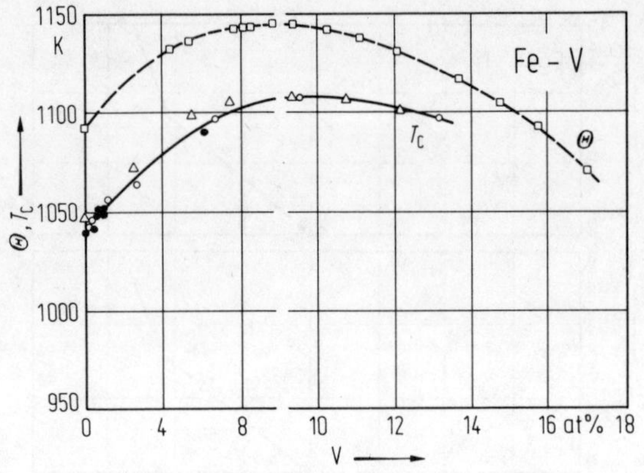

Fig. 13. Ferromagnetic and paramagnetic Curie temperatures, T_C and Θ, respectively, for Fe–V alloys. Open circles: [34 E 1], triangles: [69 S 1], squares: [62 A 1], solid circles: [76 T 1].

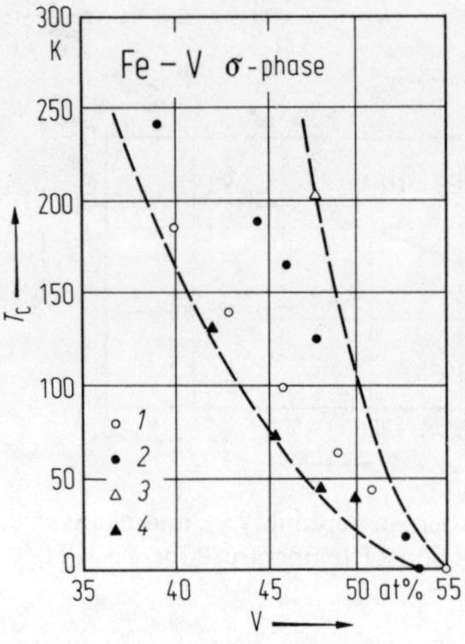

Fig. 14. Curie temperature T_C as dependent on composition and on heat treatment for σ-phase Fe–V alloys. *1*: quenched from 1000 °C [69 M 1], *2*: annealed at 650 °C for 350 h [69 M 1], *3*: annealed at 1000 °C for 120 h [63 N 2, p. 101], *4*: annealed at 1050 °C for 4 days [60 P 1].

2.3.3 Fe–Cr

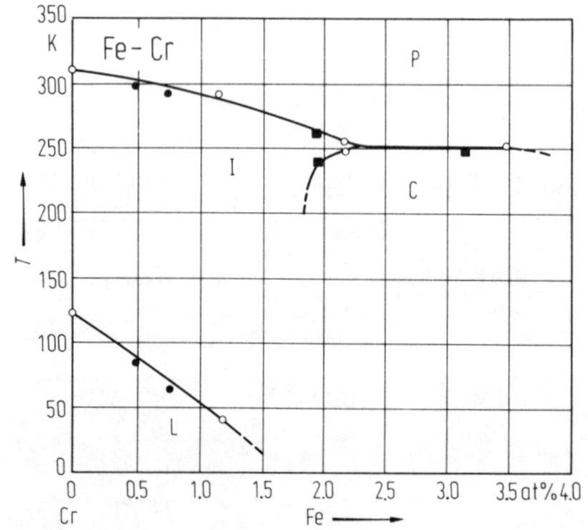

Fig. 15. Low Fe-concentration part of the magnetic phase diagram of Fe–Cr alloys [82 B 1], see also [76 M 1]. Open circles: [82 B 1], solid circles: [67 A 1], squares: [76 M 1], P: paramagnetic state, I: transverse incommensurate spin density wave state, L: longitudinal spin density wave state, C: commensurate spin density wave state.

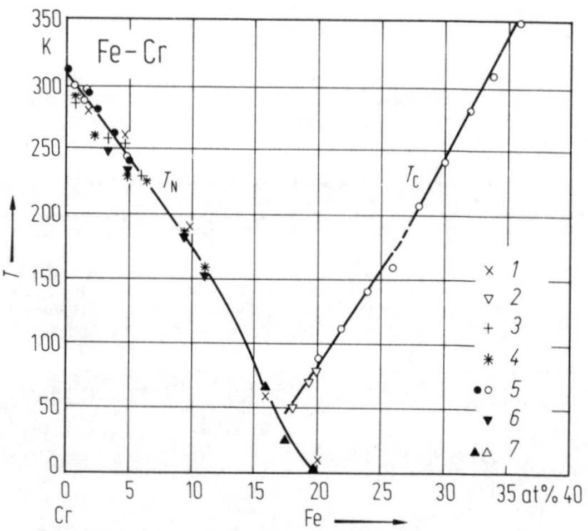

Fig. 16. Néel and Curie temperatures, T_N and T_C, respectively, as well as the extrapolated resistivity ϱ at 0 K vs. Fe concentration for Fe–Cr alloys [75 L 1]. *1*: [60 R 1], *2*: [63 N 1], *3*: [65 I 1], *4*: [66 A 1], *5*: [67 I 1], *6*: [71 A 1], *7*: [75 L 1].

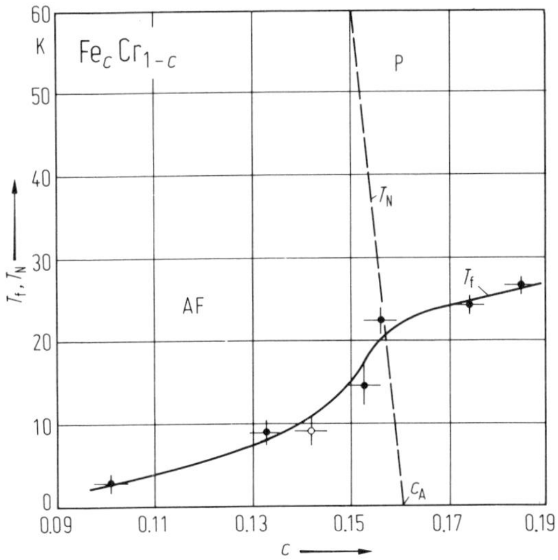

Fig. 17. Spin-glass freezing temperature T_f vs. composition for Fe–Cr alloys [83 B 1]. The broken line is the antiferromagnetic phase boundary [78 B 1]. Open circle: [79 S 1].

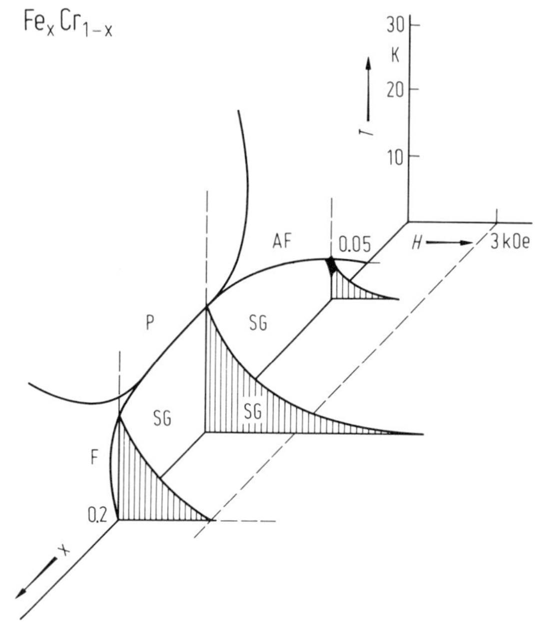

Fig. 18. Schematic temperature – magnetic fieldcomposition phase diagram for Fe–Cr alloys as derived from irreversible relaxations in the magnetization curves [83 P 1]. P: paramagnetic; F: ferromagnetic, AF: antiferromagnetic, SG: spin glass.

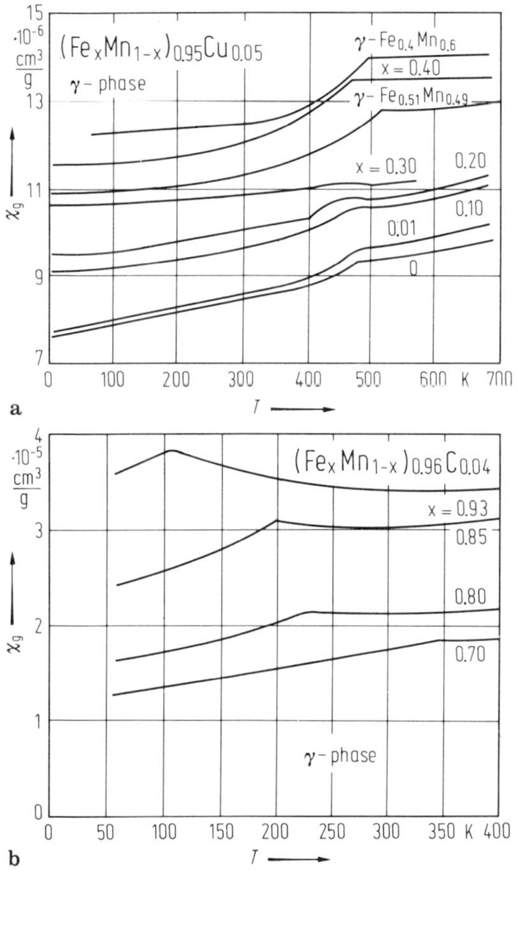

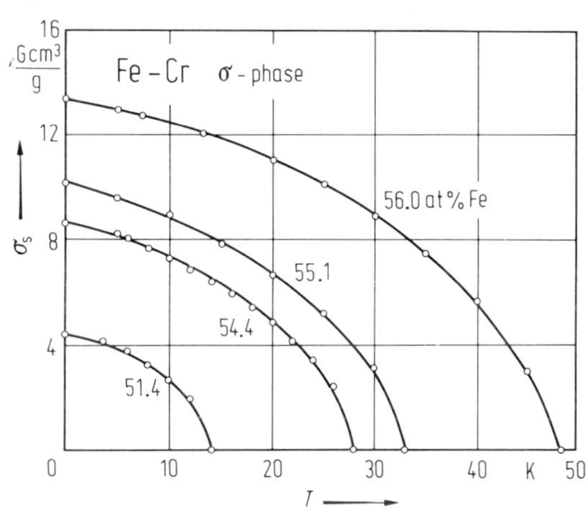

Fig. 19. Spontaneous magnetic moment per unit of mass, σ_s, as a function of temperature for σ-phase Fe–Cr alloys [66 R 1].

2.3.4 Fe–Mn

Fig. 20. Magnetic mass susceptibility χ_g vs. temperature for γ-Fe–Mn alloys [71 E 1], see also [66 U 1]. (a) $Fe_{0.51}Mn_{0.49}$, $Fe_{0.4}Mn_{0.6}$ and $(Fe_xMn_{1-x})_{0.95}Cu_{0.05}$, $0 \leq x \leq 0.4$. Cu has been added to stabilize the γ structure. (b) $(Fe_xMn_{1-x})_{0.96}C_{0.04}$. $0.7 \leq x \leq 0.93$. C has been added to stabilize the γ structure.

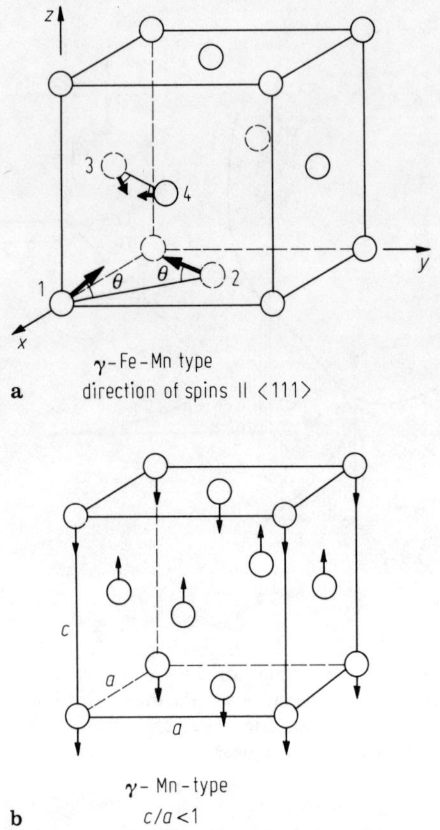

Fig. 21. Spin structure of Fe–Mn allows. (a) γ-Fe–Mn type [66 U 1], (b) γ-Mn type [57 M 1].

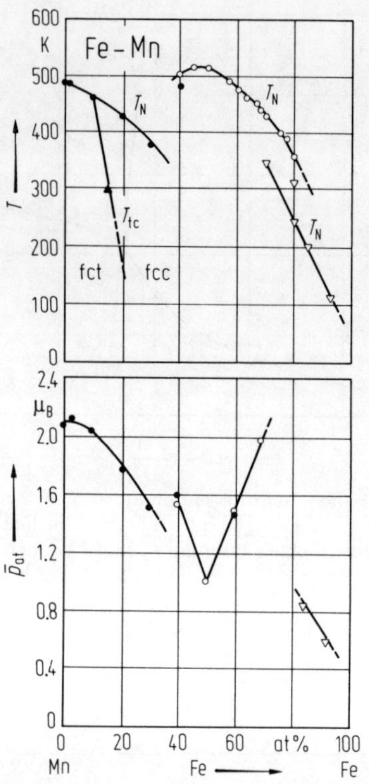

Fig. 22. Néel temperature T_N and average magnetic moment per atom, $\bar{p}_{at}$, vs. composition of Fe–Mn alloys. T_{tc} represents the fcc–fct transition temperature [68 S 1].

2.3.5 Co–Ti

Table 3. Magnetic data of the hexagonal Laves phase compounds Co–Ti. χ_0 is a temperature-independent paramagnetic susceptibility [70 N 1].

Ti at%	Structure	$\sigma_s(4\,K)$ $G\,cm^3\,g^{-1}$	T_C K	χ_0 $10^{-6}\,cm^3\,g^{-1}$	C_g $10^{-4}\,cm^3\,K\,g^{-1}$
28.0	$MgNi_2+Cu_3Au$	4.90	42	–	–
29.0	$MgNi_2$	8.50	44	12.71	16.24
30.0	$MgNi_2$	6.20	38	13.83	9.77
30.1	$MgNi_2$	5.60	33	13.09	9.77
31.3	$MgNi_2$	2.77	18	9.35	8.76
32.8	$MgNi_2+MgCu_2$	2.80	17	–	–

2.3.6 Co–Cr

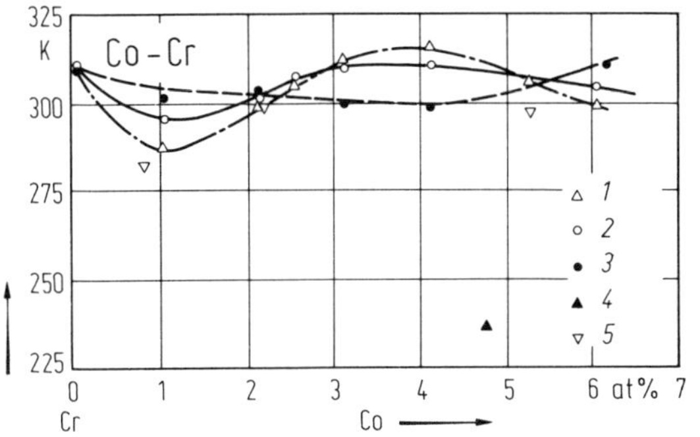

Fig. 23. Néel temperature T_N for Co–Cr alloys [83 A 1]. (*1*) Minima in resistivity vs. T, (*2*) inflection points in bulk modulus B vs. T, (*3*) inflection points in linear thermal expansion coefficient, (*4*) neutron diffraction [78 K 1]. (*5*) neutron diffraction [68 E 1].
See also Fig. 31.

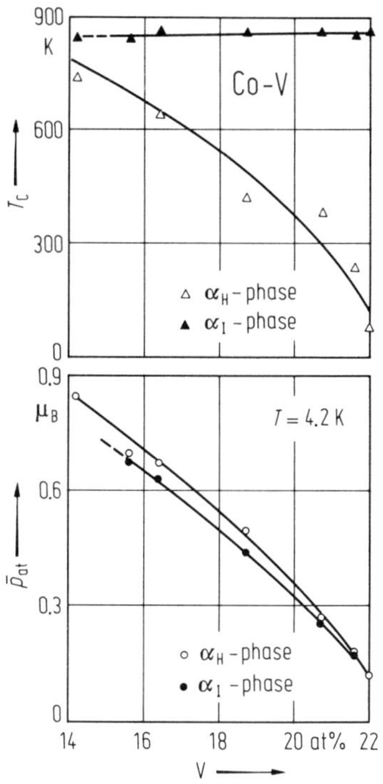

Fig. 24. Ferromagnetic Curie temperature T_C and average magnetic moment per atom, $\bar{p}_{at}$, for Co–V alloys at 4.2 K [80 A 1]. α_H-phase: high-temperature phase, quenched from 1200 °C, α_I-phase: intermediate phase, obtained by heat treatment between 700 and 860 °C.

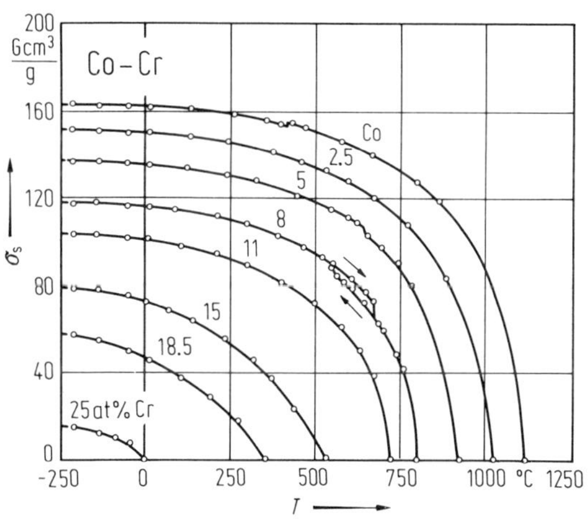

Fig. 25. Saturation magnetic moment σ_s vs. temperature T for Co–Cr alloys. The measurements were made for increasing temperatures. For Co-8 at% Cr also measurements were made when decreasing the temperature from T_C [82 G 1].

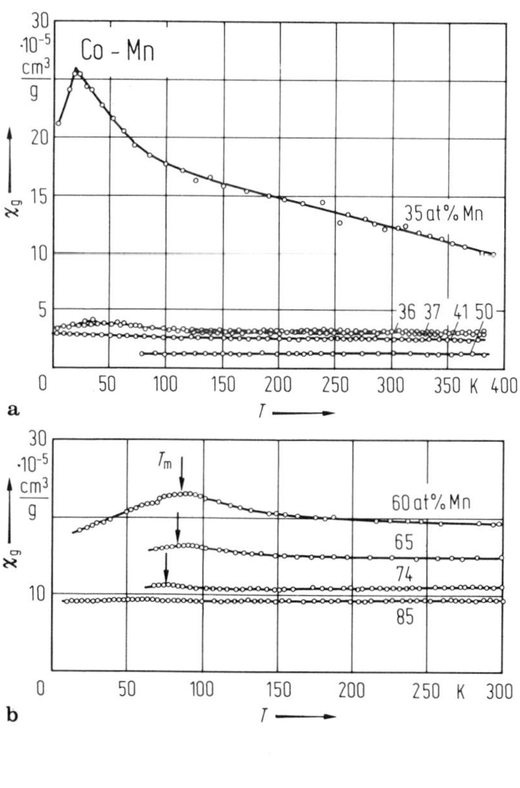

Fig. 26. Magnetic mass susceptibility χ_g vs. temperature for Co–Mn alloys. (a) 35···50 at% Mn (b) 60···85 at% Mn [75 H 1]. Arrows indicate the temperature T_m of the susceptibility maximum.

2.3.7 Co–Mn

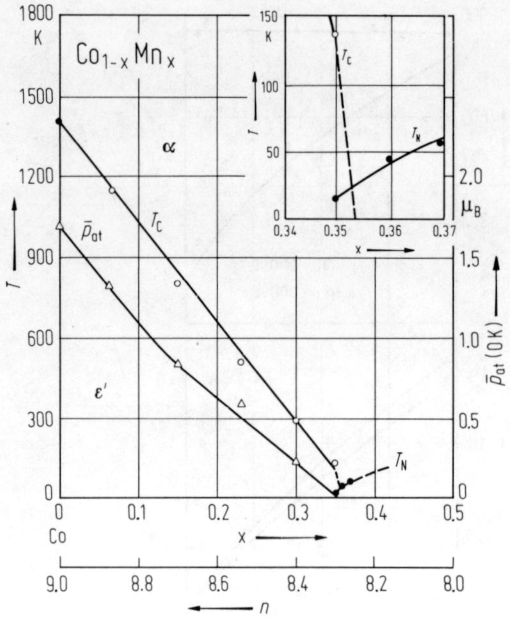

Fig. 27. Curie and Néel temperatures, T_C and T_N, and average magnetic moment per atom, $\bar{p}_{at}(0\,K)$, of $Co_{1-x}Mn_x$ alloys [70 M 1, 73 A 1, 73 M 1]. n: average number of 4s and 3d electrons per atom.

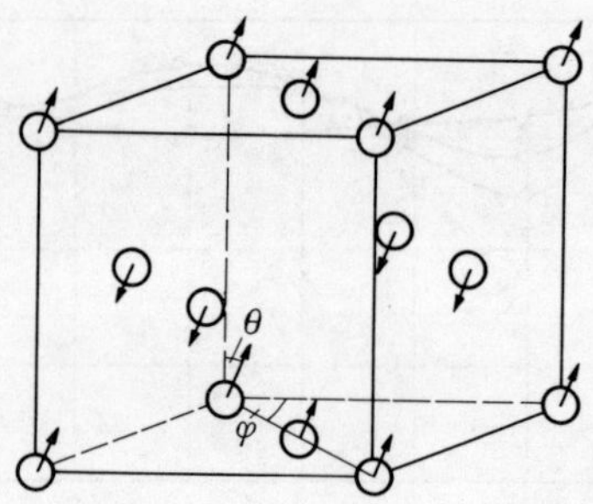

Fig. 28. Antiferromagnetic structure of Co–Mn (fcc). The spin direction is indicated by angles θ and ϕ [73 A 1].

2.3.8 Ni–Ti

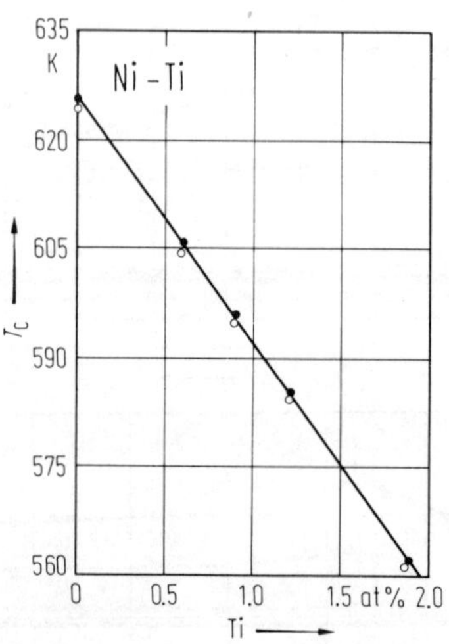

Fig. 29. Ferromagnetic Curie temperature T_C of Ni–Ti alloys, derived from resistivity measurements [78 Y 1]. Open circles: maximum in $d\varrho/dT$ and solid circles: "kink-point" technique.

2.3.9 Ni–V

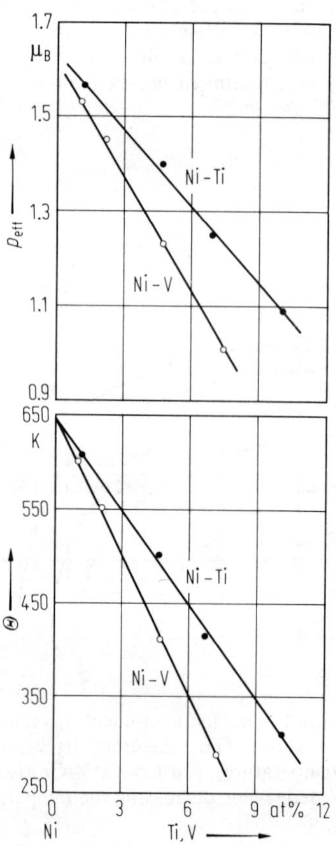

Fig. 30. Effective paramagnetic moment per atom, p_{eff}, and the paramagnetic Curie temperature Θ for Ni alloys with small V and Ti concentrations [36 M 1].

2.3.10 Ni–Cr

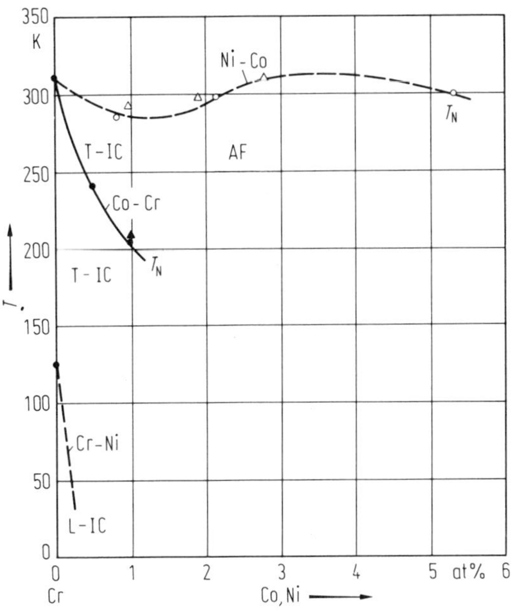

Fig. 31. Magnetic phase diagram of dilute Ni–Cr (dashed lines) and Co–Cr (solid line) alloys. T–IC and L–IC denote transverse and longitudinal spin density wave states, respectively. AF means commensurate antiferromagnetic state. Circles: [68 E 1] and triangles: [66 B 2].

Table 4. Atomic magnetic moments as determined by polarized-neutron diffuse scattering measurements on Ni–Cr and Ni–V alloys [76 C 1] and average magnetic moment per atom, $\bar{p}_{at}$ [72 B 1, 71 C 2, 32 S 1]. $T = 4.2$ K, $H = 57.3$ kOe.

Cr	V	$\bar{p}_{Ni}$	$\bar{p}_{Cr}$	$\bar{p}_V$	$\bar{p}_{at}$
at%		μ_B			
1	–	0.570	−0.20(6)	–	0.562
5	–	0.375	−0.02(12)	–	0.355
10	–	0.100	0.05(6)	–	0.095
–	5	0.346		−0.065(48)	0.325

2.3.11 Ni–Mn

Table 5. Magnetic properties of stoichiometric NiMn. p_{Mn}, p_{Ni} and H_{hyp} refer to 0 K. For spin arrangement in the antiferromagnetic phase, see Fig. 34.

Crystal structure	Magnetism	T_N K	p_{Mn}	p_{Ni}	$H_{hyp}(^{55}Mn)$
			μ_B		kOe
fct CuAu-type [1]	AF	1073(40) [68 K 1]	3.8(3) [68 K 1]	<0.6 [68 K 1]	235 [2] [67 P 1]

[1]) $a = 3.74$ Å, $c = 3.52$ Å.
[2]) Long-range order parameter $S = 0.9$.

Table 6. Magnetic propeties of stoichiometric Ni_3Mn. $\bar{p}_{at}$, p_{Mn}, p_{Ni} and H_{hyp} refer to 0 K. Order-disorder transition temperature: 510 °C.

	Crystal structure	Magnetism	T_C	$\bar{p}_{at}$	p_{Mn}	p_{Ni}	$H_{hyp}(^{55}Mn)$
			K	μ_B	μ_B		kOe
ordered	fcc Cu_3Au-type	F	753 [48 K 1]	1.1 [48 K 1]	3.83 [1]) [63 P 1]	0.47 [1]) [63 P 1]	334.5 [67 P 1]
disordered	fcc	F	132 [58 K 1]	0.73 [58 K 1]			258.7 [67 P 1]

[1]) Determined from an ordered sample having the long-range order parameter of 0.74.

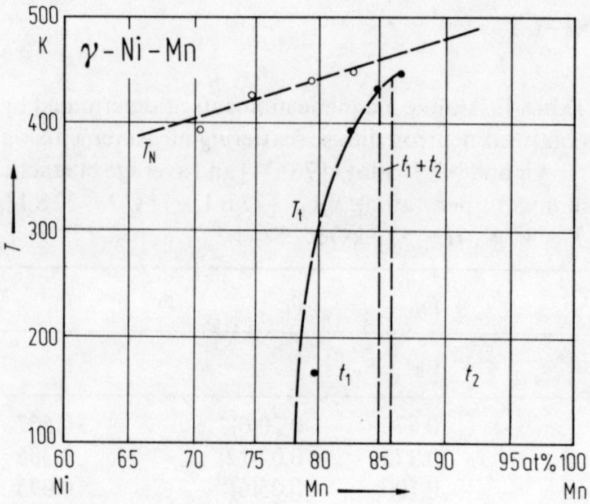

Fig. 32. Tetragonal transition temperature T_t and Néel temperature T_N of γ-phase Ni–Mn alloys. t_1: $c/a>1$, t_2: $c/a<1$ [71 U 1], see also [70 U 1].

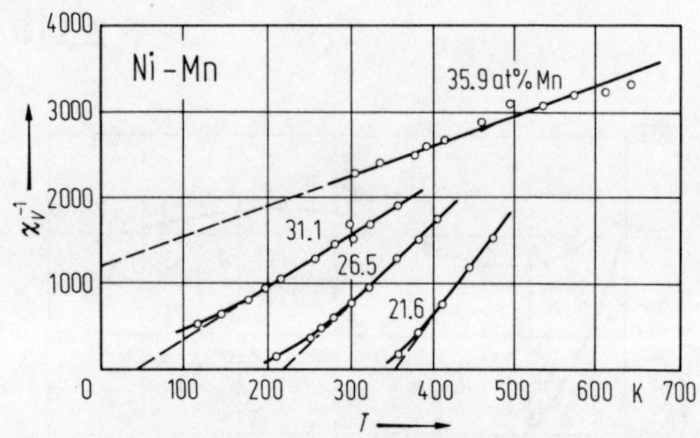

Fig. 33. Reciprocal value of the paramagnetic volume susceptibility, χ_v^{-1}, vs. temperature T for Ni–Mn alloys, 21.6⋯35.9 at % Mn [60 K 1]. See also Fig. 35.

Fig. 34. Spin structure of ordered NiMn [59 K 1]. The black and white circles denote Mn and Ni atoms, respectively. The moments are determined to be $p_{Mn}=4.0(1)$ and $p_{Ni}=0.6\,\mu_B$. The other possible spin structure is discussed in [59 K 1].

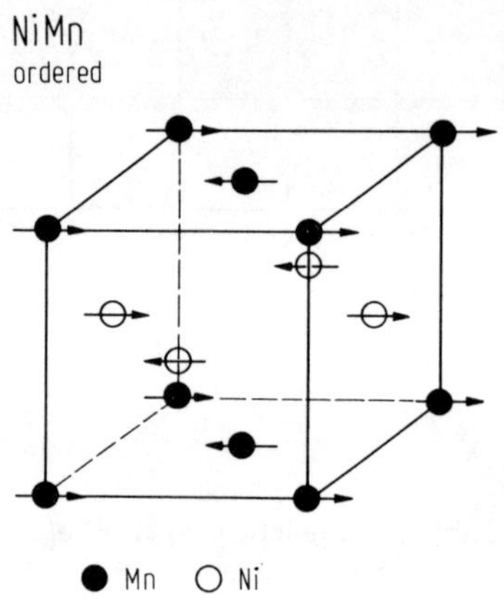

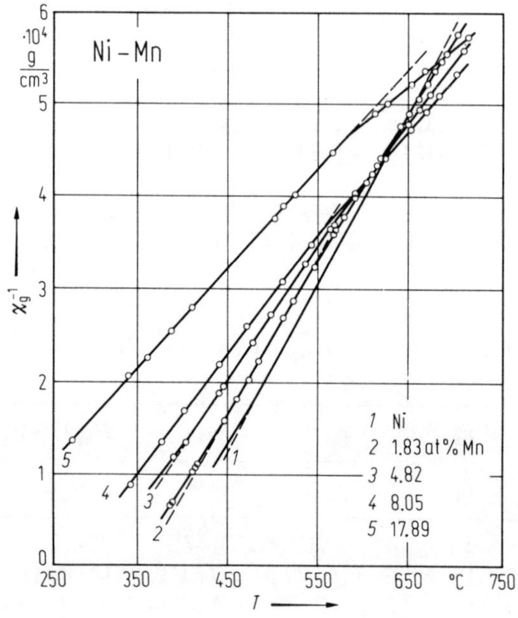

Fig. 35. Reciprocal value of the paramagnetic mass susceptibility, χ_g^{-1}, vs. temperature for Ni–Mn alloys, 0⋯17.89 at % Mn [36 M 1]. See also Fig. 33.

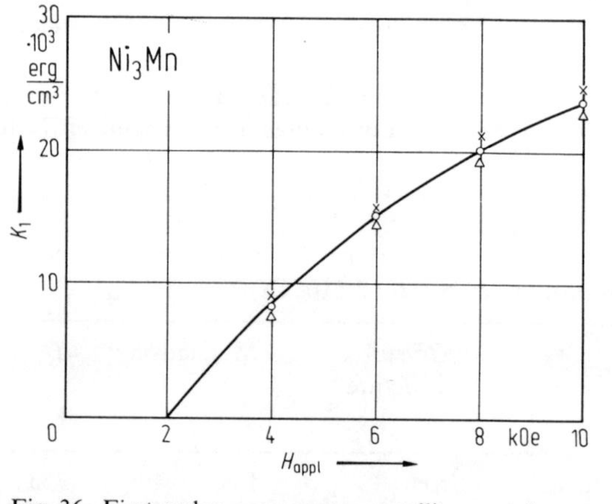

Fig. 36. First-order magnetocrystalline anisotropy constant K_1 derived from torque measurements on a single crystal of Ni_3Mn in various states of atomic order. Crosses: [100] magnetic field cooling, triangles: [010] magnetic field cooling, circles: zero-field cooling. The specimen was annealed at 420 °C for 230 h [67 S 2].

2.3.12 Mn–V–Cr

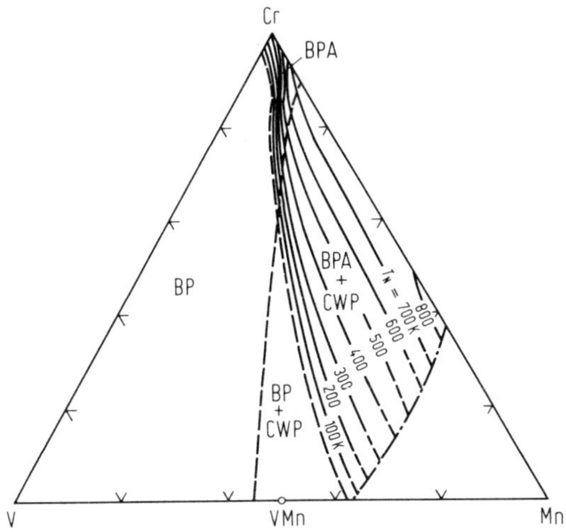

Fig. 37. Magnetic phase diagram of V–Cr–Mn alloys in the bcc phase. BP and BPA mean the band (Pauli) paramagnetic and antiferromagnetic phases, respectively. CWP is the Curie-Weiss-type paramagnetism [80 M 2]. The broken curve is the boundary for the appearance of a localized magnetic moment. The numbers give T_N in [K].

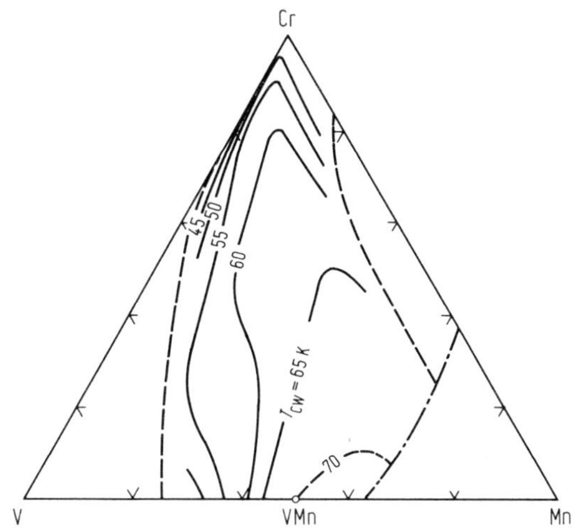

Fig. 38. Curie temperature T_{CW} of the weak ferromagnetism in V–Cr–Mn alloys (bcc). [80 M 2].

2.3.13 Fe–V–Cr

Table 7. Susceptibility data fitted to the modified Curie-Weiss law $\chi_g = A + C_g/(T - \Theta)$ for $Fe_{0.03}Cr_{0.97-x}V_x$ alloys [77 A 1].

	V at%	T_N K	A 10^{-6} cm^3 g^{-1}	C_g 10^{-6} K cm^3 g^{-1}	Θ K
$T > T_N$	0	250	3.34	692.0	−16.9
	0.5	162	3.20	695.4	−14.8
	1.0	108	3.16	693.3	−14.3
$T < T_N$	0		4.29	335.90	− 3.7
	0.5		3.98	519.2	− 6.6
	1.0		3.74	580.7	− 7.2

2.3.14 Fe–Co–Ti

Table 8. Effective paramagnetic moment p_{eff} and average magnetic moment per atom, $\bar{p}_{at}$, of $(Fe_xCo_{1-x})_{0.5}Ti_{0.5}$ alloys, as derived from, respectively, the Curie-Weiss constant and the spontaneous magnetic moment σ_s at low temperature [78 B 2, 78 B 3, 73 A 2]. Θ: paramagnetic Curie temperature, T_C: ferromagnetic Curie temperature, γ: electronic specific heat coefficient, χ_{HF}: high-field magnetic susceptibility, p: hydrostatic pressure, M_s(0 K): spontaneous magnetization at low temperature, D: spin-wave stiffness constant.

x	p_{eff} μ_B	Θ K	T_C K	σ_s(0 K) G cm³ g⁻¹	$\bar{p}_{at}$ μ_B	γ mJ mol⁻¹ K⁻²
0.00	1.87	−440				18.8
0.10	1.16	−140				20.2
0.20	1.14	− 42				22.5
0.30	1.00	8				22.1
0.40	0.97	38	16.5	3.60	0.069	20.4
0.50	1.06	56	46.0	8.52	0.16	17.9
0.60	0.78	58	57.0	6.00	0.11	14.8
0.70	0.42	37	36.0	3.16	0.059	10.3
0.80	0.28	14	10.5	1.38	0.026	7.42
0.90	0.33	− 13				4.22
1.00	0.35	− 22				0.92

x	T_C [1] K	dT_C/dp K kbar⁻¹	$\sigma_{m,s}$(0 K) G cm³ mol⁻¹	dM_s(0 K)/dp G kbar⁻¹	χ_{HF} 10⁻³ cm³ mol⁻¹	D meV Å²
0.405	13.9	−1.92	317	−13.3	2.6	42(5)
0.43	33.9	–	686	–	2.2	66(7)
0.5	55	−1.21	993	−13.7	2.1	94(5)
0.55	–	−0.66	683	–	1.6	–
0.67	39.2	−0.63	435	− 3.75	1.0	132(15)
0.73	–	–	178	− 1.6	0.74	53(5)
0.78	8.4	−0.16	113	− 1.3	–	–

[1]) From Arrot plots.

2.3.15 Fe–Co–Cr

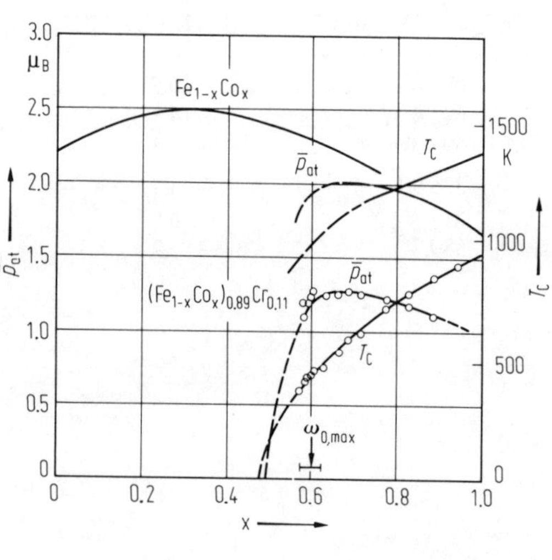

Fig. 39. Composition dependence of the average magnetic moment per atom $\bar{p}_{at}$ and the Curie temperature T_C for $(Fe_{1-x}Co_x)_{0.89}Cr_{0.11}$ and $Fe_{1-x}Co_x$ alloys [70 F 1]. The arrow below $\omega_{0,max}$ indicates the composition giving the maximum volume magnetostriction.

2.3.16 Fe–Co–Mn

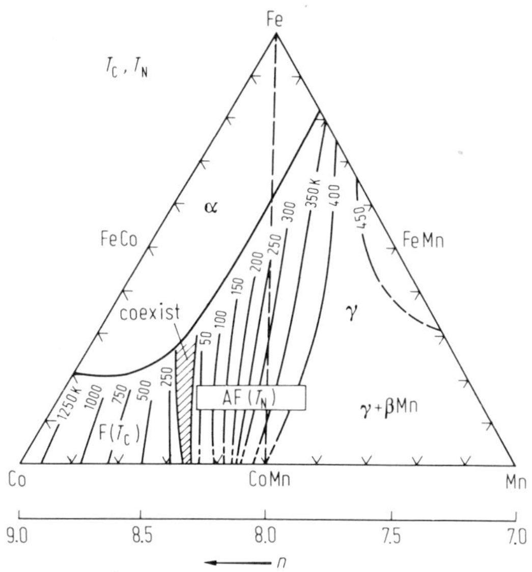

Fig. 40. Ferromagnetic Curie temperature T_C and Néel temperature T_N, in [K], for the γ-phases of Fe–Co–Mn alloys [73 M 1]. n: number of 4s and 3d electrons per atom.

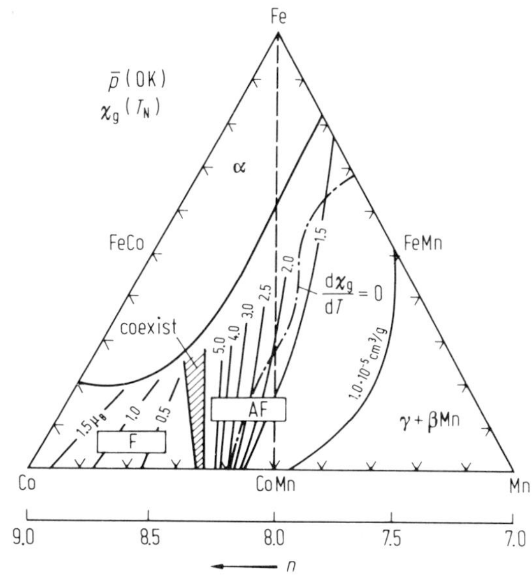

Fig. 41. Saturation magnetic moment at 0 K, $\bar{p}$(0 K), in the ferromagnetic state and magnetic mass susceptibility χ_g at the Néel temperature T_N in the antiferromagnetic state of Fe–Co–Mn alloys. Dashed-dotted curve: composition at which $d\chi_g/dT$ changes sign. The shadowed area is the region for the coexistence of ferro- and antiferromagnetic phases [73 M 1]. n: number of 4s and 3d electrons per atom.

2.3.17 Fe–Ni–V

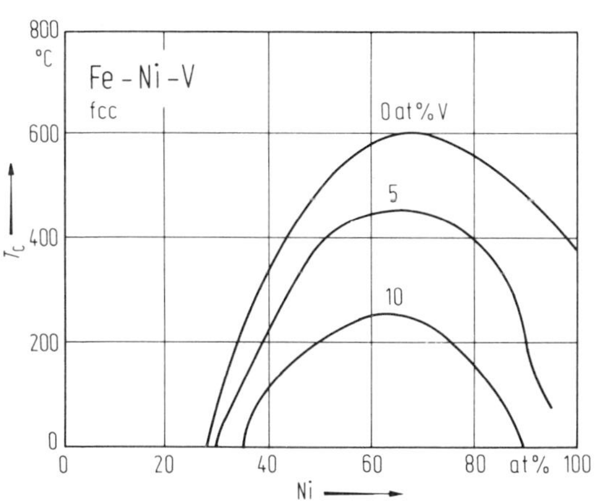

Fig. 42. Curie temperature T_C of Fe–Ni–V alloys in the fcc phase [34 K 1, 51 B 1, p. 186].

2.3.18 Fe–Ni–Cr

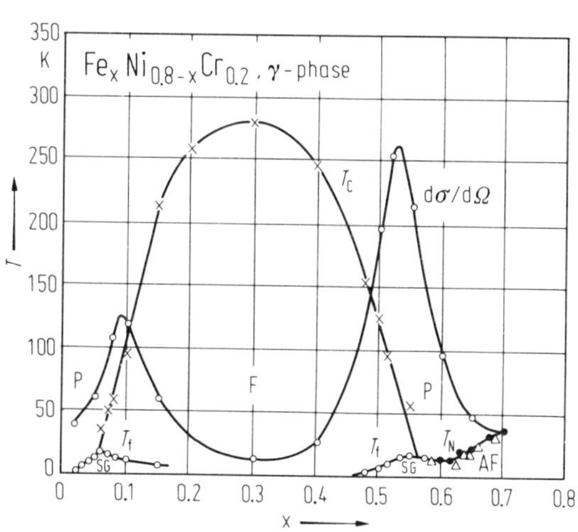

Fig. 43. Magnetic phase diagram and differential cross section of critical neutron scattering for γ-$Fe_xNi_{0.8-x}Cr_{0.2}$. P, F, AF and SG are the paramagnetic, ferromagnetic, antiferromagnetic and spin glass phases, respectively, and dσ/dΩ represents the differential cross section for neutron scattering in relative units. Open circles and crosses: [82 M 1], triangles: [76 W 1] and solid circles: [70 K 1].

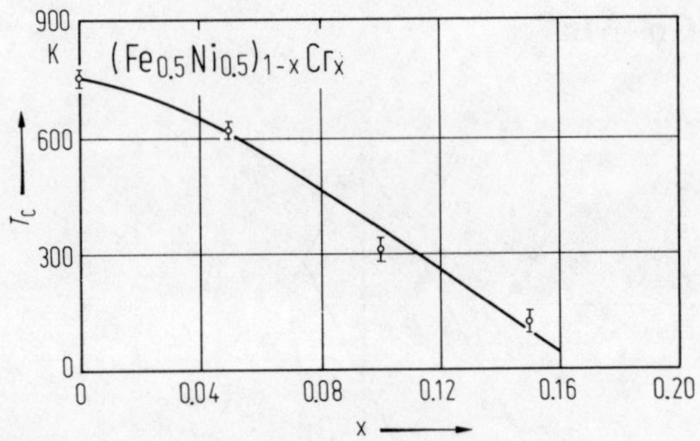

Fig. 44. Ferromagnetic Curie temperature T_C derived from Mössbauer spectra for stable disordered fcc $(Fe_{0.5}Ni_{0.5})_{1-x}Cr_x$ alloys, $x \leqq 0.15$ [76 B 1].

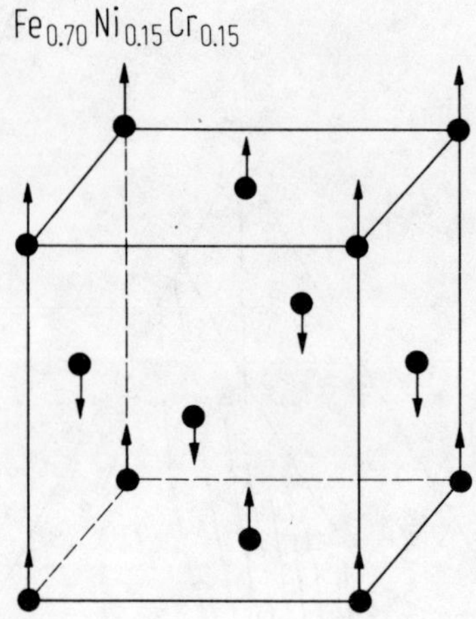

Fig. 45. Magnetic structure of fcc $Fe_{0.70}Ni_{0.15}Cr_{0.15}$ derived from neutron scattering. The long-range antiferromagnetic order is shown. The sublattice magnetic moment, $\bar{p}_{at} = 0.3(1)\,\mu_B$ at 4.2 K is aligned along the [001] direction. The spin components perpendicular to [001] have only a short-range correlation [75 I 1], see also [70 I 1].

2.3.19 Fe–Ni–Mn

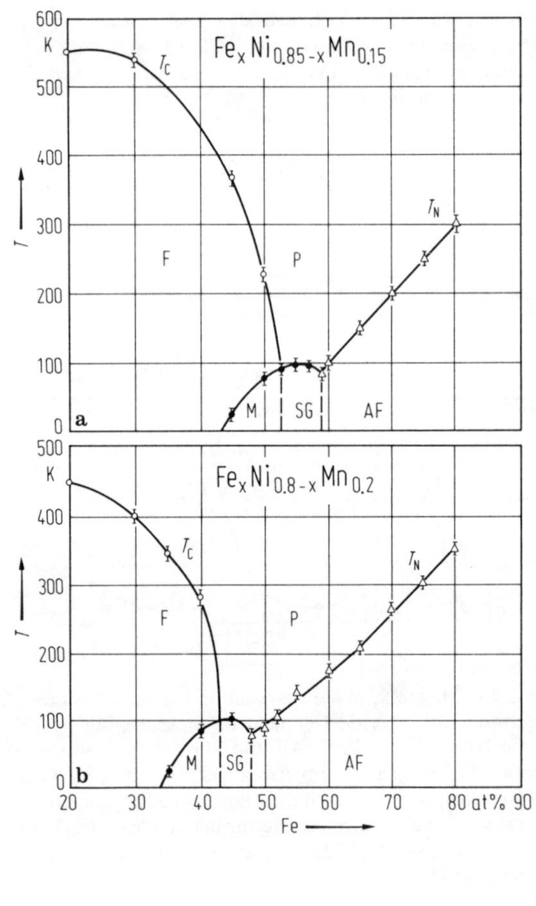

◀ Fig. 46. Magnetic phase diagrams for (a) $Fe_xNi_{0.85-x}Mn_{0.15}$ and (b) $Fe_xNi_{0.8-x}Mn_{0.2}$ alloys [81 M 1], see also [80 M 1]. P: paramagnetic, F: ferromagnetic, AF: antiferromagnetic, M: metamagnetic, SG: spin-glass. T_C: Curie temperature, T_N: Néel temperature.

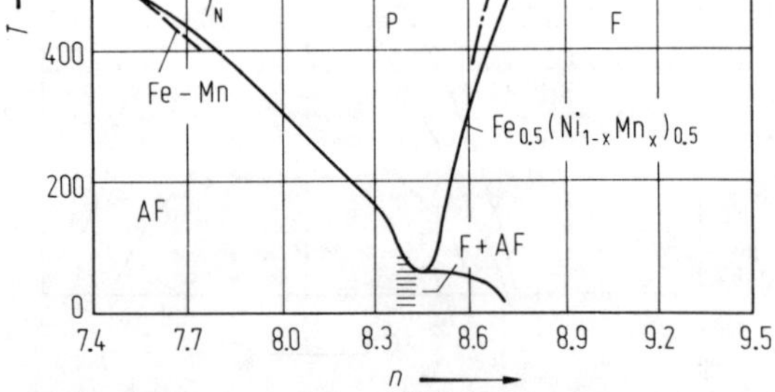

Fig. 47. Magnetic phase diagram for $Fe_{0.5}(Ni_{1-x}Mn_x)_{0.5}$ alloys. The compositions are indicated by the average number n of 4s and 3d electrons per atom [78 B 4]. P: paramagnetic, F: ferromagnetic, AF: antiferromagnetic. The shaded area indicates coexistence of ferro- and antiferromagnetism. T_C: Curie temperature, T_N: Néel temperature.

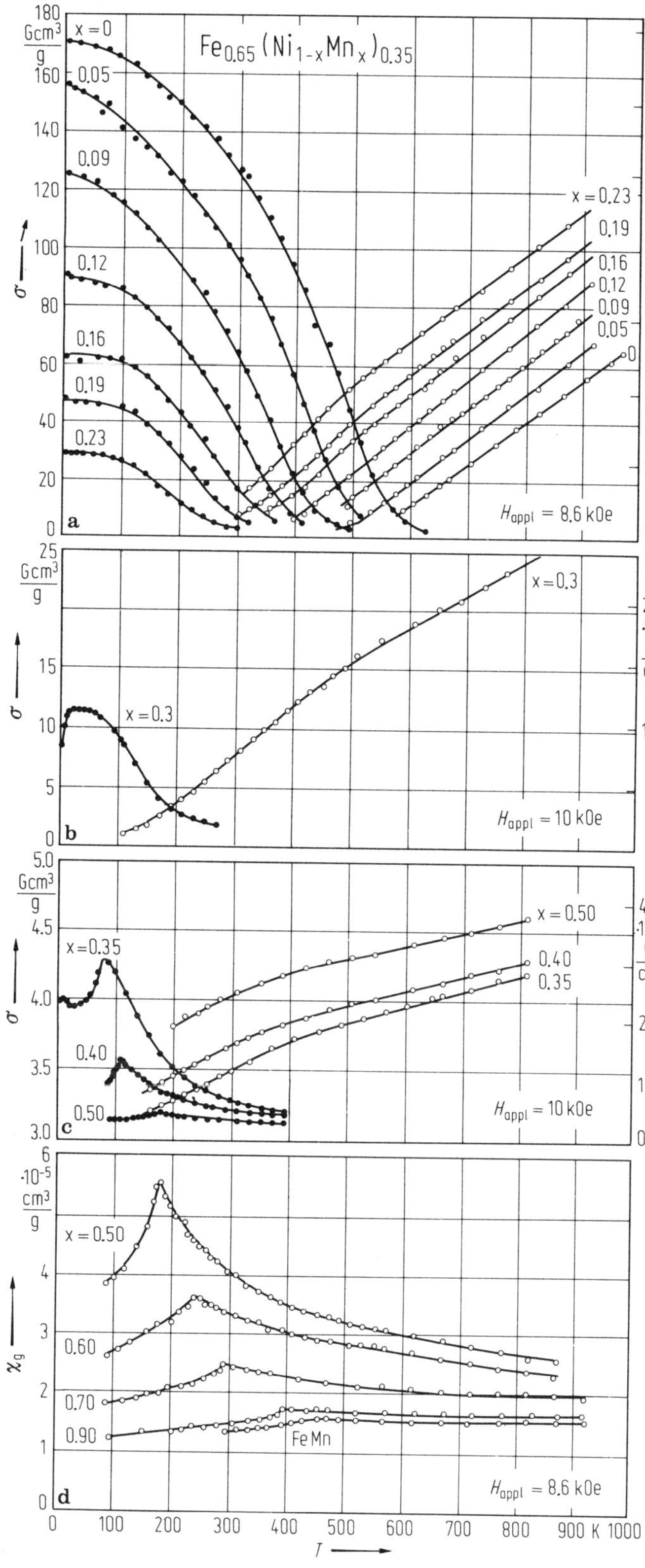

Fig. 48. Inverse magnetic mass susceptibility, χ_g^{-1}, and the magnetic moment σ vs. temperature for $Fe_{0.65}(Ni_{1-x}Mn_x)_{0.35}$ alloys.
(a) $0 \leqq x \leqq 0.23$, $H_{appl} = 8.6$ kOe,
(b) $x = 0.3$, $H_{appl} = 10$ kOe,
(c) $0.35 \leqq x \leqq 0.5$, $H_{appl} = 10$ kOe and
(d) $0.5 \leqq x \leqq 0.9$, $H_{appl} = 8.6$ kOe
(χ_g only); for comparison χ_g of FeMn is also shown [67 S 1].

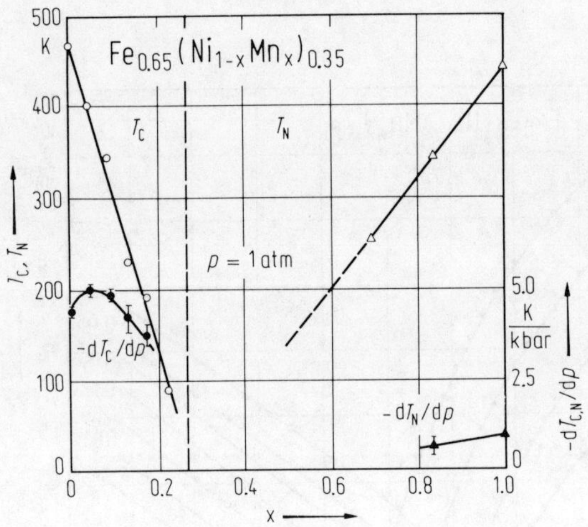

Fig. 49. Magnetic phase diagram for $Fe_{0.65}(Ni_{1-x}Mn_x)_{0.35}$ alloys at atmospheric pressure and the variation with pressure of the Curie and the Néel temperature, dT_C/dp and dT_N/dp, respectively, as measured for pressures up to 25 kbar [75 D 1], see also [71 B 1, 71 N 1].

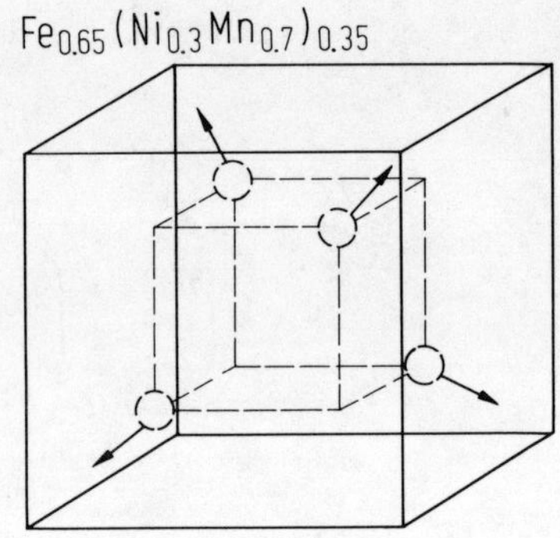

Fig. 50. Proposed spin configuration of antiferromagnetic $Fe_{0.65}(Ni_{0.3}Mn_{0.7})_{0.35}$ [69 N 1].

2.3.20 Co–Ni–Mn

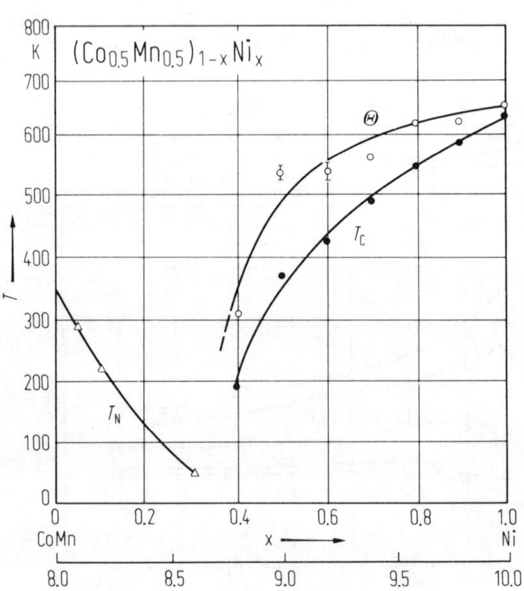

Fig. 51. Magnetic phase diagram of $(Co_{0.5}Mn_{0.5})_{1-x}Ni_x$ alloys as a function of composition and average number n of 4s and 3d electrons per atom. T_C, Θ and T_N are ferromagnetic Curie temperature, paramagnetic Curie temperature and Néel temperature, respectively [74 A 1].

2.3.21 References for 2.2 and 2.3

32 S 1 Sadron, C.: Ann. Phys. **17** (1932) 371.
34 E 1 Edlund, D.L.: Ph.D. Thesis, Massachusetts Inst. of Techn., Cambridge, Mass. **1934**.
34 K 1 Kuhlewein, H.: Z. Anorg. Allg. Chem. **218** (1934) 65.
36 M 1 Manders, C.: Ann. Phys. **5** (1936) 167.
48 K 1 Köster, W., Rauscher, W.: Z. Metallkd. **39** (1948) 178.
51 B 1 Bozorth, R.M.: Ferromagnetism, New York: D. Van Nostrand **1951**.
57 M 1 Meneghetti, D., Sidhu, S.S.: Phys. Rev. **105** (1957) 130.
58 K 1 Kouvel, J.S., Graham, C.D., Jr., Becker, J.J.: J. Appl. Phys. **29** (1958) 518.
59 K 1 Kasper, J.S., Kouvel, J.S.: J. Phys. Chem. Solids **11** (1959) 586.
60 K 1 Kouvel, J.S.: J. Phys. Chem. Solids **16** (1960) 107.
60 P 1 Parsons, D.: **185** (1960) 840.
60 R 1 Rajan, N.S., Waterstrat, R.M., Beck, P.A.: J. Appl. Phys. **31** (1960) 731.
62 A 1 Arajs, S., Colvin, R.V., Chessin, H., Peck, J.M.: J. Appl. Phys. **33** (1962) 1353.
62 T 1 Taniguchi, S., Tebble, R.S., Williams, D.E.G.: Proc. R. Soc. London Ser. A **265** (1962) 502.
62 W 1 Wilkinson, M.K., Wollan, E.O., Koehler, W.C., Cable, J.W.: Phys. Rev. **127** (1962) 2080.
63 N 1 Nevitt, M.V., Aldred, A.T.: J. Appl. Phys. **34** (1963) 463.
63 N 2 Nevitt, M.V.: Electronic Structure and Alloy Chemistry of the Transition Elements (Beck, P.A., ed.), New York: Interscience **1963**, p. 101.
63 P 1 Paoletti, A., Ricci, F.P.: J. Appl. Phys. **34** (1963) 1571.
65 H 1 Hamaguchi, Y., Wollan, E.O., Koehler, W.C.: Phys. Rev. **138** (1965) A 737.
65 I 1 Ishikawa, Y., Tournier, R., Fillipi, J.: J. Phys. Chem. Solids **26** (1965) 1727.
66 A 1 Arajs, S., Durmyre, G.R.: J. Appl. Phys. **37** (1966) 1017.
66 B 1 Bastow, T.J.: Proc. Phys. Soc. **88** (1966) 935.
66 B 2 Booth, J.G.: J. Phys. Chem. Solids **27** (1966) 1639.
66 K 1 Koehler, W.C., Moon, R.M., Trego, A.L., Mackintosh, A.R.: Phys. Rev. **151** (1966) 405.
66 R 1 Read, D.A., Thomas, E.H.: IEEE Trans. Magn. **MAG-2** (1966) 415.
66 S 1 Suzuki, T.: J. Phys. Soc. Jpn. **21** (1966) 442.
66 U 1 Umebayashi, H., Ishikawa, Y.: J. Phys. Soc. Jpn. **21** (1966) 1281.
67 A 1 Arrott, A., Werner, S.A.: Phys. Rev. **153** (1967) 624.
67 I 1 Ishikawa, Y., Endoh, Y.: J. Phys. Soc. Jpn. **23** (1967) 205.
67 K 1 Komura, S., Kunitomi, N., Hamaguchi, Y.: J. Phys. Soc. Jpn. **23** (1967) 171.
67 P 1 Proctor, W., Scurlock, R.G., Wray, E.M.: Proc. Phys. Soc. **90** (1967) 697.
67 S 1 Shiga, M.: J. Phys. Soc. Jpn. **22** (1967) 539.
67 S 2 Satoh, T., Yokoyama, Y., Nagashima, T.: J. Phys. Soc. Jpn. **22** (1967) 1296.
67 W 1 West. G.W.: Philos Mag. **15** (1967) 855.
68 E 1 Endoh, Y., Ishikawa, Y., Ohno, H.: J. Phys. Soc. Jpn. **24** (1968) 263.
68 K 1 Krén, E., Nagy, E.,Nagy, I., Pál, L., Szabó, P.: J. Phys. Chem. Solids **29** (1968) 101.
68 S 1 Smith, J.H.: J. Appl. Phys. **39** (1968) 675
69 M 1 Mori, N., Mitsui, T.: J. Phys. Soc. Jpn. **26** (1969) 1087.
69 N 1 Nakamura, Y., Shiga, M., Takeda, Y.: J. Phys. Soc. Jpn. **27** (1969) 1470.
69 S 1 Sueda, N., Fujiwara, Y., Fujiwara, H.: J. Sci. Hiroshima Univ. Ser. A-II, **33** (1969) 267.
70 F 1 Fujimori, H., Saito, H.: Trans. Jpn. Inst. Met. **11** (1970) 72.
70 I 1 Ishikawa Y., Endoh, Y., Takimoto, T.: J. Phys. Chem. Solids **31** (1970) 1225.
70 K 1 Kohlhaas. R., Raible, A.A., Rocker, W.: Z. Angew. Phys. **30** (1970) 254.
70 M 1 Matsui, M., Ido, T., Sato, K., Adachi, K.: J. Phys. Soc. Jpn. **28** (1970) 791.
70 N 1 Nakamichi, T., Aoki, Y., Yamamoto, M.: J. Phys. Soc. Jpn. **28** (1970) 590.
70 U 1 Uchishiba, H., Hori, T., Nakagawa, Y.: J. Phys. Soc. Jpn. **28** (1970) 792.
71 A 1 Arajs, S., Anderson, E.E.: Physica **54** (1971) 617.
71 B 1 Bartel, L.C., Edwards, L.R., Samara, G.A.: AIP Conf. Proc. **5** (1971) 482.
71 C 1 Chiu, C.H., Jericho, M.H., March, R.H.: Can. J. Phys. **49** (1971) 3010.
71 C 2 Chiffey, R., Hicks, T.J.: Phys. Lett. A **34** (1971) 267
71 E 1 Endoh, Y., Ishikawa, Y.: J. Phys. Soc. Jpn. **30** (1971) 1614.
71 N 1 Nakamura, Y., Hayase, M., Shiga, M., Miyamoto, Y., Kawai, N.: J. Phys. Soc. Jpn. **30** (1971) 720.
71 O 1 Okazaki, M.: J. Phys. (Paris) **32** (1971) C 1–874.

71U1 Uchishiba, H.: J. Phys. Soc. Jpn. **31** (1971) 436.
71V1 Von Meerwall, E., Schreiber, D.S.: Phys. Rev. B **3** (1971) 1.
72B1 Besnus, M.J., Gottehrer, Y., Munschy, G.: Phys. Status Solidi B **49** (1972) 597.
73A1 Adachi, K., Matsui, M., Mitani, S.: J. Phys. Soc. Jpn. **35** (1973) 426.
73A2 Asada, Y., Nosé, H.: J. Phys. Soc. Jpn. **35** (1973) 409.
73M1 Matsui, M., Sato, K., Adachi, K.: J. Phys. Soc. Jpn. **35** (1973) 419.
74A1 Adachi, K., Sato, K., Katata, A., Mori, M.: Proc. Int. Conf. Magn. Moscow, 1973, **1(2)** (1974) 256.
75D1 Dubovka, G.T., Ponyatovskii, E.G., Georgieva, I.Ya., Antonov, V.E.: Phys. Status Solidi A **32** (1975) 301.
75H1 Hori, T.: L. Phys. Soc. Jpn. **38** (1975) 1780.
75I1 Ishikawa, Y., Kohgi, M., Noda, Y.: J. Phys. Soc. Jpn. **39** (1975) 675.
75L1 Loegel, B.: J. Phys. F**5** (1975) 497.
76B1 Bansal, C., Chandra, G.: Solid State Commun. **19** (1976) 107.
76C1 Cable, J.W., Medina, R.A.: Phys. Rev. B **13** (1976) 4868.
76M1 Mori, M., Tsunoda, Y., Kunitomi, N.: Solid State Commun. **18** (1976) 1103.
76T1 Teoh, W., Arajs, S., Abukay, D., Anderson, E.E.: J. Mag. Magn. Mater. **3** (1976) 260.
76W1 Warnes, L.A.A., King, H.W.: Cryogenics **16** (1976) 473, 659.
77A1 Astrom, H.U., Gudmundsson, H., Hedman, L., Rao, K.V.: Physica B + C **86–88** (1977) 332.
77H1 Hedgcock, F.T., Strom-Olsen, J.O., Wilford, D.F.: J. Phys. F **7** (1977) 855.
78B1 Burke, S.K., Rainford, B.D.: J. Phys. F **8** (1978) L239.
78B2 Beille, J., Bloch, D., Towfig, F.: Solid State Commun. **25** (1978) 57.
78B3 Beille, J., Towfig, F.: J. Phys. F **8** (1978) 1999.
78B4 Bendick, W., Ettwig, H.H., Pepperhoff, W.: J. Phys. F**8** (1978) 2525.
78K1 Kondorskii, E.I., Kostina, T.I., Galkin, Y.Yu.: Conf. Ser. – Inst. Phys. **39** (Transition Met., 1977) (1978) 611.
78Y1 Yao, Y., Arajs, S.: Phys. Status Solidi (b) **89** (1978) 201
79M1 Maki, S., Adachi, K.: J. Phys. Soc. Jpn. **46** (1979) 1131.
79S1 Strom-Olsen, J.O., Wilford, D.F., Burke, S.K., Rainford, B.D.: J. Phys. F **9** (1979) L 95.
80A1 Aoki, Y., Hiroyoshi, H., Kawakami, M.: J. Mag. Magn. Mater **15–18** (1980) 1179.
80H1 Hornreich, R.M.: J. Mag. Magn. Mater **15–18** (1980) 387.
80M1 Men'shikov, A.Z., Kuzmin, N.N., Dorofeev, Yu.A., Kazantsev, V.A., Sidorov, S.K.: J. Mag. Magn. Mater. **20** (1980) 134.
80M2 Maki, S., Adachi, K.: Trans. Jpn. Inst. Met. **22** (1980) 182.
81M1 Men'shikov, A.Z., Burlet, P., Chamberod, A., Tholence, J.L.: Solid State Commun. **39** (1981) 1093.
81P1 Peretto, P., Venegas, R., Rao, G.N.: Phys. Rev. B **23** (1981) 6544.
82B1 Benediktsson, G., Hedman, L., Aastroem, H.U., Rao, K.V.: J. Phys. F **12** (1982) 1439.
82G1 Gavoille, G., Durupt, S., Hubsch, J.: J. Phys. (Les. Ulis, Fr.) **43** (1982) 773.
82M1 Men'shikov, A.Z., Takezei, G.A., Teplykh, A.E.: Fiz. Met. Metalloved. **54** (1982) 465 (Russ).
83A1 Alberts, H.L., Lourens, J.A.J.: J. Phys. F **13** (1983) 873.
83B1 Burke, S.K., Rainford, B.D.: J. Phys. F **13** (1983) 441.
83P1 Palumbo, A.C., Parks, R.D., Yeshurun, Y.: J. Mag. Magn. Mater. **36** (1983) 66.

3 4d and 5d elements

3.1 Introduction to the paramagnetism of 4d and 5d transition metals

To estimate the Stoner (Landau) enhancement factor F for 4d and 5d transition metals, one should single out χ_{spin} from the observed susceptibility at 0 K, χ (0 K). For that purpose one needs either to evaluate theoretically or to determine experimentally χ_{orb} and $\chi_{sp\text{-}orb}$. Such a procedure is not quite well defined except for Knight-shift experiments and causes ambiguity. Here we adopt, as a very crude estimate, χ (0 K) for χ_{spin}; the results for F calculated from the values of γ_{ob}, χ (0 K) and λ are listed in Table 1. It is concluded that, within the above-mentioned uncertainty, Os and Ru are in the weakly paramagnetic region, Pd is in the nearly ferromagnetic region and others are intermediate.

The 4d and 5d transition metals and their mutual alloys and compounds generally exhibit paramagnetism; the major part of their susceptibility arises from the spin paramagnetism of itinerant d electrons; paramagnetic susceptibilities χ_{orb} and $\chi_{sp\text{-}orb}$, which arise, respectively, from the terms of the orbital angular momentum and the spin-orbit interaction, are known also to be important [81 S 1].

Table 1. Basic constants and susceptibility data for 4d and 5d transition metals.
$\chi_m(0\,K)$ and $\chi_m(20\,°C)$: observed magnetic susceptibility at 0 K and 20 °C, respectively
γ: observed electronic specific heat coefficient
λ: electron-phonon interaction constant
F: Stoner (Landau) enhancement factor.
Reference to each value of $\chi_m(20\,°C)$ is found in the corresponding place in the column of references. See [82 V 1, 81 S 1] for references to the values of γ and λ

	Crystal structure	$\chi_m(0\,K)$ $10^{-4}\frac{cm^2}{mol}$	γ $\frac{mJ}{mol\,K^2}$	λ	F	$\chi_m(20\,°C)$ $10^{-4}\frac{cm^3}{mol}$			Ref.		
Zr	hcp	1.06	2.77	0.4	3.9	1.14	1.17 [1]	1.18	55 K 1	71 C 1	41 S 1
						1.18 [1]	1.21	1.21	68 V 1	61 K 1	62 T 1
						1.26	1.29		65 S 1	65 V 1	
Nb	bcc	2.27	7.80	0.9	4.0	2.02	2.04	2.05	65 S 1	53 K 1	76 H 1
						2.09	2.14	2.14	54 A 1	61 K 1	71 K 1
Mo	bcc	0.82	1.83	0.4	4.6	0.65	0.67	0.72	61 K 1	77 P 1	71 K 1
						0.73	0.83		57 A 1	53 K 1	
Tc	hcp	1.07	6.28	–	–	1.07	1.23		80 R 1	70 I 1	
Ru	hcp	0.39	2.8	0.4	1.4	0.34	0.41	0.43	61 K 1	70 I 1	31 G 1
Rh	fcc	0.95	4.64	–	–	1.01	1.02	1.02	81 A 1	51 H 1	60 B 1
						1.06			61 K 1		
Pd	fcc	7.0	9.36	0.7	9.3	5.25	5.29	5.31	51 H 1	60 B 1	63 M 1
						5.4	5.8		61 K 1	70 T 1	
Hf	hcp	0.71	2.21	0.3	3.1	0.68	0.70	0.70	61 K 1		65 V 1
						0.74 [1]	0.75		71 C 1	55 K 1	
Ta	bcc	1.54	6.15	0.7	3.1	1.50	1.52	1.54	53 K 1	54 A 1	51 H 1
						1.54	1.62		61 K 1	71 K 1	
W	bcc	0.52	0.90	0.25	5.3	0.52	0.53	0.53	75 A 1	61 K 1	71 K 1
						0.54	0.59		57 A 1	53 K 1	
Re	hcp	0.63	2.35	0.4	2.7	0.56	0.65	0.68	54 A 1	61 K 1	71 K 1
						0.68	0.71 [1]		52 W 1	69 V 1	
Os	hcp	0.09	2.35	0.4	0.4	0.10	0.10 [1]	0.13	31 G 1	73 G 1	67 W 1
Ir	fcc	1.98	3.19	0.3	5.9	0.23	0.27		60 B 1	61 K 1	
Pt	fcc	2.06	6.48	0.6	3.7	1.87	1.89	1.89	78 I 1	51 H 1	61 K 1
						1.91	2.0	2.05	60 B 1	70 T 1	74 W 1

[1]) Single-crystal specimens, $(\chi_{m\parallel}(20\,°C)+2\chi_{m\perp}(20\,°C))/3$ is listed.

3.2 Magnetic susceptibility

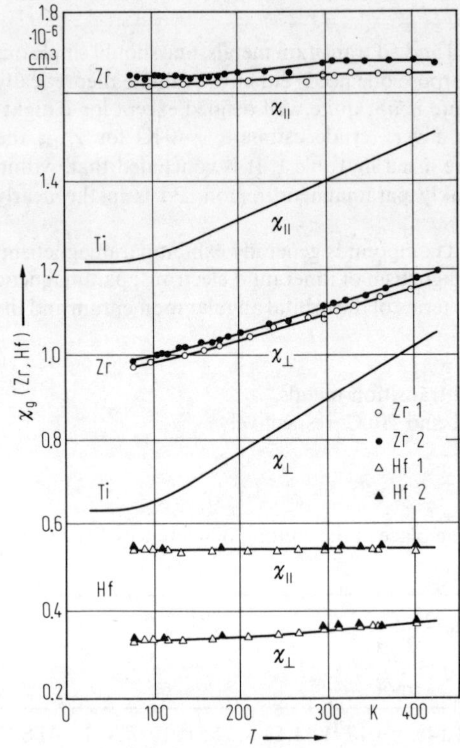

Fig. 1. Temperature dependence of the susceptibility of Zr and Hf for, respectively, two different specimens; for reference the data on Ti is included [71 C 1].

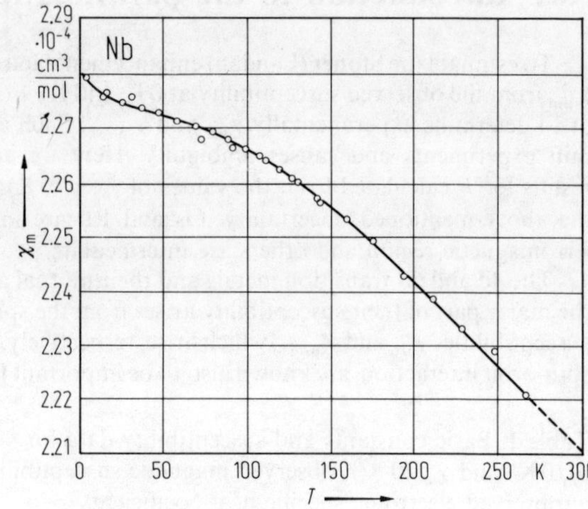

Fig. 2. Temperature dependence of the susceptibility of Nb [76 H 1].

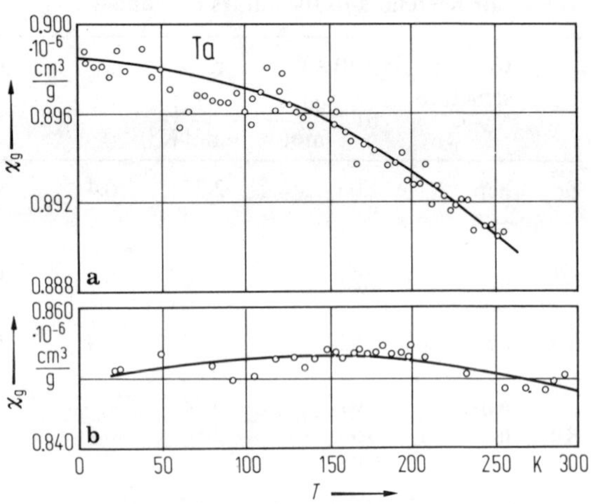

Fig. 3. Temperature dependence of the susceptibility of Ta. (a) [78 K 1], (b) [54 H 1].

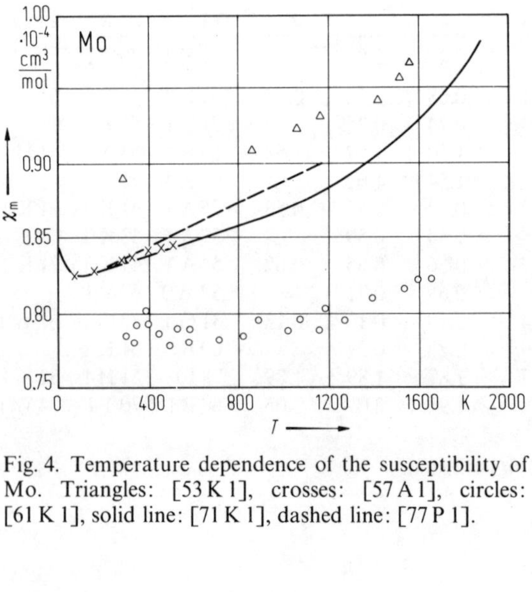

Fig. 4. Temperature dependence of the susceptibility of Mo. Triangles: [53 K 1], crosses: [57 A 1], circles: [61 K 1], solid line: [71 K 1], dashed line: [77 P 1].

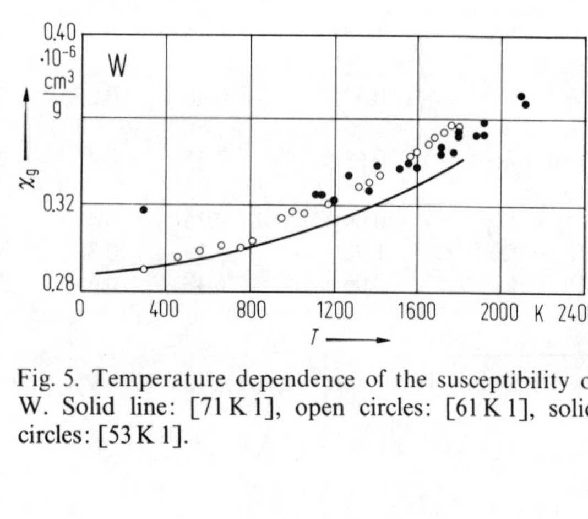

Fig. 5. Temperature dependence of the susceptibility of W. Solid line: [71 K 1], open circles: [61 K 1], solid circles: [53 K 1].

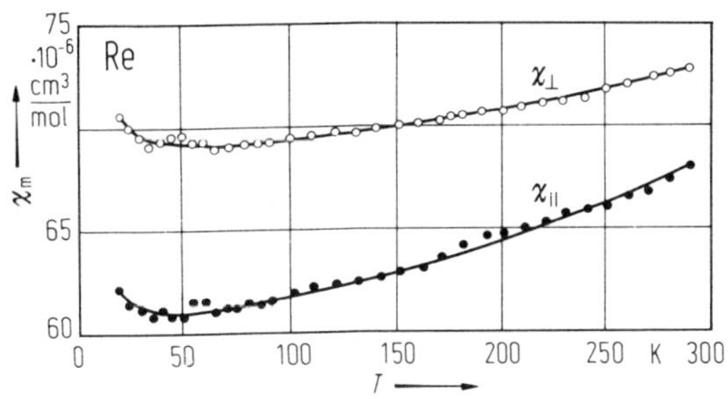

Fig. 6. Temperature dependence of the susceptibility of Re in two crystallographic directions. $\chi_\perp$: $H_{appl} \perp c$, $\chi_\parallel$: $H_{appl} \parallel c$ [69 V 1].

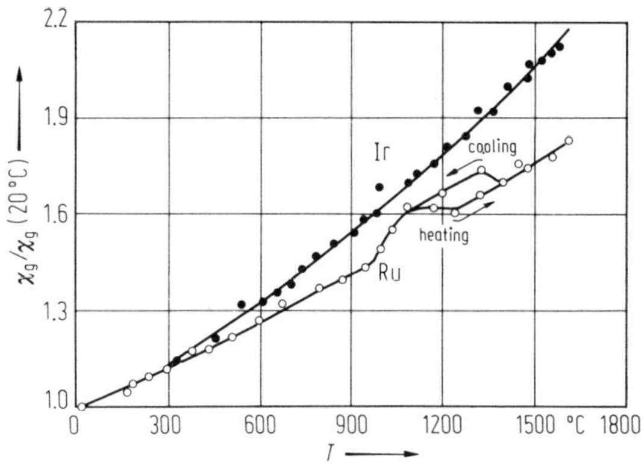

Fig. 7. Temperature dependence of the relative susceptibility of Ru and Ir. χ_g (20 °C) is 0.34 and 0.14 in units of $[10^{-6} cm^3/g]$ for Ru and Ir, respectively [61 K 1].

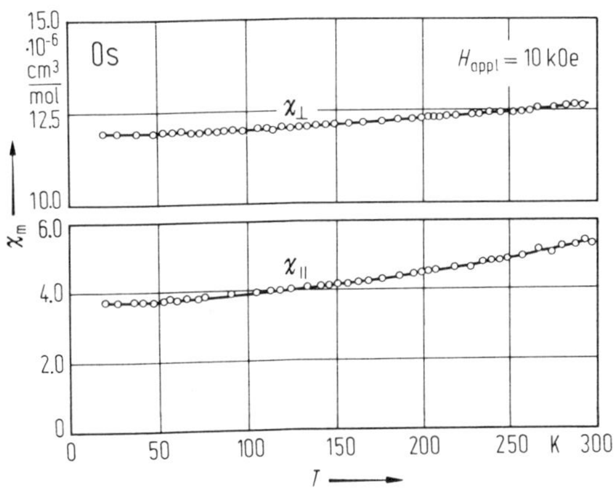

Fig. 8. Temperature dependence of the susceptiblity of Os single crystal for two directions of the magnetic field at 10 kOe; $\chi_\parallel$: $H_{appl} \parallel c$ and $\chi_\perp$: $H_{appl} \perp c$ [73 G 1].

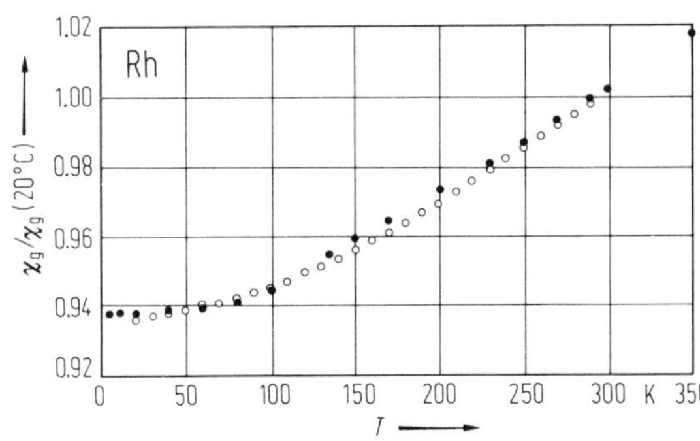

Fig. 9. Temperature dependence of the relative susceptibility of Rh. Solid circles: [81 A 1], open circles: [60 B 1].

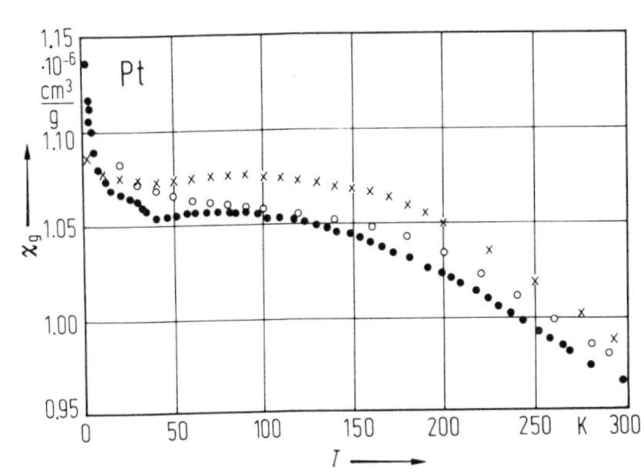

Fig. 10. Temperature dependence of the susceptiblity of Pt. Crosses: [73 D 1], open circles: [60 B 1], solid circles: [78 I 1].

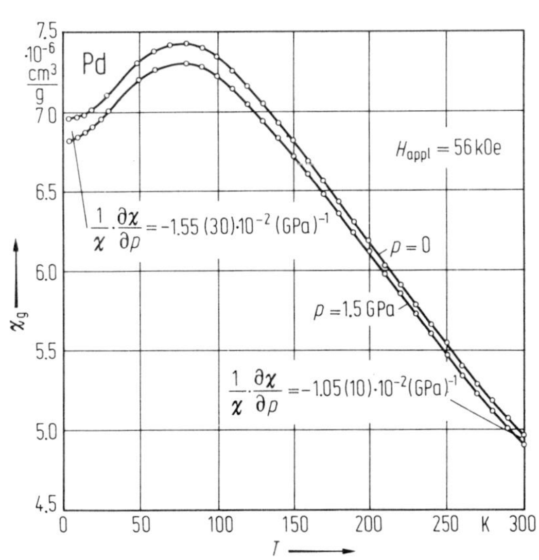

Fig. 11. Temperature dependence of the susceptiblity of Pd at hydrostatic pressure of 0 and 1.5 GPa in a magnetic field of 56 kOe [81 G 1].

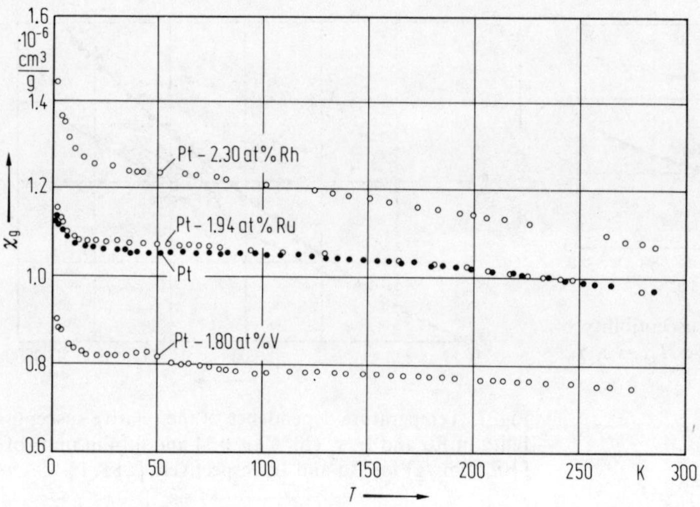

Fig. 12. Temperature dependence of the susceptibility of Pt-based dilute alloys Pt–Ru, Pt–Rh, Pt–V [78 I 1].

3.3 References for 3

31 G 1 Guthrie, A.N., Bourland, L.T.: Phys. Rev. **37** (1931) 303.
41 S 1 Squire, C.F., Kaufmann, A.R.: J. Chem. Phys. **9** (1941) 673.
51 H 1 Hoare, F.E., Walling, J.C.: Proc. Phys. Soc. (London) Sect. B **64** (1951) 337.
52 H 1 Hoare, F.E., Matthews, J.C.: Proc. R. Soc. London Ser. A **212** (1952) 137.
52 W 1 Wucher, J., Perakis, N.: C. R. Acad. Sci. **235** (1952) 419.
53 K 1 Kriessman, C.J.: Rev. Mod. Phys. **25** (1953) 122.
54 A 1 Asmussen, R.W., Soling, H.: Acta Chem. Scand. **8** (1954) 563.
54 H 1 Hoare, F.E. Kouvelites, J.S., Matthews, J.C., Preston, J.: Proc. Phys. Soc. (London) Sect. B **67** (1954) 728.
55 K 1 Kriessman, C.J., McGuire, T.R.: Phys. Rev. **98** (1955) 936.
57 A 1 Asmussen, R.W., Potts-Jensen, J.: Acta Chem. Scand. **11** (1957) 1271.
60 B 1 Budworth, D.W., Hoare, F.E., Preston, J.: Proc. R. Soc. London Ser. A **257** (1960) 250.
61 K 1 Kojima, H., Tebble, R.S., Williams, D.E.G.: Proc. R. Soc. London Ser. A **260** (1961) 237.
62 T 1 Taniguchi, S., Tebble, R.S., Williams, D.E.G.: Proc. R. Soc. London Ser. A **265** (1962) 502.
63 M 1 Manuel, A.J.. St. Quinton, J.M.P.: Proc. R. Soc. London Ser. A **273** (1963) 412.
65 S 1 Suzuki, H., Miyahara, S.: J. Phys. Soc. Jpn. **20** (1965) 2102.
65 V 1 Volkenshtein, N.V., Galoshina, E.V.: Phys. Met. Metallogr. USSR (English Transl.) **20** (1965) No. 3, 48.
67 W 1 Weiss, W.D., Kohlhaas, R.: Z. Angew. Phys. **23** (1967) 175.
68 V 1 Volkenshtein, N.V., Galoshina, E.V., Shchegolikhina, N.I.: Phys. Met. Metallogr. USSR (English Transl.) **25** (1968) No. 1, 166.
69 V 1 Volkenshtein, N.V., Galoshina, E.V., Shchegolikhina, N.I.: Sov. Phys. JETP (English Transl.) **29** (1969) 79.
70 I 1 Isaacs, L.L., Lam, D.J.: J. Phys. Chem. Solids **31** (1970) 2581.
70 T 1 Treutmann, W.: Z. Angew. Phys. **30** (1970) 5.
71 C 1 Collings, E.W., Ho, J.C.: Phys. Rev. B **4** (1971) 349.
71 K 1 Kohlhaas, R., Wunsch, K.M.: Z. Angew. Phys. **32** (1971) 158.
73 D 1 van Dam, J.E.: Thesis, University of Leiden **1973**.
73 G 1 Galoshina, E.V., Gorina, N.B., Polyakova, V.P., Savitskii, E.M., Shchegolikhina, N.I., Volkenshtein, N.V.: Phys. Status Solidi (b) **58** (1973) K 45.
74 W 1 Weaver, H.T., Quinn, R.K.: Phys. Rev. B **10** (1974) 1816.
75 A 1 Alekseyeva, L.I., Budagovskiy, S.S., Bykov, V.N., Kondakhchan, I.G., Povarova, K.P., Podolyan, N.I., Savitskiy, Ye.M.: Phys. Met. Metallogr. USSR (English Transl.) **40** (1975) No. 5, 87.
76 H 1 Hechtfischer, D.: Z. Phys. B **23** (1976) 255.
77 P 1 Ploumbidis, D.: Z. Phys. B **28** (1977) 61.

78 I 1 Inoue, N., Sugawara, T.: J. Phys. Soc. Jpn. **44** (1978) 440.
78 K 1 Köbler, U., Schober, T.: J. Less-Common Met. **60** (1978) 101.
80 R 1 Radhakrishna, P., Brown, P.J.: J. Phys. F **10** (1980) 489.
81 A 1 Abart, J., Voitländer, J.: Solid State Commun. **40** (1981) 277.
81 G 1 Gerhardt, W., Razavi, F., Schilling, J.S., Hüser, D., Mydosh, J.A.: Phys. Rev. B **24** (1981) 6744.
81 S 1 Shimizu, M.: Rep. Prog. Phys. **44** (1981) 329.
82 V 1 Vonsovsky, S.V., Izyumov, Yu.A., Kurmaev, E.Z.: Superconductivity in Transition Metals, Berlin, Heidelberg, New York: Springer **1982**, ch. 4.

4 Alloys and compounds of 3d elements and 4d or 5d elements

4.1 3d elements and Zr, Nb, Mo or Hf, Ta, W, Re

4.1.1 Introduction

a) Phase diagram and crystal structure

Solubility and intermetallic compounds in binary systems are shown in Table 1. As seen in the table, most of the 4d and 5d elements are not soluble in the 3d elements, but they can form intermetallic compounds. Among them the Laves phase compound, AB_2, is the most important one. In particular, Laves phase compounds containing Fe or Co are extensively studied in respect of magnetism.

Laves phases have one of the three following structure types:
(i) C15 (cubic, $MgCu_2$ type),
(ii) C14 (hexagonal, $MgZn_2$ type),
(iii) C36 (hexagonal, $MgNi_2$ type).

In the present alloy systems, Mg is replaced by one of the 4d or 5d elements and a 3d element takes place of Cu or Zn site. The C36 type rarely appears.

For Table 1, see next page.

b) Magnetic properties

Most of the alloys and the intermetallic compounds are Pauli paramagnetic except for some of the Fe alloys and intermetallic compounds which show ferro- or antiferromagnetism. However, they usually show a temperature-dependent susceptibility. In some cases, this looks like a Curie-Weiss law. This behavior may be ascribed to a high density of states of the d band at the Fermi level. For these cases, we use a term "nonmagnetic", meaning a magnetic state in which no local-moment alignment, including ferro- and antiferromagnetism, takes place at the lowest temperature.

4.1.2 Ti and V alloys and compounds

No alloys and intermetallic compounds other than nonmagnetic ones have been found so far. Most of them become a superconductor at low temperatures. Fairly extensive studies including magnetic properties have been done because of the interest in superconductors, in particular, for Laves phase compounds of V_2M ($M=Zr$, Hf, etc.).

Table 2. Magnetic and related properties of V_2M ($M=Zr$, Hf, and Ta). All these materials are nonmagnetic and become a superconductor at low temperatures.

	V_2Zr	V_2Hf	V_2Ta
Structure	C15	C15	C15
χ_m(RT) [$10^{-4}\,cm^3\,mol^{-1}$]	10.5	10.6	6.0
χ_g(RT) [$10^{-6}\,cm^3\,g^{-1}$]	5.4	3.8	2.1
	[80 M 1]	[81 V 1]	[78 H 1]
T_s [K]	7.7	8.5	3.6
	[80 M 1]	[80 M 1]	[78 H 1]
K (^{51}V) at RT [%]	0.57	0.51	
	[78 P 1]	[81 V 1]	
e^2qQ (^{51}V) at RT [MHz]		1.5	
		[81 V 1]	

Table 1. Equilibrium phases and solubility limits of binary alloys between 3d and 4d or 5d elements. See [58 h 1] if not otherwise noted. For intermetallic compounds, the crystal structure is given in parentheses.

	Zr	Hf	Nb	Ta	Mo	W	Re
Ti	Solid solution [1]) Low-temperature phase α (hcp) High-temperature phase β (bcc)	Solid solution [1])	Solid solution Martensitic transformation (β→α) on Ti-rich side	Solid solution	Solid solution	Not soluble [1]) Ti in W 10 at%	Not soluble [1]) Ti_5Re_{24}
V	V_2Zr(C15) Zr in V 3 at%	V_2Hf(C15)[2])	Solid solution	Miscibility gap V in Ta 36 at% Ta in V 10 at% [1])	?	?	V in Re 3 at% Re in V 65 at% VRe_3 [1])
Cr	Not soluble $Cr_2Zr\binom{C15}{C14}$	Not soluble $Cr_2Hf\binom{C15}{C14}$	?	Ta in Cr 1.5 at% Cr in Ta 9 at% Cr_2Ta(C15, C14)	Miscibility gap	Cr in W 5 at% W in Cr 2 at%	Cr in Re 5 at% Re in Cr 35 at% Cr_2Re_3σ-phase [1])
Mn	Not soluble[1]) Mn_2Zr(C14)	Mn_2Hf(C14)	?	Mn_2Ta(C14)	σ-phase	?	$Mn_{14}Re_{16}$ [1]) σ-phase
Fe	Not soluble Fe_2Zr(C15)	Not soluble [1]) Fe_2Hf(C14+C15)	Fe_2Nb(C14)	Fe_2Ta(C14)	Mo in Fe 4 at% $Fe_7Mo_6(D8_5)$	Fe_2W(C14) Fe_3W_2 $Fe_7W_6(D8_5)$	Re in Fe 11 at% Fe_3Re_2 σ-phase
Co	Not soluble CoZr(CsCl) $Co_{11}Zr_2$, Co_4Zr $CoZr_2(CuAl_2)$ Co_2Zr(C15) [1])	Co_6Hf Co_2Hf(C15) CoHf(CsCl) $CoHf_2(Ti_2Ni)$ [1])	Co_2Nb(C15)	Co_2Ta [1])	$Co_3Mo(D0_{19})$ $Co_7Mo_6(D8_5)$	$Co_3W(D0_{19})$ $Co_7W_6(D8_5)$	Solid solution
Ni	$Ni_5Zr(AuBe_5)$ Ni_5Zr_2, $Ni_{10}Zr_7$ $Ni_{11}Zr_9$, NiZr $NiZr_2(CuAl_2)$[1])	Ni_5Hf Ni_7Hf_2, Ni_5Hf_2 $Ni_{10}Hf_7$ $Ni_{11}Hf_9$, NiHf $NiHf_2(AlCu_2)$[1])	Ni_3Nb NiNb	$Ni_3Ta(TiCu_3)$ Ni_2Ta_3	Mo in Ni 13 at% Ni_4Mo Ni_3Mo	W in Ni 12 at% Ni_4W	Ni in Re 45 at% Re in Ni 5 at% [1])

[1]) [65 e 1].
[2]) [69 s 1].

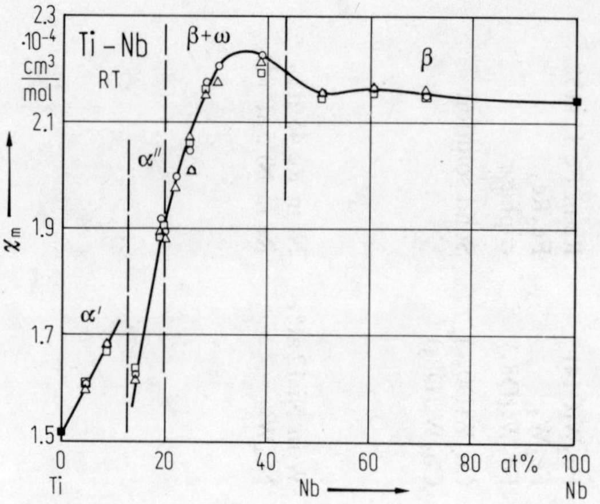

Fig. 1. Composition dependence of the room-temperature susceptibility of Ti–Nb alloys. Triangles: as cast, open circles: quenched from 1000 °C, open squares: quenched from 1350 °C, solid squares: pure Ti and Nb. Dashed lines indicate the boundary of metastable phases obtained by quenching from high temperatures. The crystal structures are as follows: α′: hcp, α″: tetragonal martensite, ω: complex, β: bcc. [76 C 1].

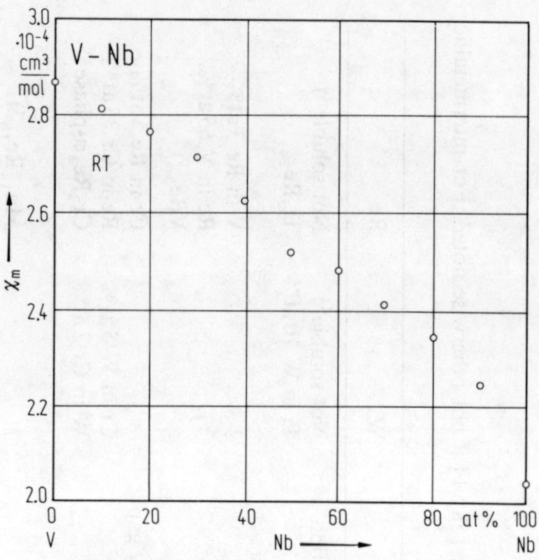

Fig. 2. Composition dependence of the magnetic susceptibility at room temperature of V–Nb alloys [67 L 1]. The V–Nb system forms a solid solution with the bcc structure over whole concentration range [54 W 1]. The specimens were homogenized at 1050 °C for one week and subsequently water quenched.

4.1.3 Cr alloys and compounds

Cr-rich alloys are in the spin density wave state or become antiferromagnetic. Intermetallic compounds of Cr seem to be nonmagnetic.

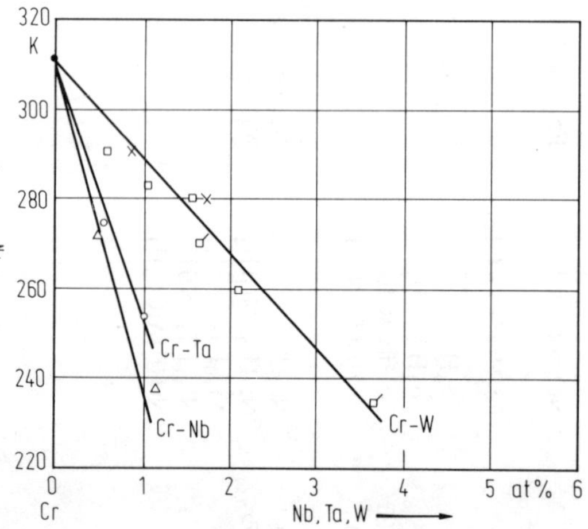

Fig. 3. Variation of the Néel temperature T_N in Cr–Nb [76 F 1]. Cr–Ta [76 F 1] and Cr–W, squares: [76 F 1], crosses: [66 K 1].

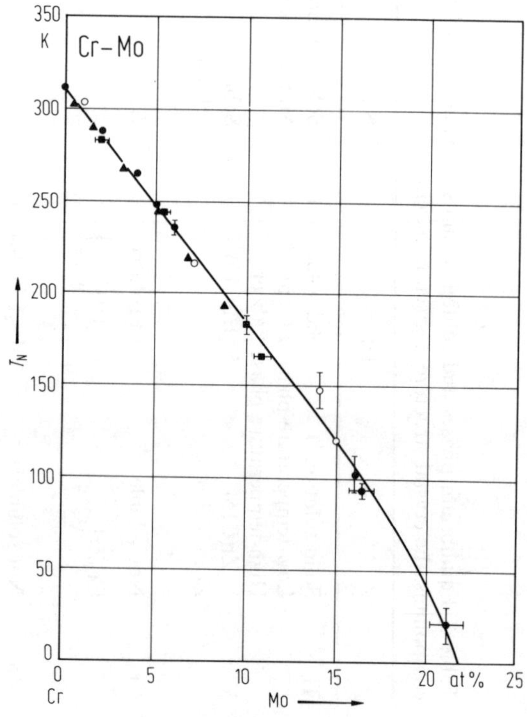

Fig. 4. Variation of the Néel temperature T_N in Cr–Mo alloys. Solid circles: [66 H 1], triangles: [68 A 1], open circles: [70 B 1] and squares: [80 S 1].

Table 3. Magnetic properties of Cr_2Zr and Cr_2Hf [83 H 1].

	Cr_2Zr	Cr_2Hf
Structure	C15	C14
Magnetism	Pauli paramagnetism	
χ_m (300 K) [10^{-4} cm^3 mol^{-1}]	9.2	4.0
Ref.	[83 H 1]	[68 A 2]

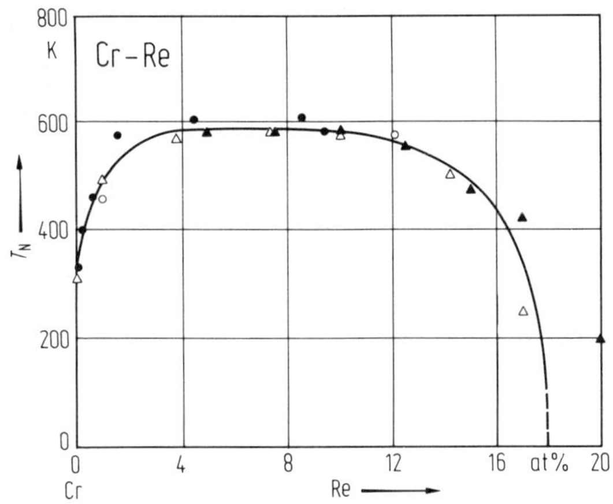

Fig. 5. Concentration dependence of the Néel temperature of Cr–Re alloys. Solid circles: [65 B 1], open circles: [70 B 1]; determined by electric conductivity measurements. Solid triangles: [64 B 2], open triangles: [70 B 1]; determined by susceptibility measurements.

4.1.4 Mn alloys and compounds

Mn-rich alloys are polymorphic and show complicated phase diagrams whose magnetic properties are not well understood. Neither ferromagnetic nor antiferromagnetic phases have been found in intermetallic compound except their hydrides.

4.1.5 Fe alloys and compounds

The solubility limit of 4d and 5d elements in Fe is very small. Therefore, interests in magnetism of Fe-rich single-phase alloys are reduced to the effects of impurities on ferromagnetism of bcc Fe, which are treated in sect. 1.1. On the other hand, the solubility of Fe in 4d and 5d transition metals is also low. The magnetic properties of Fe impurities in 4d and 5d metals and alloys are studied in relation to the Kondo effect, which are described in Landolt-Börnstein, NS, vol. III/15a [82 f 1].

Laves phase Fe_2Zr and Fe_2Hf are the only ferromagnetic binary stoichiometric compounds treated in this section. In particular Fe_2Zr and related alloys are most extensively studied. Distinct magneto-volume effects are observed in these materials. For instance, the $Fe_2Zr_{0.7}Nb_{0.3}$ alloy exhibits an Invar-like thermal expansion curve.

The $Fe_2Hf_{1-x}Ta_x$ system has an interesting magnetic phase diagram. At $x=0.2$, ferromagnetic to antiferromagnetic and antiferromagnetic to paramagnetic transitions take place with increasing temperature and metamagnetic properties are observed in the antiferromagnetic region.

The $(Fe_{1-x}Co_x)_2Zr$ system exhibits spin-glass characteristics above $x=0.5$.

Table 4. Magnetic and related properties of $FeHf_2$ and $FeHf_2H_3$ hydride [79 B 1].

	$FeHf_2$	$FeHf_2H_3$
Structure type	Ti_2Ni	Ti_2Ni
a at RT [Å]	12.04	12.87
Magnetism	Paramagnetic	Ferromagnetic
σ at 4.2 K under 9 kOe [G cm^3 g^{-1}]		10.8
T_C [K]		73
IS (^{57}Fe) relative to α-Fe at RT [mm s^{-1}]	−0.12	0.285
$\frac{1}{2}e^2qQ$ (^{57}Fe) at RT [mm s^{-1}]	0.47	

Table 5. Magnetic and related properties of Fe_2Zr. θ: angle between hyperfine field and [111] direction.

	4.2 K	RT	Ref.
Structure	C15		
a [Å]		7.073	79 M 1
Magnetism	Ferromagnetism		
σ_s [$G\,cm^3\,g^{-1}$]	88.2	80.9	79 M 1
p_{Fe} [μ_B]	1.6		79 M 1
T_C [K]	630		79 M 1
Θ [K]	643		79 M 1
C_m [$cm^3\,mol^{-1}\,K^{-1}$]	1.18		79 M 1
Axis of easy magnetization	[111]		64 W 1
dT_C/dp [$K\,kbar^{-1}$]	−2		73 B 1
$d\omega/dH$ [$10^{-10}\,Oe^{-1}$]	4.6	7.0	80 M 2
χ_{HF} [$10^{-5}\,cm^3\,cm^{-3}$]	2		79 M 1
γ [$erg\,mol^{-1}\,K^{-2}$]	13		79 M 1
H_{hyp} (^{57}Fe) [kOe]			
$\theta = 0°$ site	206		64 W 1
$\theta = 70.3°$ site	223	200	64 W 1
H_{hyp} (^{91}Zr) [kOe]	125		64 B 1
IS (^{57}Fe) relative to α-Fe [$mm\,s^{-1}$]	−0.057	−0.155	64 N 1
$\frac{1}{2}e^2qQ$ (^{57}Fe) [$mm\,s^{-1}$]	0.5		64 W 1

Table 6. Magnetic and related properties of Fe_2Hf. θ: angle between hyperfine field and [111] direction.

		Ref.
Structure	Mixture of C14 and C15 (fraction depends on heat treatment)	
a at RT [Å]	7.028 (C15)	70 N 1
Magnetism	Ferromagnetic	
σ (80 K) [$G\,cm^3\,g^{-1}$]	52.8	70 N 1
T_C [K]	591	
Axis of easy magnetization	[111] (C15)	76 L 1
H_{hyp} (^{57}Fe) at 300 K [kOe]		
$\theta = 0$ site	176	76 L 1
$\theta = 70.3°$ site	184	
H_{hyp} (^{181}Ta on Hf site) at RT [kOe]	130 (C15) 85 (C14)	76 L 1
$\frac{1}{2}e^2qQ$ (^{57}Fe) at RT [$mm\,s^{-1}$]	−0.41	76 L 1

Table 7. Magnetic and related properties of Fe_2M (M = Nb, Ta, W).

	Fe_2Nb	Fe_2Ta	Fe_2W	Ref.
Structure	C14	C14	C14	
a at RT [Å]	4.831	4.827	4.736	58 p 1
c at RT [Å]	7.882	7.838	7.719	58 p 1
Magnetism	Pauli paramagnetic			
χ_m (RT) [$10^{-3}\,cm^3\,mol^{-1}$]	2.3	2.5		69 K 1
IS (^{57}Fe) relative to α-Fe [$mm\,s^{-1}$]				
at 4.2 K	−0.13	−0.08		64 N 1
at RT	−0.23	−0.23	−0.25	64 N 1
$\frac{1}{2}e^2qQ$ (^{57}Fe) [$mm\,s^{-1}$]				
at 4.2 K	0.37	0.30		64 N 1
at RT	0.33	0.29	0.24	64 N 1

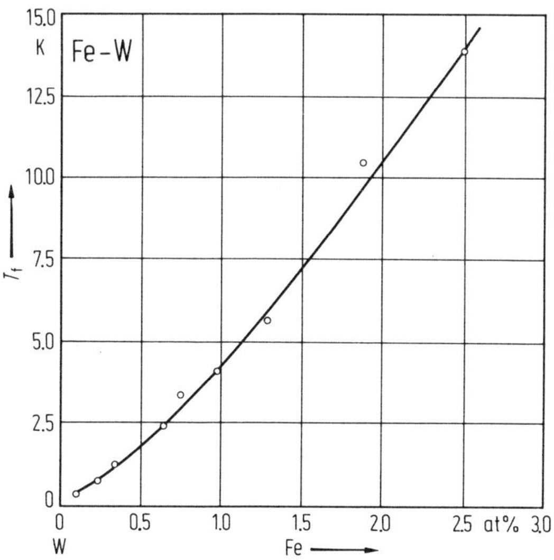

◀ Fig. 6. Composition dependence of the spin-glass freezing temperature T_f of Fe–W alloys deduced from the maximum of $\chi - T$ curves. The specimens were heat treated at 2280 °C for 3 h [80 K 1].

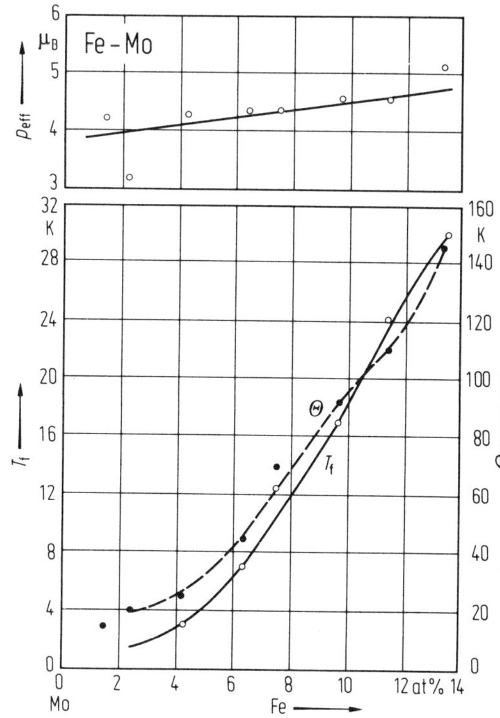

Fig. 7. Composition dependence of the spin-glass freezing temperature T_f, the paramagnetic Curie temperature Θ and the effective moment per Fe atom, p_{eff}, deduced from the Curie constants of Fe–Mo alloys. T_f was obtained from the maximum of the susceptibility vs. temperature curves under the applied field of 60 Oe. The samples were prepared by sintering of powder constituents under hydrogen atmosphere at 100 K below the melting temperature and then rapidly cooled [76 A 1].

4.1.6 Co and Ni alloys and compounds

Many types of intermetallic compounds are found but magnetic properties are scarcely known. Supposedly, most of them will be nonmagnetic. Among them, Laves phase compounds of Co are fairly well studied. Only off-stoichiometric $Co_{2+x}Zr$ is weakly ferromagnetic.

Table 8. Magnetic and related properties of certain intermetallic compounds of Co–Hf, Co–Ta, and Co–Nb [82 B 1]. For the compounds not showing magnetic ordering, the Pauli susceptibility is given instead of p_{Co} and T_C.

Compound	Structure	a [3]) Å	T_C K	p_{Co} μ_B	χ_g $\cdot 10^{-6}\,cm^3\,g^{-1}$
Co_2Hf	C15	6.896	–		7.8
Co_3Hf	C15	6.833	40	0.12	
$Co_{23}Hf_6$	Th_6Mn_{23}	11.502	499	0.66	
Co_7Hf	hexagonal	5.477 [1]) 8.070 [2])	600	1.14	
Co_2Ta	C15	6.761	–		12
Co_3Ta	Cu_3Au	6.788	–		10
Co_2Nb	C15	6.773	–		7
Co_3Nb	C15	6.717		0.17	

[1]) a, in hexagonal plane.
[2]) c, along hexagonal axis.
[3]) RT.

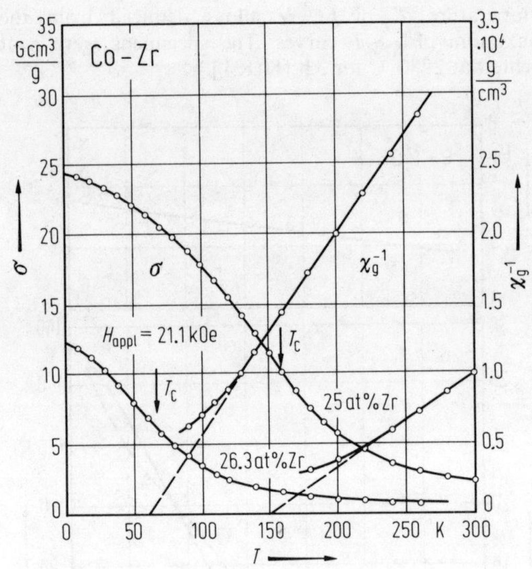

Fig. 8. Temperature dependence of the magnetization at 21.1 kOe and the inverse susceptibility for $ZrCo_x$ with $x = 2.8$ (26.3 at% Zr) and $x = 3.0$ (25.0 at% Zr) [81 F 2].

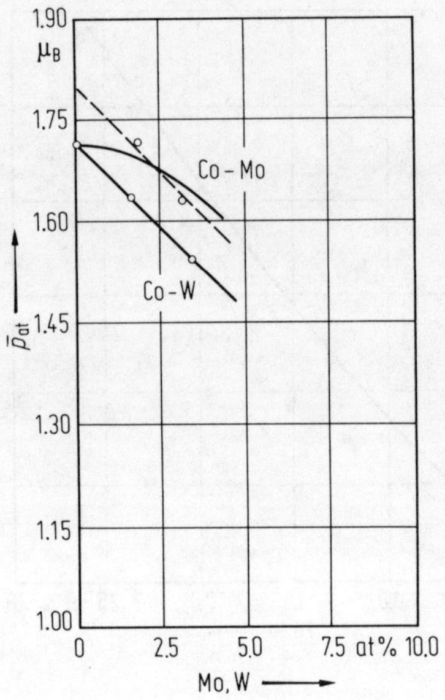

Fig. 9. Composition dependence of the average atomic moment of Co–Mo and Co–W alloys [32 S 1, 37 F 1].

Table 9. Magnetic and related properties of some intermetallic compounds of Ni–Zr [82 A 1]. Except for Ni_5Zr the susceptibility is temperature-independent.

Compound	Structure type	χ_m $\cdot 10^{-6}$ cm³ mol⁻¹	γ mJ mol⁻¹ K⁻²	Θ_D K
Zr	bcc	137	2.80	291
$NiZr_2$	$CuAl_2$	95	4.85	216
NiZr	CrB	72	2.0	270
$Ni_{11}Zr_9$	$Ni_{11}Zr_9$	79		
$Ni_{10}Zr_7$	$Ni_{10}Zr_7$	73		
$Ni_{21}Zr_8$	$Ni_{21}Zr_8$	79	2.5	351
Ni_3Zr	Ni_3Sn	89		
Ni_7Zr_2	Ni_7Zr_2	98		
Ni_5Zr	$AuBe_5$	[1])	5.75	485

[1]) Weak ferromagnet.

4.1.7 References for 4.1

58 h 1 Hansen, M.: Constitution of Binary Alloys, New York: McGraw-Hill Inc. **1958**.

58 p 1 Pearson, W.B.: A Handbook of Lattice Spacings and Structures of Metals and Alloys, London: Pergamon Press **1958**.

65 e 1 Elliott, R.P.: Constitution of Binary Alloys, 1st. suppl., New York: McGraw-Hill Inc. **1965**.

69 s 1 Shunk, F.A.: Constitution of Binary Alloys, 2nd. suppl., New York: McGraw-Hill Inc. **1969**.

82 f 1 Fischer, K.H., in: Landolt-Börnstein, New Series (Hellwege, K.H., Olsen, J.L., eds.), Berlin, Heidelberg, New York: Springer, vol. **15a** (1982) 289.

32 S 1 Sadron, C.: Ann. Phys. Ser. 10, **17** (1932) 371.

37 F 1 Farcas, T.: Ann. Phys. Ser. 11, **8** (1937) 146.

54 W 1 Wilhelm, H.A., Carlson, O.N., Dickinson, J.M.: Trans. AIME **200** (1954) 915.

64 B 1 Betsuyaku, H., Komura, S., Betsuyaku, Y.: J. Phys. Soc. Jpn. **19** (1964) 1262.
64 B 2 Booth, J.G.: Phys. Status Solidi **7** (1964) K 157.
64 N 1 Nevitt, M.V., Kimball, C.W., Preston, R.S.: Proc. ICM, Nottingham **1964**, p. 137.
64 W 1 Wertheim, G.K., Jaccarino, V., Wernick, J.H.: Phys. Rev. A **135** (1964) 151.
65 B 1 Butylenko, A.K., Gridnev, V.N.: Fiz. Met. Metalloved. **19** (1965) 205.
66 H 1 Heiniger, F.: Phys. Kondens. Mater. **5** (1966) 285.
66 K 1 Koehler, W.C., Moon, R.M., Trego, A.L., Mackintosh, A.R.: Phys. Rev. **151** (1966) 405.
67 L 1 Lam, D.J., Spokas, J.J., Van Ostenburg, D.O.: Phys. Rev. **156** (1967) 735.
68 A 2 Abel, A.W., Craig, R.S.: J. Less-Common Met..**16** (1968) 77.
68 A 1 Arajs, S.: J. Appl. Phys. **39** (1968) 673.
69 K 1 Kai, K.: Dr. Thesis, Tohoku Univ. **1969**.
70 B 1 Bender, D. Muller, J.: Phys. Kondens Mater. **10** (1970) 342.
70 N 1 Nakamichi, T., Kai, K., Aoki, Y., Ikeda, K., Yamamoto, M.: J. Phys. Soc. Jpn. **29** (1970) 794.
73 B 1 Brouha, M., Buschow, K.H.J.: J. Appl. Phys. **44** (1973) 1813.
76 A 1 Amamou, A., Caudron, R., Costa, P., Friedt, J.M., Gautier, F., Loegel, B.: J. Phys. F **6** (1976) 2371.
76 C 1 Collings, E.W., Smith, R.D.: J. Less-Common Met. **48** (1976) 187.
76 F 1 Fukamichi, K., Saito, H.: J. Jpn. Inst. Metals (in Japanese) **40** (1976) 22.
76 L 1 Livi, F.P., Rogers, J.D., Viccaro, P.J.: Phys. Status Solidi (a) **37** (1976) 133.
78 H 1 Hafstrom, J.W., Knapp, G.S., Aldred, A.T.: Phys. Rev. B **17** (1978) 2892.
78 P 1 Pop, I., Coldea, M., Rao, V.U.S.: Phys. Status Solidi (a) **49** (1978) 207.
79 B 1 Buschow, K.H.J., van Diepen, A.M.: Solid State Commun. **31** (1979) 469.
79 M 1 Muraoka, Y., Shiga, M., Nakamura, Y.: J. Phys. F **9** (1979) 1889.
80 K 1 Krischel, D., Thomas, L.K.: J. Phys. F **10** (1980) 115.
80 M 1 Marchenko, V.A., Polovov, V.M.: Zh. Eksp. Teor. Fiz. **78** (1980) 1062 [Sov. Phys. JETP **51** (1980) 535].
80 M 2 Muraoka, Y., Shiga, M., Nakamura, Y.: J. Phys. F **10** (1980) 127.
80 S 1 Ström-Olsen, J.O., Wilford, D.F.: J. Phys. F **10** (1980) 1467.
81 F 2 Fujii, H., Pourarian, F., Wallace, W.E.: J. Mag. Magn. Mater. **24** (1981) 93.
81 V 1 Verkhovskiy, S.V., Kozhanov, V.N., Stepanov, A.P., Shevchenko, A.D., Pan, V.M., Bulakh, I.Ye.: Fiz. Met. Metalloved. **49** (1981) 553 [Phys. Met. Metallogr. (USSR) 49 No. 3 (1981) 91].
82 A 1 Amamou, A., Kuentzler, R., Dossmann, Y., Forey, P., Glimois, J.L., Feron, J.L.: J. Phys. F **12** (1982) 2509
82 B 1 Buschow, K.H.J.: J. Appl. Phys. **53** (1982) 7713.
83 H 1 Hirosawa, S., Pourarian, F., Sinha, V.K., Wallace, W.E.: J. Mag. Magn. Mater. **38** (1983) 159.

4.2 3d elements and Ru, Rh, Pd or Os, Ir, Pt

4.2.1 3d-rich alloys

Hardly any information is available for the 3d-rich alloys based on the paramagnetic 3d elements Sc, Ti, and V.

For antiferromagnetic Cr, the Néel temperature starts to increase with Ru, Rh, Os, Ir, and Pt content, reaching maximum values at a few at%. For Cr–Pd alloys T_N shows a different composition dependence (Fig. 1). The transition from incommensurate to commensurate state has been studied for Cr–Ru alloys. Magnetic phase diagrams showing the commensurate, incommensurate and paramagnetic states as a function of temperature and pressure are known for Cr + 0.3 at% Ru and Cr + 0.6 at% Ru (Fig. 2).

The γ-Mn phase can be stabilized by alloying Mn with small amounts of other elements (e.g. Pd) and quenching from high temperatures. In dilute Fe-(Ru, Rh, Os, Ir) alloys the Curie temperature decreases linearly with composition; the magnetic moment per atom initially increases with composition for Fe–Ir, Fe–Rh, and Fe–Pt, remains nearly constant for Fe–Ru and Fe–Pd, and decreases for Fe–Os (Fig. 8). The magnetic moment per atom and the Curie temperature both decrease with composition in Co-(Ru, Rh, Pd, Os, Ir, Pt) alloys (Figs. 7 and 9).

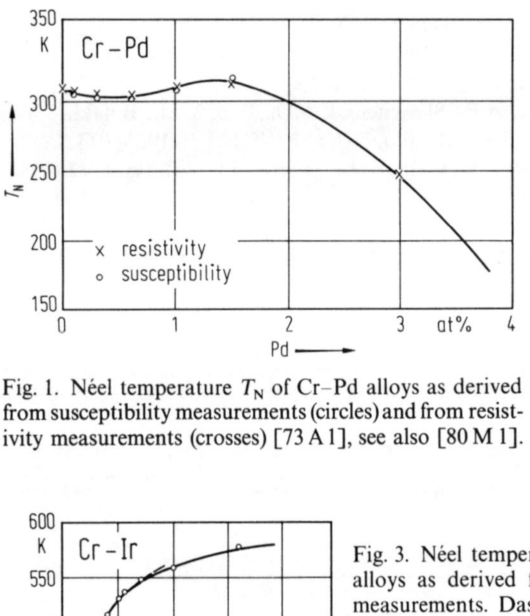

Fig. 1. Néel temperature T_N of Cr–Pd alloys as derived from susceptibility measurements (circles) and from resistivity measurements (crosses) [73 A 1], see also [80 M 1].

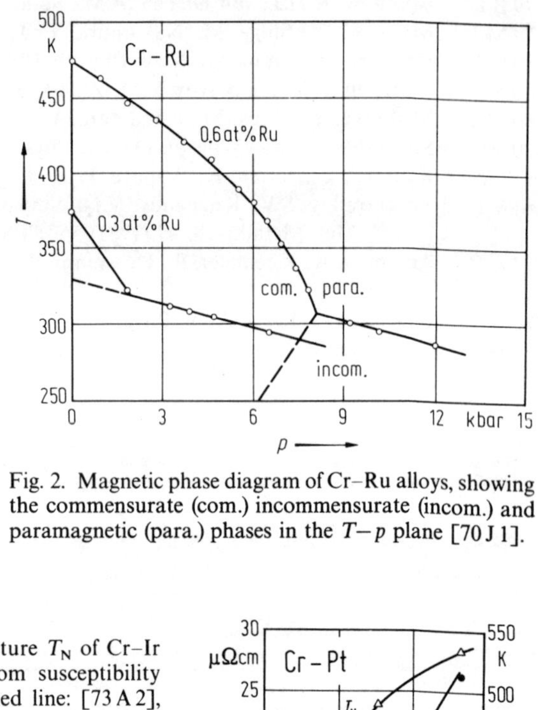

Fig. 2. Magnetic phase diagram of Cr–Ru alloys, showing the commensurate (com.) incommensurate (incom.) and paramagnetic (para.) phases in the $T-p$ plane [70 J 1].

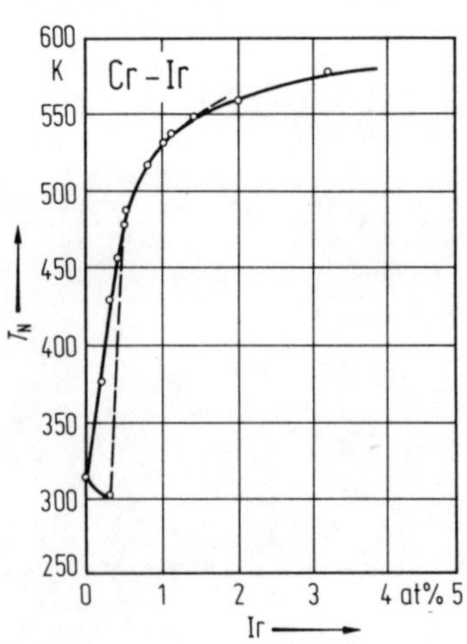

Fig. 3. Néel temperature T_N of Cr–Ir alloys as derived from susceptibility measurements. Dashed line: [73 A 2], solid line: [75 F 1].

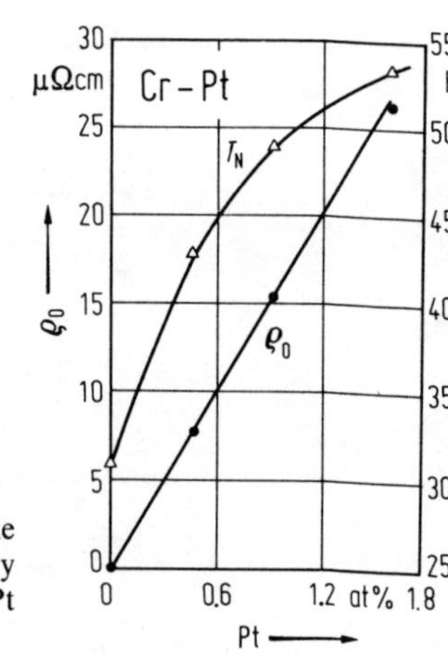

Fig. 4. Néel temperature T_N and the low-temperature electrical resistivity ϱ_0, determined at 4.2 K, of dilute Cr–Pt alloys [75 A 1].

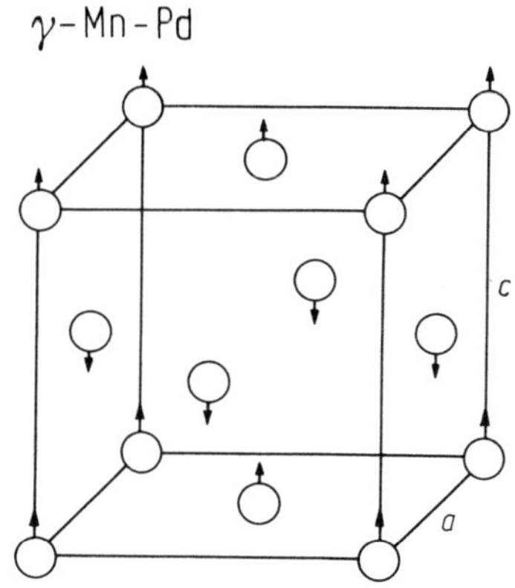

Fig. 5. Magnetic structure of γ-phase Mn–Pd alloys [68 H 1].

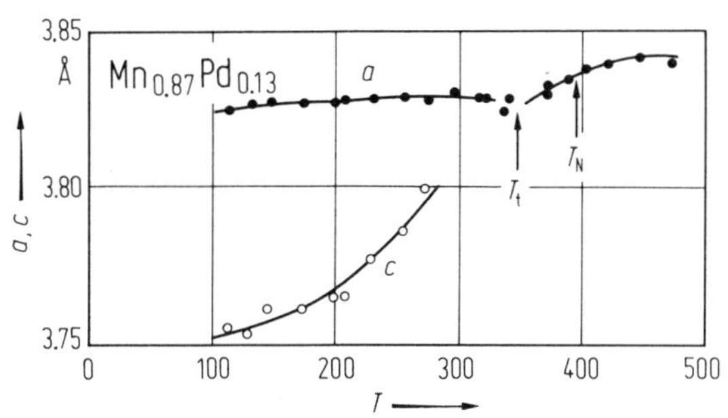

Fig. 6. Lattice parameters a and c vs. temperature for $Mn_{0.87}Pd_{0.13}$, T_N and T_t are the magnetic and structural transformation temperatures, respectively [68 H 1].

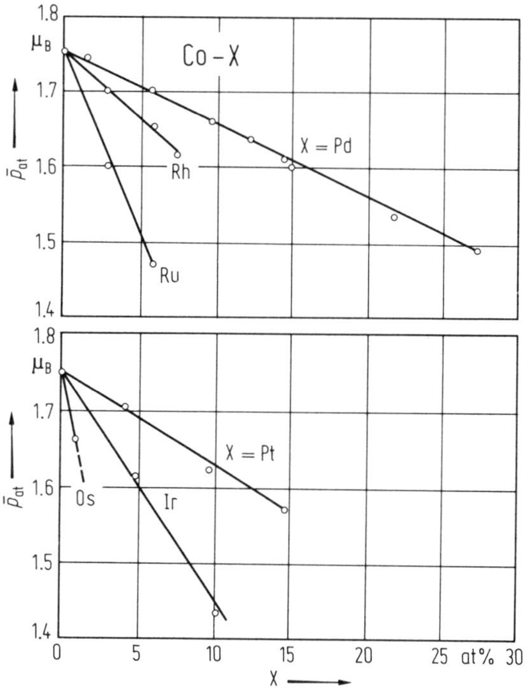

Fig. 7. Average magnetic moment per atom $\bar{p}_{at}$ of cobalt-rich fcc Co-(Ru, Rh, Pd, Os, Ir, Pt) alloys as a function of composition [60 C 1].

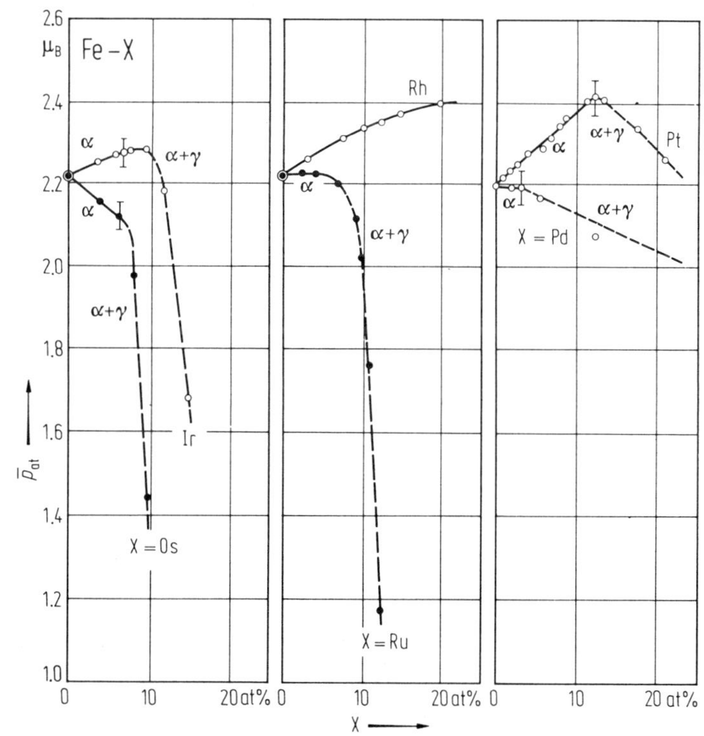

Fig. 8. Average magnetic moment per atom $\bar{p}_{at}$ in iron-rich Fe-(Ru, Rh, Pd, Os, Ir, Pt) alloys as a function of solute concentration [36 F 1].

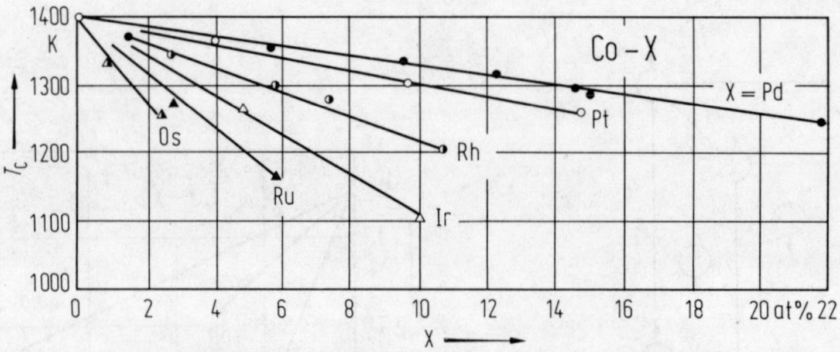

Fig. 9. Composition dependence of the Curie temperature of cobalt-rich fcc Co-(Ru, Rh, Pd, Os, Ir, Pt) alloys [60 C 1], see also [52 K 1].

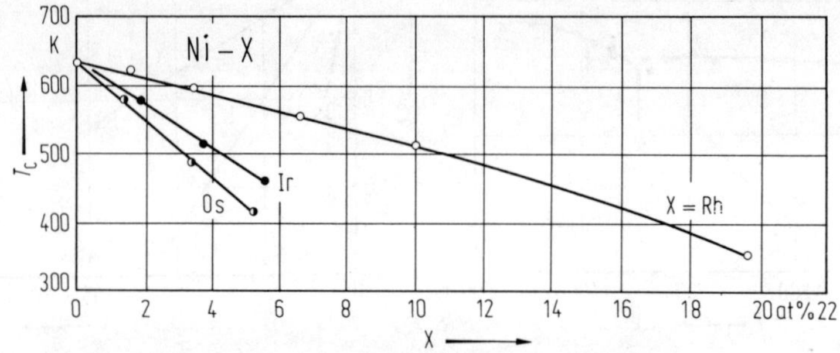

Fig. 10. Composition dependence of the Curie temperature of nickel-rich Ni–(Rh, Ir, Os) alloys [60 C 1].

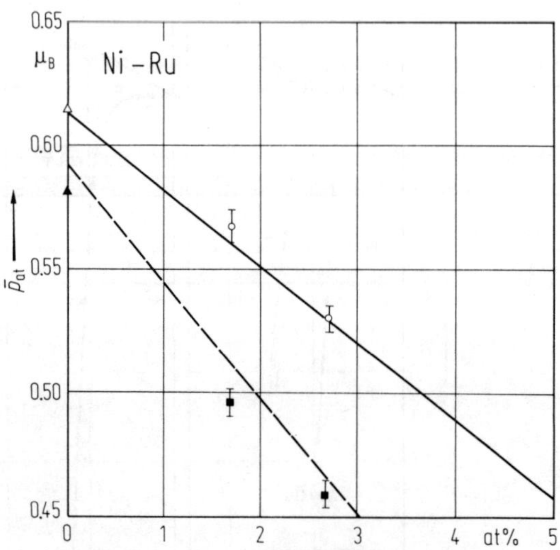

Fig. 11. Average magnetic moment per atom $\bar{p}_{at}$ for Ni–Ru alloys. Solid line: 0K [32 S 1], dashed line: 290 K [32 S 1], solid triangle: [71 C 1], open triangle: [71 M 1], circles: 4.2 K [79 P 1], squares: 290 K [79 P 1].

4.2.2 Concentrated alloys and intermetallic compounds

4.2.2.0 Introduction

Detailed phase diagrams for the binary constitutional alloys T–M, in which T is a 3d transition element and M a group-VIII 4d or 5d element, are available for most of the systems. The T–M alloys show a wide variety of ordered structures at room temperature. In many cases the experimentally determined magnetic parameters strongly depend on the state of atomic ordering. For the system T–Pt this is indicated in Table 1. The different types of magnetic order and the magnetic parameters of the ordered T–Pt compounds are summarized in Table 2.

Table 1. Crystallographic structure and type of magnetic order in atomically ordered (o) and disordered (d) T–Pt compounds [81 K 1]. T: 3d transition elements Ti···Ni; P: paramagnetic, F: ferromagnetic, AF: antiferromagnetic.

		T						
		Ti	V	Cr	Mn	Fe	Co	Ni
T_3Pt	o	A15 P	A15 P	A15 P	Cu_3Au AF	Cu_3Au F	–	Cu_3Au F
	d	–	–	–	fcc AF	fcc F	–	fcc F
TPt	o	AuCd	AuCd CuAuI P P	CuAuI AF	CuAuI AF	CuAuI F	CuAuI F	CuAuI P
	d	–	–	fcc P	fcc P?F	fcc F	fcc F	fcc F
TPt_3	o	Cu_3Au P	$TiAl_3$ Cu_3Au F F	Cu_3Au F	Cu_3Au F	Cu_3Au AF	Cu_3Au F	–
	d	–	fcc P	fcc P	fcc P?	fcc F	fcc F	–

Table 2. Magnetic moment per atom and magnetic ordering temperature of some ordered T–Pt compounds T: 3d transition elements V···Ni; P: paramagnetic, F: ferromagnetic, AF: antiferromagnetic, FI: ferrimagnetic; T_s: temperature at which a transition between two different types of magnetic order occurs. These quantities strongly depend on the atomic state of ordering; data from various authors generally disagree. The values given in this table have therefore qualitative significance only.

	T					
	V	Cr	Mn	Fe	Co	Ni
T_3Pt	P	P	AF, $T_N=485$ K $p_{Mn}=3.0\,\mu_B$ 2) $T_s=365$ K	F, $T_C=400$ K $p_{Fe}=2.7\,\mu_B$ 3) $p_{Pt}=0.5\,\mu_B$ 3)		F
TPt	P	AF $p_{Cr}=2.24\,\mu_B$ 1)	AF, $T_N=975$ K $p_{Mn}=4.3\,\mu_B$ 1) $T_s=715$ K	FI, $T_C=670$ K $p_{Fe}=2.8\,\mu_B$ 1) $p_{Pt}=-0.25\,\mu_B$	F, $T_C=750$ K $p_{Co}=1.6\,\mu_B$ $p_{Pt}=0.25\,\mu_B$	P
TPt_3	FI, $T_C=200\cdots240$ K $p_V=1\,\mu_B$ 4) $p_{Pt}=-0.3\,\mu_B$ 4)	FI, $T_C=481$ K $p_{Cr}=2.33\,\mu_B$ 1) $p_{Pt}=-0.27\,\mu_B$ 1)	F, $T_C=395$ K $p_{Mn}=3.60\,\mu_B$ 2) $p_{Pt}=0.17\,\mu_B$ 2)	AF, $T_N=170$ K $p_{Fe}=3.3\,\mu_B$	F, $T_C=290$ K $p_{Co}=1.64\,\mu_B$ $p_{Pt}\approx0.26\,\mu_B$	

1) At 300 K. 2) At 77 K. 3) At 4.2 K. 4) At 1.7 K.

4.2.2.1 Sc, Ti, and V alloys and compounds

Incidental experimental work has been reported for the CsCl-type of compounds ScRu and TiRu. Intermetallic compounds of Ti or V with Pd or Pt have been studied more frequently, especially VPt_3, which compound orders magnetically below 240 K in the atomically ordered Cu_3Au structure, the spontaneous magnetic moment per cell, $\bar{p}_{cell}$, being as small as 0.101 μ_B. For the tetragonal, $TiAl_3$-type of structure, compound VPt_3 different results are obtained: $T_C = 210$ K; $\bar{p}_{cell} = 0.075\,\mu_B$. NMR experiments on $V(Ir_{1-x}Pt_x)_3$ indicate that VPt_3 has a ferrimagnetic structure with $\bar{p}_V = 1\,\mu_B$ and $\bar{p}_{Pt} = -0.3\,\mu_B$. Disordered VPt_3 has been reported to be paramagnetic. Experiments on VPd_3 down to 1.6 K do not give any evidence for magnetic order in this compound.

4.2.2.2 Cr alloys and compounds

Thermomagnetic measurements have been performed on concentrated Cr–M alloys (M: group VIII 4d or 5d element) over the whole composition range (Figs. 12 and 13).

Cr–Pd and Cr–Pt have been studied in more detail. The fcc Cr–Pd alloys in the composition range 25⋯35 at % Cr show mictomagnetic behavior (Fig. 14).

Neutron diffraction measurements reveal that the magnetic structure of ordered $CrPt_3$ is ferrimagnetic with two sublattices: the Cr atoms occupy one sublattice, the Pt atoms the other one. An antiferromagnetic structure in which Cr spins are antiferromagnetic within each layer was detected in the ordered CrPt compound (CuAuI-structure) at room temperature (Fig. 15). For the disordered systems superparamagnetic behavior is suggested [73 B 2].

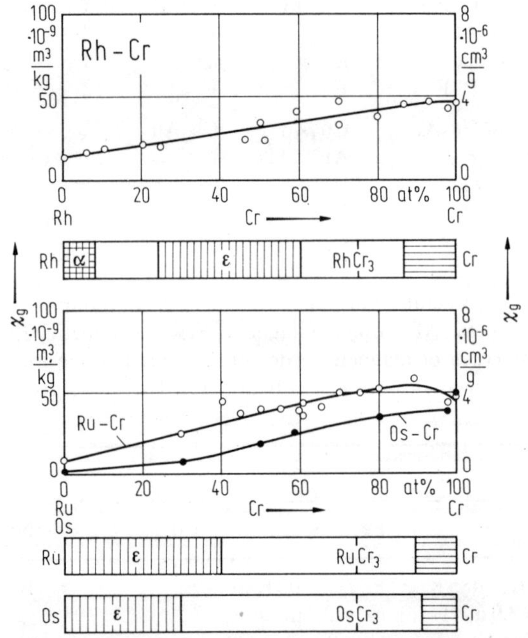

Fig. 12. Mass susceptibility χ_g as a function of composition in the systems Cr–Ru, Cr–Os, and Cr–Rh [68 K 1].

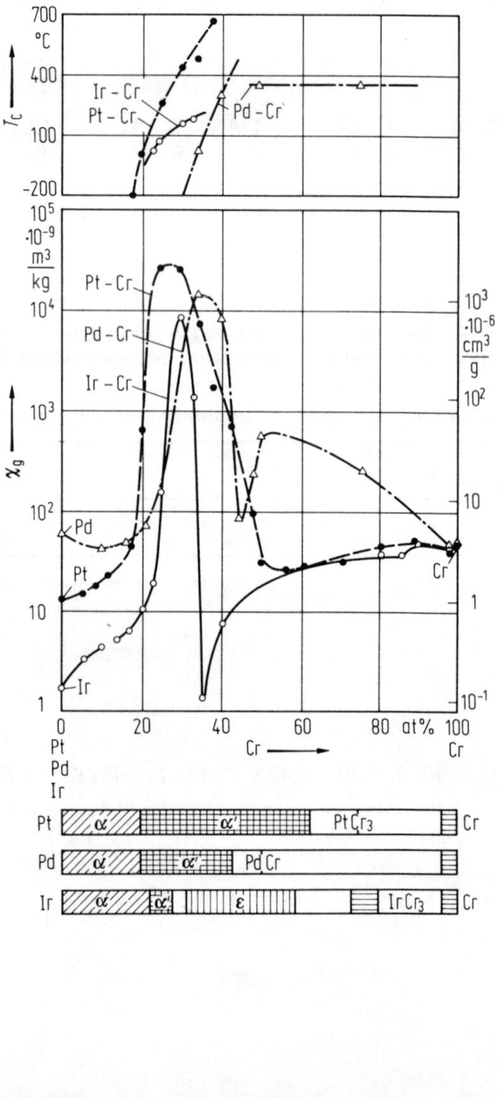

Fig. 13. Mass susceptibility χ_g and Curie temperature T_C of annealed Cr–Pd, Cr–Ir, and Cr–Pt alloys (for ferromagnetic samples χ_g refers to a field of $4\cdot10^5$ A/m ≙ 5.03 kOe [68 K 1].

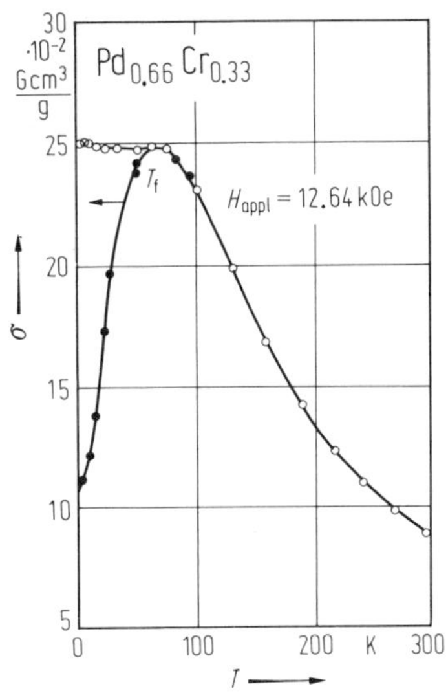

Fig. 14. Magnetic moment per unit of mass, σ, vs. temperature for quenched $Pd_{0.66}Cr_{0.33}$ at $H_{appl}=12.64$ kOe; open circles: field-cooled state, solid circles: cooled in zero field [71 C 2].

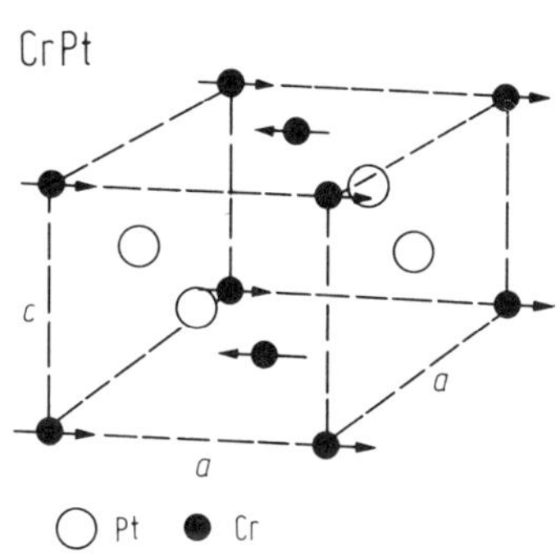

Fig. 15. Spin structure of CrPt (CuAuI-type of structure) [63 P 1].

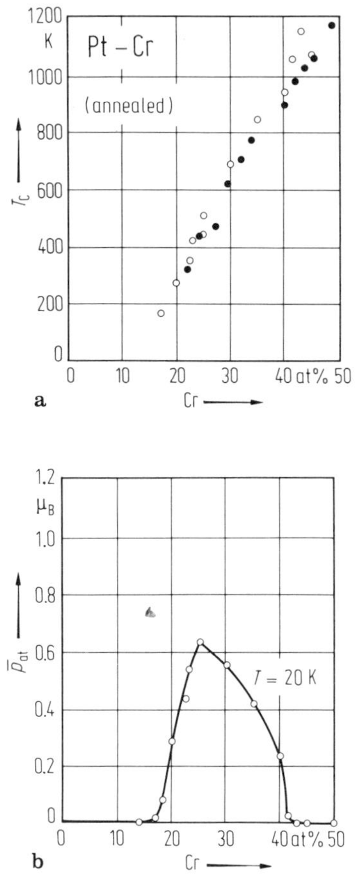

Fig. 16. Variation with composition of the Curie temperature T_C (a) and the mean magnetic moment (b) at 20 K for annealed Pt–Cr samples [73 B 1].

4.2.2.3 Mn alloys and compounds

Different types of antiferromagnetic order have been observed in the ordered Mn–M compounds. Susceptibility and resistivity versus temperature curves for the ordered compound MnRh show a large temperature hysteresis, indicating a gradual transition from the high-temperature CsCl-type structure to the low-temperature CuAuI-type structure Fig. 17. Tetragonal MnRh is a strong antiferromagnet with a Néel temperature well above 200 K. Neutron diffraction experiments on Mn_3Rh reveal a triangular antiferromagnetic structure below $T_N=850(10)$ K with a value for the magnetic moment per Mn atom of $3.6(4)\mu_B$ (Fig. 18).

The face-centered tetragonal form of Mn can be stabilized by addition of small amounts of other elements. Neutron cross sections of 5, 10, and 16 at% Pd in antiferromagnetic γ-Mn have been measured at 4.2 K. In MnPd (CuAuI-type structure) an antiferromagnetic structure with $\bar{p}_{Mn}=4.4(4)\,\mu_B$ and $T_N=810(10)$ K has been established by neutron diffraction. The possible magnetic structures are given in Fig. 23. Above 870 K the tetragonal structure of MnPd transforms into a CsCl-type of structure. In the stoichiometric $MnPd_3$ compound a one-dimensional, long-period superlattice based on the Cu_3Au-type of order with periodic antiphase domains has been observed. The magnetic moments in antiferromagnetic $MnPd_3$ change their direction with composition, being parallel to the c axis below 25 at% Mn and perpendicular above (Fig. 19).

A neutron diffraction study in disordered Mn–Ir alloys with face-centered cubic and face-centered tetragonal lattices reveals an increase in T_N with increasing Ir content in the composition range 10···30 at% Ir. By ordering in the Cu_3Au-type of structure, T_N is further increased. The magnetic transition is not associated with the crystallographic transition fcc-fct, but occurs at higher temperatures. A magnetic phase diagram of Mn–Ir is shown in Fig. 20; the magnetic structure is given in Fig. 21. Fig. 22 represents the lattice parameter together with the average magnetic moment and the Néel temperature as a function of composition.

In the Mn–Pt system ferromagnetic order has been reported for ordered $MnPt_3$ and antiferromagnetic order for ordered MnPt and Mn_3Pt. The results of magnetization studies on Mn–Pt alloys near the $MnPt_3$ composition are shown in Fig. 25). Values for the magnetic moments in $MnPt_3$ at 77 K are: $\bar{p}_{Mn} = 3.60(9)\mu_B$, $\bar{p}_{Pt} = 0.17(4)\mu_B$ [62 C 1]; a slightly different value for $\bar{p}_{Pt}$ of $0.26(3)\mu_B$ is reported in [69 A 1]. Ordered MnPt is antiferromagnetic with $\bar{p}_{Mn} = 4.3(2)\mu_B$ and $T_N = 970(10)$ K; its possible magnetic structures are given by Fig. 24a. Between 600 K and 800 K a transition from structure II to structure III occurs for the stoichiometric compound. In antiferromagnetic Mn_3Pt two different magnetic structures are found; the possible magnetic structures for this compound are shown in Fig. 24b. The values for the magnetic moment at 77 K and for the Néel temperature of the stoichiometric compound are: $\bar{p}_{Mn} = 3.0(3)\mu_B$ and $T_N = 475(10)$ K; the transition between the two structures D and F occurs at 365(10) K. A magnetic phase diagram of the whole Mn–Pt system is given in Fig. 24c.

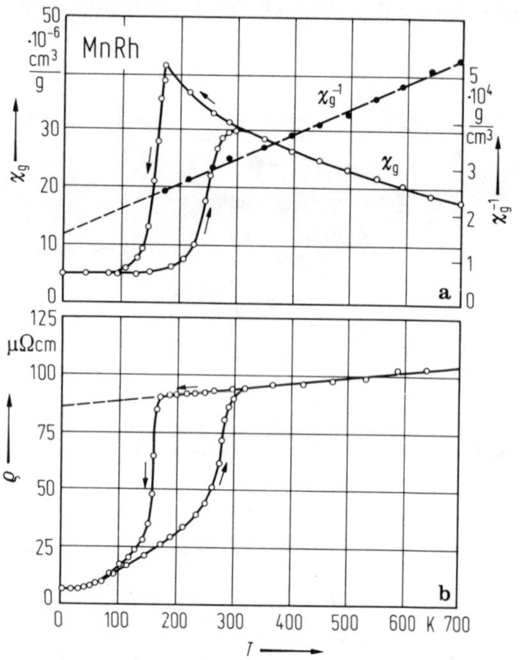

Fig. 17. Mass susceptibility χ_g (and its reciprocal) and ▶ electrical resistivity ϱ of equiatomic MnRh during a complete temperature cycle between 4.2 K and 700 K, indicated by arrows. In the ordered high-temperature cubic phase, the susceptibility is of a Curie-Weiss type. The susceptibility and resistivity of the low-temperature tetragonal (CuAuI-type) phase suggest that MnRh in its tetragonal form is a strong antiferromagnet [63 K 1].

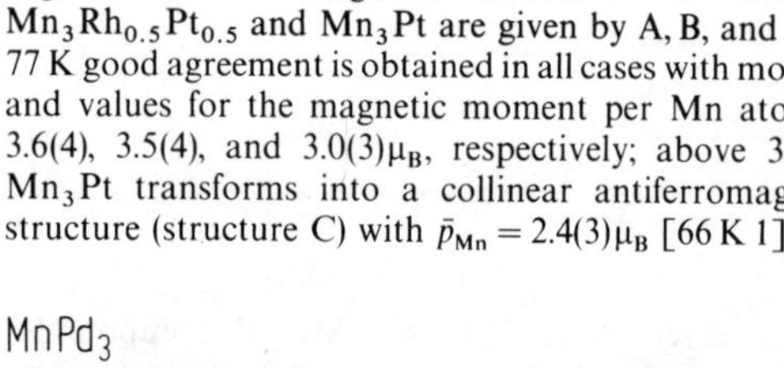

Fig. 18. Possible magnetic structures of Mn_3Rh, $Mn_3Rh_{0.5}Pt_{0.5}$ and Mn_3Pt are given by A, B, and C; at 77 K good agreement is obtained in all cases with model A and values for the magnetic moment per Mn atom of 3.6(4), 3.5(4), and $3.0(3)\mu_B$, respectively; above 380 K, Mn_3Pt transforms into a collinear antiferromagnetic structure (structure C) with $\bar{p}_{Mn} = 2.4(3)\mu_B$ [66 K 1].

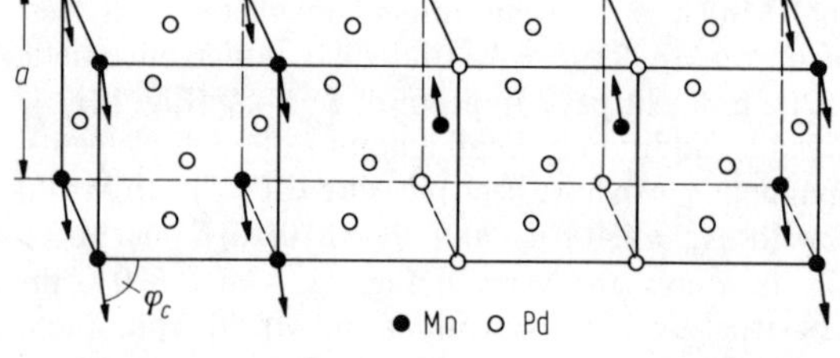

Fig. 19. Crystallographic and magnetic structures of $MnPd_3$; the crystallographic structure can be considered as an antiphase domain structure based on a nonstoichiometric CuAuI-type of order, indicating a continuous transition to the CuAuI-type phase existing at higher Mn concentration [69 K 1].

Table 3. Crystal and magnetic structure data for Mn–Pd alloys, the angle φ_c is defined in Fig. 19; S is the atomic long-range order parameter. The magnetic moments have been obtained at 77 K assuming $p_{Pd} = 0$ [69 K 1].

	at% Pd			
	80.5	77.3	74.7	69.6
a [Å]	3.87	3.87	3.87	3.93
c [Å]	–	15.48	15.48	15.05
c/a	1	4	4	3.83
S	–	0.85(5)	0.90(5)	1.00(5)
$\bar{p}_{Mn}$ [μ_B]	–	4.2(3)	4.1(3)	4.1(3)
φ_c [deg]	–	0(2)	8(2)	90(2)
T_N [K]	–	205(15)	220(10)	235(10)

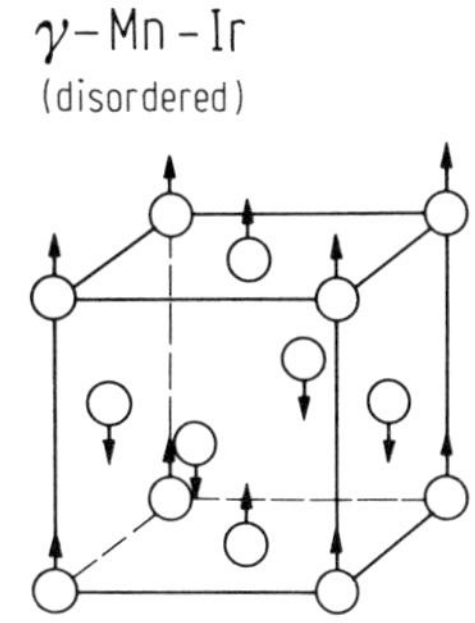

Fig. 21. The antiferromagnetic two-sublattice spin structure of the disordered γ-phase Mn–Ir alloys [74 Y 2].

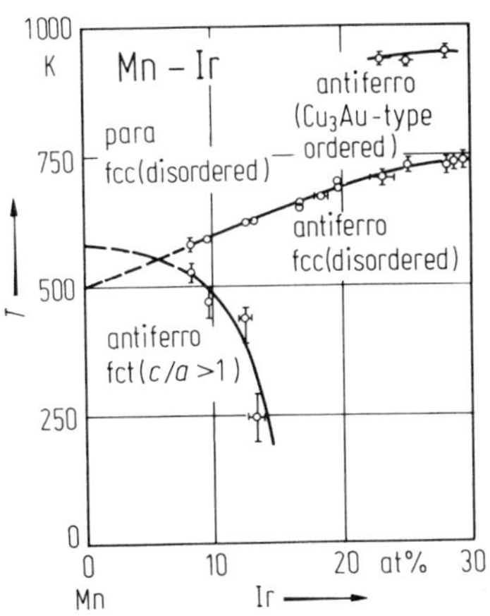

Fig. 20. Magnetic phase diagram of γ-phase $Mn_{1-x}Ir_x$ alloys [74 Y 1].

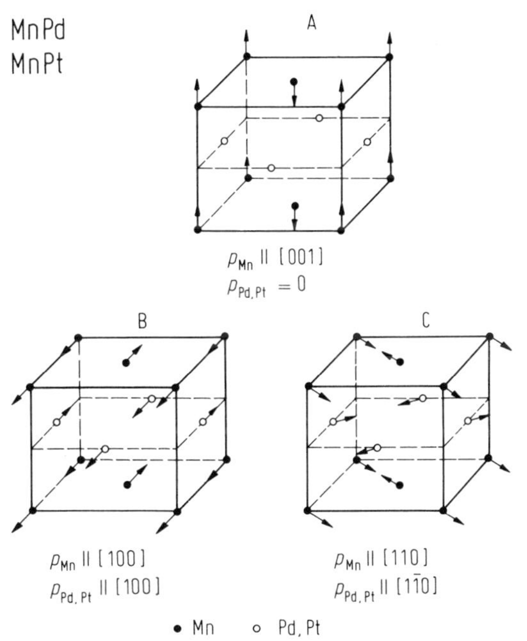

Fig. 23. Allowed magnetic structures of equiatomic MnX (CuAuI-type of structure) from neutron diffraction experiments. The Mn and X atoms are represented by open and solid circles, respectively; X = Pd, Pt; for MnPd the structure is either B or C; for MnPt structure A is found below T_s and structure B between T_s and T_N, see Table 4 [68 P 1].

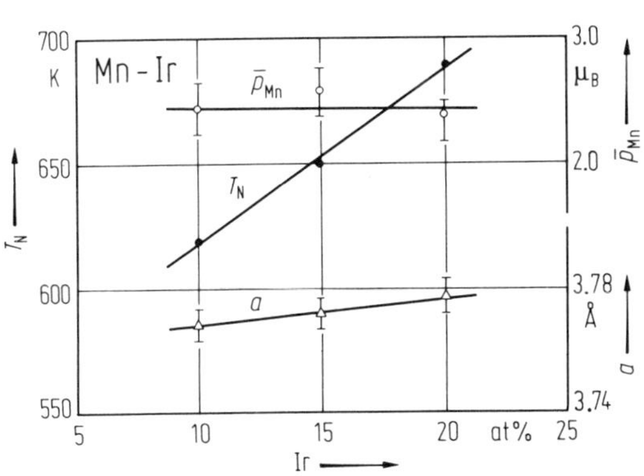

Fig. 22. Antiferromagnetic transition temperatures T_N, average magnetic moments per Mn atom, $\bar{p}_{Mn}$, and lattice parameters a of fcc Mn–Ir alloys [71 Y 1].

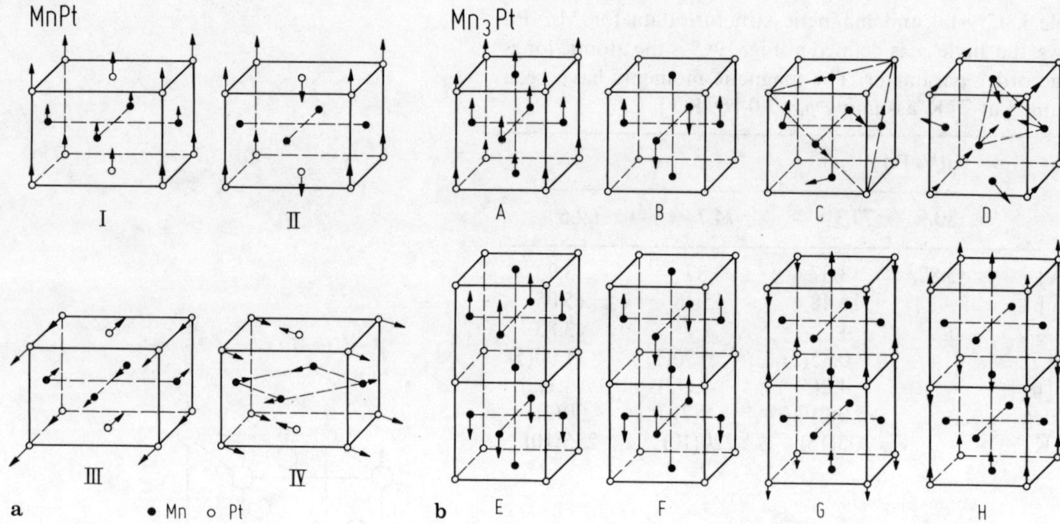

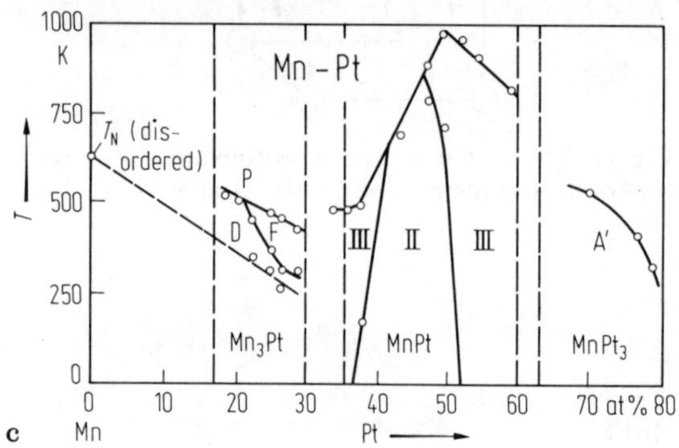

Fig. 24. (a) Possible magnetic structures of MnPt (CuAuI-type of structure); the Mn and Pt atoms are represented by solid and open circles, respectively; structures II ··· IV correspond to, respectively, structures A ··· C in Fig. 23, (b) possible magnetic structures of Mn_3Pt (Cu_3Au-type of structure), (c) magnetic phase diagram of the Mn–Pt system; the letters denote, for the ordered alloys, the stability regions of the various magnetic structures given in (a) and (b) [68 K 2]. A′: structure A in (b) with Mn on Pt-sites and vice versa. The dashed line represents T_N for disordered Mn–Pt.

Table 4. Crystal and magnetic structural data at room temperature for Pd–Mn and Pt–Mn alloys; values of the transition temperature T_s between different magnetic structures, and the Néel point T_N [68 P 1]. For type of magnetic structure, see Fig. 23. The magnetic moments of Mn are room-temperature values, with vanishing p_{Pd} and p_{Pt}.

Specimen	at% Mn	a Å	c Å	Type	$\bar{p}_{Mn}$ μ_B	T_s °C	T_N °C
$Mn_{0.90}Pt_{1.10}$	45.3	3.97	3.73	B	4.2(2)		630(20)
$Mn_{0.96}Pt_{1.04}$	48.0	3.99	3.67	B	4.2(2)	< -268	680(10)
MnPt	50.3	4.00	3.67	A	4.3(2)	≈440	700(10)
$Mn_{1.04}Pt_{0.96}$	52.3	3.99	3.67	A	4.2(2)	≈510	610(20)
$Mn_{1.13}Pt_{0.87}$	56.4	3.97	3.71	A	3.9(2)		420(20)
$Mn_{0.80}Pd_{1.20}$	40.1	4.05	3.64	B	4.2(2)		340(10)
$Mn_{0.88}Pd_{1.12}$	44.0	4.06	3.61	B	4.4(3)		560(20)
MnPd	50.0	4.07	3.58	B	4.4(4)		540(10)
$Mn_{1.03}Pd_{0.97}$	51.6	4.07	3.59	B	4.3(2)		490(10)

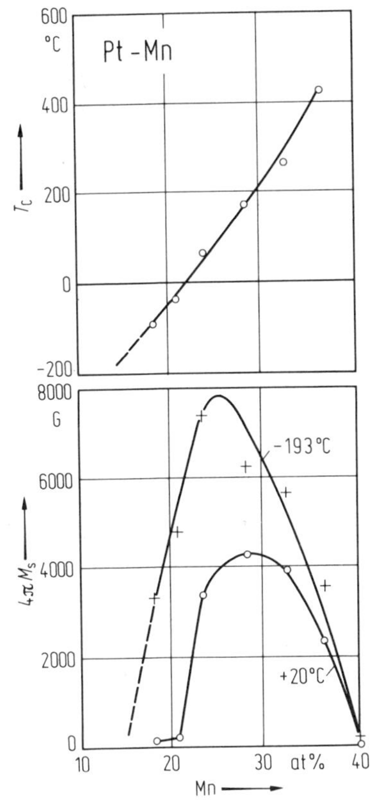

Fig. 25. Saturation magnetization M_s and Curie temperature T_C of Pt–Mn alloys near the 3:1 composition [50 A 1].

4.2.2.4 Fe alloys and compounds

Fe–Rh alloys near the equiatomic composition have the CsCl-type of structure. At low temperatures the alloys are antiferromagnetic with magnetic moment of 3.3 μ_B antiparallel on neighboring iron atoms. Above 320 K the magnetic structure transforms into a ferromagnetic state with magnetic moments of 3.17 μ_B and 0.97 μ_B on the Fe and Rh sites, repectively. The magnetic moment distribution for the nonstoichiometric Fe-rich compounds is shown in Fig. 27. The antiferromagnetic-ferromagnetic and ferromagnetic-paramagnetic transition temperatures have been studied as a function of pressure for equiatomic FeRh and $Fe_{0.515}Rh_{0.485}$.

The phase diagram of the Fe–Pd system reveals two ordered structures: FePd(CuAuI-type) and $FePd_3$; both order ferromagnetically. For FePd the iron and palladium magnetic moments amout to 2.85 μ_B and 0.35 μ_B, respectively [65 C 1]. Different results for the atomic magnetic moments have been reported for ordered $FePd_3$. At 300 K: $\bar{p}_{Fe} = 2.37(13)\mu_B$, $\bar{p}_{Pd} = 0.51(4)\mu_B$ [62 P 1]; in reference [65 C 1]: $\bar{p}_{Fe} = 3.10\mu_B$, $\bar{p}_{Pd} = 0.42\,\mu_B$.

In the Fe–Pt system several types of magnetic order have been observed. The Curie temperature as a function of composition is given in Fig. 29. In Fig. 32 the basic types of magnetic structures for this system are shown. In the composition range 24 ··· 36 at% Fe, a change in the type of antiferromagnetic order occurs from parallel iron moments in the (110) planes to parallel moments in the (100) or (010) planes. A value of 3.3 μ_B has been reported for the magnetic moment per iron atom in $FePt_3$ [63 B 1]. Above 32 at% Fe ferromagnetism becomes predominant. The saturation magnetization for ordered and disordered alloys is shown in Fig. 31.

In the pseudobinary series of $Fe(Pd_xPt_{1-x})_3$ a transition from a ferromagnetic state as in $FePd_3$ to an antiferromagnetic state as in $FePt_3$ occurs near x = 0.5. In the intermediate region an additional magnetic transition from a high-temperature ferromagnetic to a low-temperature canted-ferrimagnetic structure is found. The ordered equiatomic FePt alloy (CuAuI-type of structure) is hard to magnetize. A value for the mean atomic magnetic moment of 0.77 μ_B has been reported. By disordering FePt the bulk magnetization increases considerably. From these observations ferrimagnetic order was concluded for the ordered compound. Neutron diffraction studies, however, point to Fe magnetic moments at 300 K of 2.8(1) μ_B, parallel to the *c* axis, whereas Pt moments could not be determined [73 K 1]. In [74 M 1] the hypothetical values $\bar{p}_{Fe} = 2.75\,\mu_B$ and $\bar{p}_{Pt} = -0.25\,\mu_B$ are mentioned for the ordered FePt compound. The Fe–Pt alloys near the 25 at% Pt composition show invar characteristics as well in the Cu_3Au-type ordered state as in the disordered state. At low temperature these alloys undergo a martensitic transformation. The results of magnetic moment and Curie temperature studies on ordered and disordered alloys with 20 ··· 30 at% Pt are given in Figs. 33, 34. Anomalies in the thermal expansion are of a similar type as those in Fe–ni invar alloys.

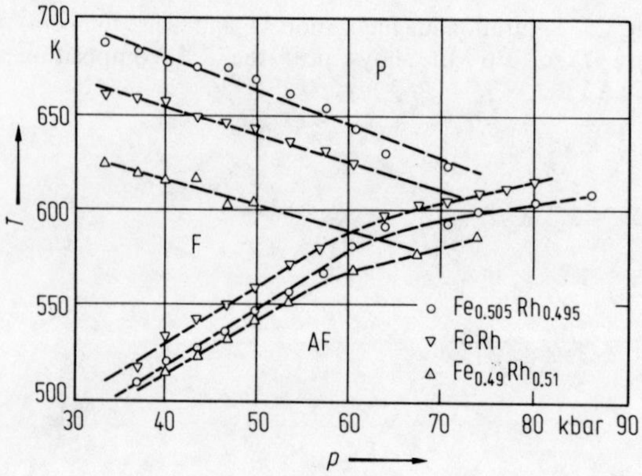

Fig. 26. $p-T$ magnetic phase diagram for Fe–Rh alloys of various composition [76 V 1], see also [81 V 1]. (The dashed lines serve only to guide the eye).

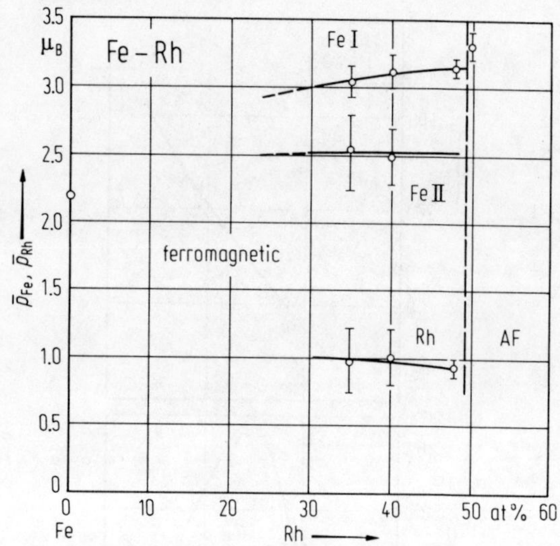

Fig. 27. Magnetic moment distribution in Fe–Rh alloys at room temperature, derived from neutron diffraction experiments: FeI and FeII denote iron atoms on the Fe and Rh sublattice, respectively [64 S 1].

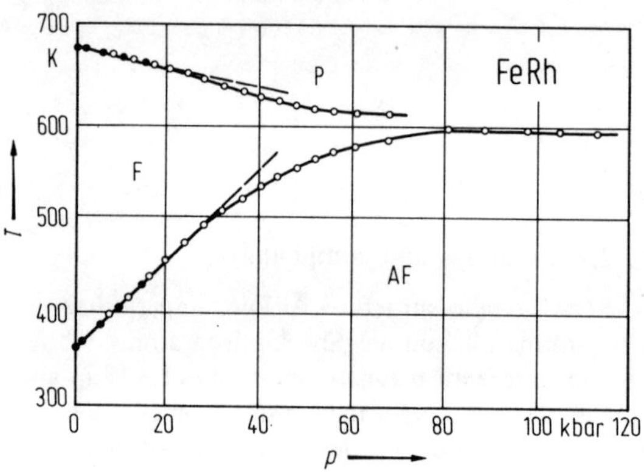

Fig. 28. $p-T$ phase diagram for FeRh. Open circles: [74 D 1], solid circles: [68 P 2]; see also [78 P 1].

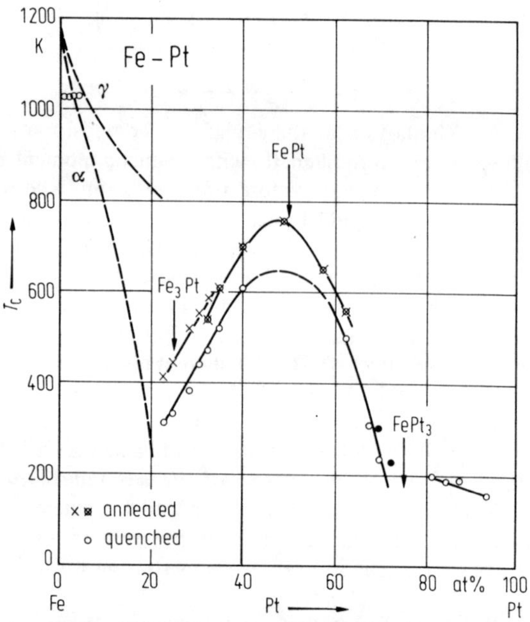

Fig. 29. Curie temperature of Fe–Pt alloys; open circles: quenched, crosses: annealed samples [50 K 1]; solid circles: [59 C 1].

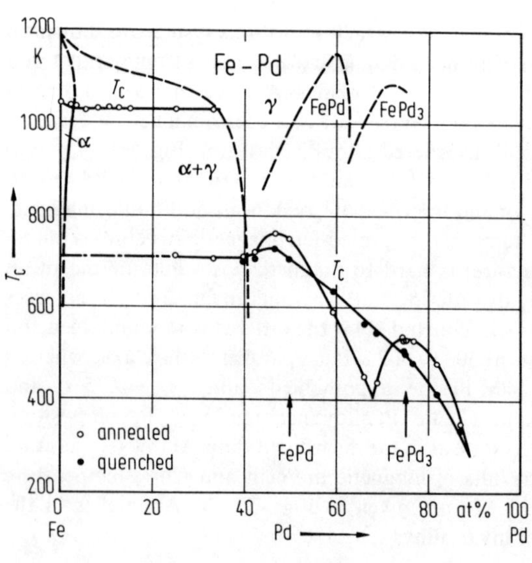

Fig. 30. Curie temperature of Fe–Pd alloys; open circles: annealed, solid circles: quenched samples; the dashed curve represents the phase diagram [36 F 1].

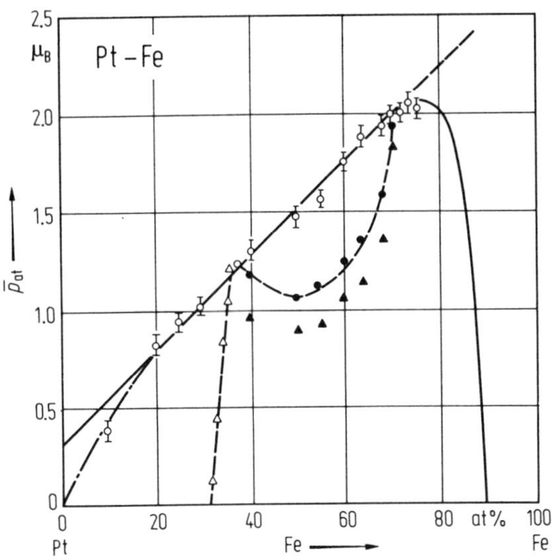

Fig. 31. Concentration dependence of the mean atomic magnetic moment of Pt–Fe alloys in the disordered (open circles) and ordered (solid circles, triangles) states [74 M 1]. Solid curve: theoretical.

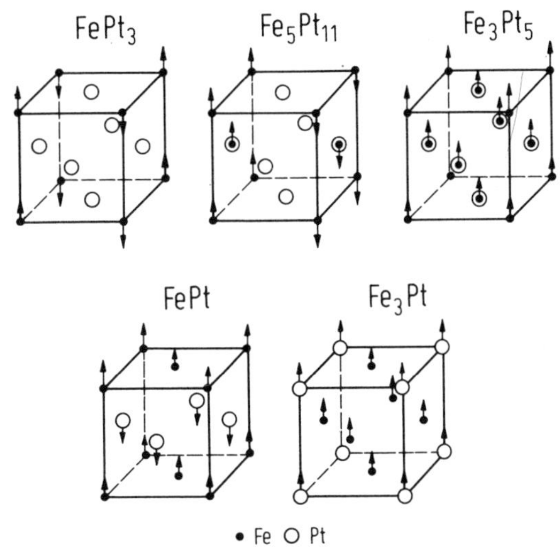

Fig. 32. The basic types of magnetic structures in ordered Fe–Pt compounds; solid circles indicate Fe, open circles Pt atoms [74 M 1].

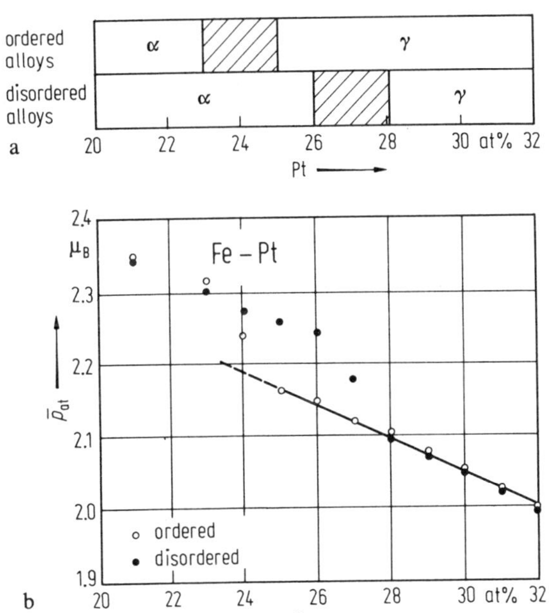

Fig. 33. (a) Phase boundary of Fe–Pt alloys with 20⋯30 at% Pt at 4.2 K, estimated from magnetization measurements; the hatched region indicates a two-phase region; (b) concentration dependence of the magnetic moment for Fe–Pt alloys; open circles: ordered solid circles: disordered alloys [83 S 1].

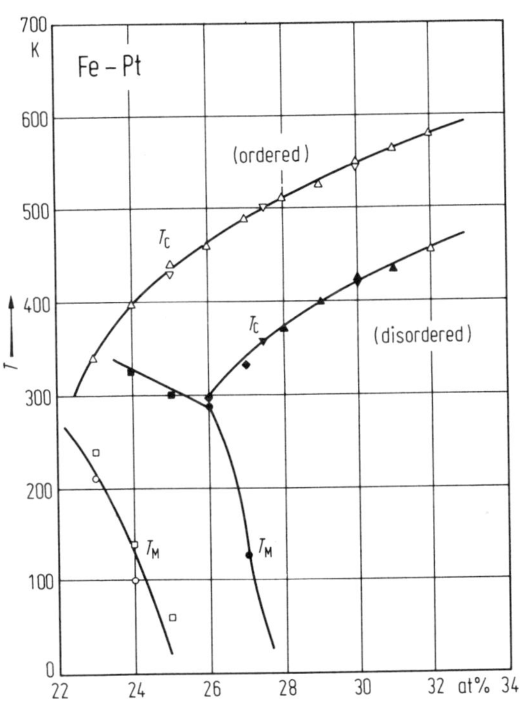

Fig. 34. Curie temperature T_C and martensitic transition temperature T_M for Fe–Pt alloys as a function of Pt content near the $FePt_3$ composition; open symbols: ordered alloys; solid symbols: disordered alloys. Different symbols refer to the results of different authors. [83 S 1].

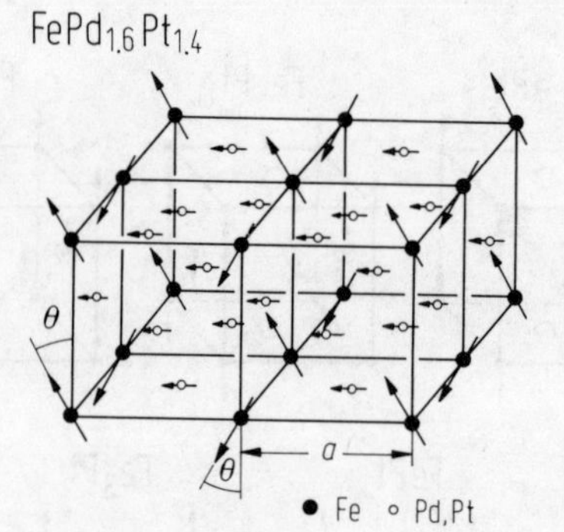

Fig. 35. Low-temperature canted-ferrimagnetic structure of $FePd_{1.6}Pt_{1.4}$ [69 K 2].

4.2.2.5 Co alloys and compounds

Alloys of the systems Co–Ru, Co–Rh, Co–Os, and Co–Ir have a critical concentration for the beginning of ferromagnetism in the intermediate composition range.

Spin-glass behavior has been investigated in Co–Rh alloys near the critical composition for ferromagnetism (Fig. 36). Co–Pd alloys show ferromagnetic order over nearly the whole range of concentration, as is the case for Co–Pt alloys (Fig. 37). Annealed samples of CoPd and $CoPd_3$ exhibit short-range atomic order. Magnetic moments from neutron diffraction experiments are: CoPd: $\bar{p}_{Co}=1.97(7)\,\mu_B$, $\bar{p}_{Pd}=0.35(7)\,\mu_B$; $CoPd_3$: $\bar{p}_{Co}=1.92(11)\,\mu_B$, $\bar{p}_{Pd}=0.48(3)\,\mu_B$. Two ordered phases occur in the Co–Pt system: CoPt and $CoPt_3$. Both are ferromagnetic. In ordered CoPt (CuAuI-type of structure) the ferromagnetically coupled moments $\bar{p}_{Co}=1.6\,\mu_B$ and $\bar{p}_{Pt}=0.25\,\mu_B$ are aligned parallel to the hexagonal axis. Co–Pt alloys near the equiatomic composition have large values for the energy product $(BH)_{max}$. By substitution of Co by Fe or Ni the energy product $(BH)_{max}$ is strongly reduced.

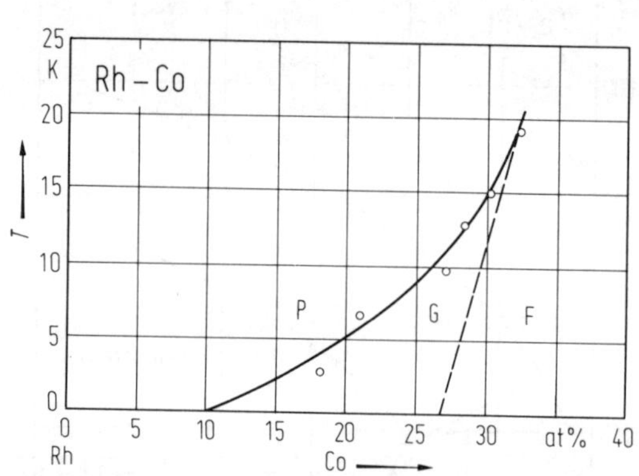

Fig. 36. Magnetic phase diagram of Rh–Co alloys near the critical Co concentration for ferromagnetism; P, G, and F designate the paramagnetic, the magnetic glass and the ferromagnetic regions, respectively [75 J 1].

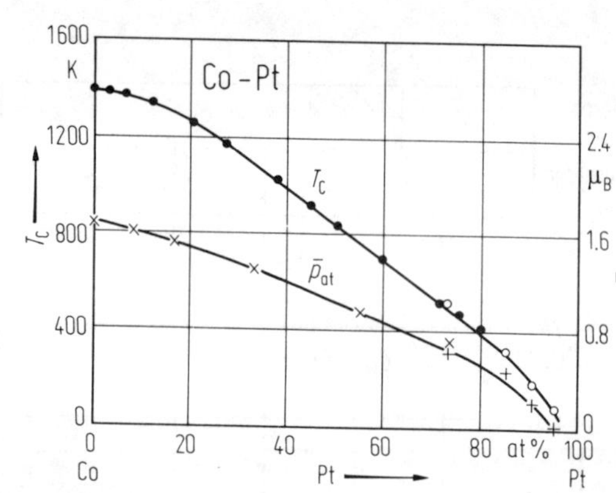

Fig. 37. Magnetic moment per atom $\bar{p}_{at}$ and Curie temperature for Co–Pt alloys as a function of composition. ×: [32 S 1]; solid circles: [40 G 1]; open circles and +: [30 C 1].

Table 5. Atomic magnetic moments in Pd_3Co and PdCo [65 C 1].

Alloy	$\bar{p}_{at}$	$\bar{p}_{Co}$ [1])	$\bar{p}_{Co}-\bar{p}_{Pd}$	$\bar{p}_{Co}$	$\bar{p}_{Pd}$
	μ_B				
Pd_3Co	0.84(2)	1.9(2)	1.45(12)	1.92(11)	0.48(3)
PdCo	1.16(3)	2.0(1)	1.63(10)	1.97(7)	0.35(7)

[1]) From the large-angle scattering data assuming no Pd contribution and the metallic Co form factor.

4.2.2.6 Ni alloys and compounds

Results of magnetic moment and Curie temperature studies of Ni–Ru, –Rh, –Pd, and Ni–Os, –Ir, –Pt alloys have been collected in Figs. 38 and 40. The onset of ferromagnetism in binary nickel alloys has extensively been investigated, in particular for Ni–Pt and Ni–Pd alloys. Important parameters for the onset of ferromagnetism are the local environment of nickel atoms and the polarization clouds around local magnetic moments.

The magnetic anisotropy is strongly reduced in single crystals of nickel with a few at% Ru. Rh. Pt.

The onset of ferromagnetic order in Ni–Pd alloys near 2.3 at% Ni is ascribed to the interplay of a strongly enhanced paramagnetic matrix and the presence of triads of nickel atoms that carry giant magnetic moments.

For the Ni–Ir system a critical concentration for ferromagnetism has been observed at 81 at% Ni.

The Ni–Pt system is subject to order-disorder transformations in two composition regions: near NiPt and near Ni_3Pt. NiPt in the ordered state has a tetragonal CuAuI-type of structure whereas the disordered alloys have a fcc structure. The disordered alloys show ferromagnetic order above 42.5 at% Ni. The ordered NiPt compound is paramagnetic, stressing the important role of local environment effects. The magnetic properties (including magnetovolumetric effects) of the disordered alloys near the critical composition for ferromagnetism have frequently been discussed in the collective model of itinerant magnetism [74 A 1, 76 K 1, 77 F 1].

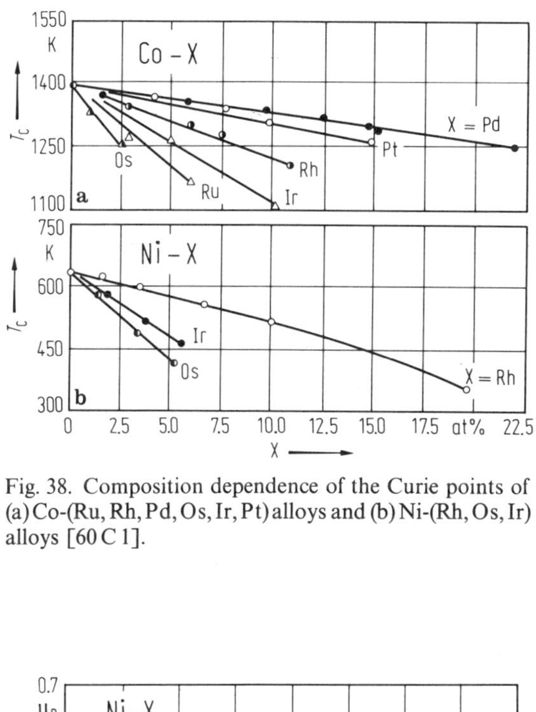

Fig. 38. Composition dependence of the Curie points of (a) Co-(Ru, Rh, Pd, Os, Ir, Pt) alloys and (b) Ni-(Rh, Os, Ir) alloys [60 C 1].

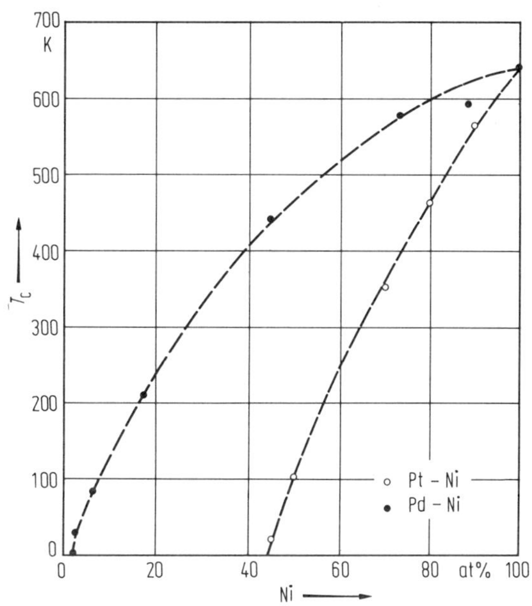

Fig. 39. Plot of the curie temperature vs. concentration for Pd–Ni and disordered Pt–Ni [72 F 1].

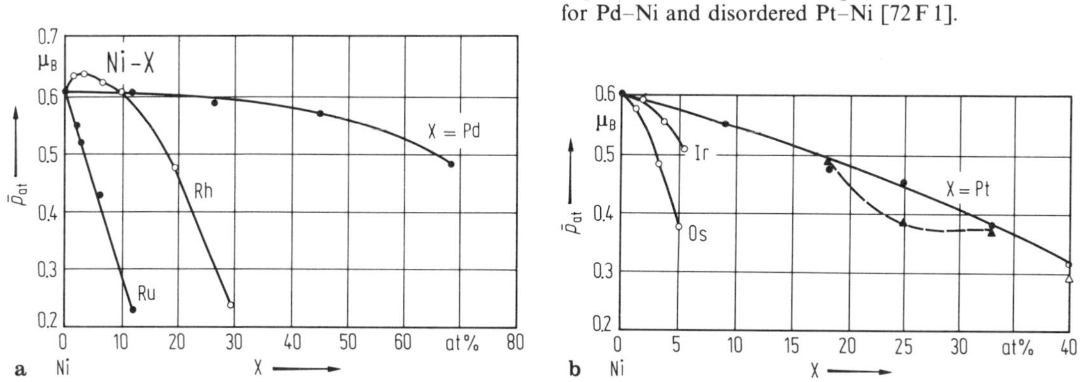

Fig. 40. Variation of the magnetic moment per atom with composition for Ni-(Ru, Rh, Pd, Os, Ir, Pt) alloys. Open circles: [60 C 1]; solid symbols: [32 S 1], [37 M 1]; see also [82 T 1] and [83 T 1].

4.2.3 4d- and 5d-rich alloys

In the dilute, 4d- and 5d-rich alloys of 3d transition metals a variety of interesting magnetic phenomena can be encountered: formation of magnetic moments, Kondo effect, local spin fluctuations, giant magnetic moments, spin-glass phenomena, etc. An introduction into these phenomena and a collection of resistivity, susceptibility, specific heat and thermopower data can be found in Landolt-Börnstein, NS, vol. III/15, and in the review by Mydosh and Nieuwenhuys [80 M 2]. In the 5d series the critical region where iron magnetic moments appear occurs just at or to the right of Ir in the periodic table [66 G 1]. Values for the Kondo (spin fluctuation) temperatures for Cr, Mn, Fe, and Co impurities in Rh, Pd, and Pt are given in Table 6. The tendency of showing spin-glass or giant moment behavior in these combinations of 3d and 4d, 5d elements is indicated in Table 7.

The appearance of ferromagnetism in dilute solutions of Fe and Co in Pd and Pt and of Ni in Pd is illustrated in Fig. 42. The critical concentrations for ferromagnetism are of the order of 0.1 and 0.2 at% Fe, Co in Pd and Pt, respectively. In these low-concentration regions giant magnetic moments are formed around the Fe, Co atoms by polarization of the surrounding Pd, Pt matrix (Fig. 43). Below the critical concentration of ferromagnetism the interaction between the polarization clouds becomes possibly antiferromagnetic, pointing to a RKKY-type of interaction [71 C 3]. In dilute solutions of Mn in Pd a more complex situation is encountered. Besides the ferromagnetic interaction between the Mn moments by means of the polarized Pd matrix a direct antiferromagnetic interaction between neighboring Mn moments occurs. These two competing interactions give rise to spin-glass behavior in the composition range 2 ⋯ 10 at% Mn (see Fig. 41). By adding small amounts of Fe to Pd the transition from ferromagnetic to spin-glass behavior can be shifted to higher Mn concentrations (see Fig. 44). The critical concentration for ferromagnetism in dilute Pd–Ni alloys is around 2.3 at% Ni and is believed to be connected with a percolation among interacting magnetic clusters, nucleated by triads of Ni atoms [81 C 1].

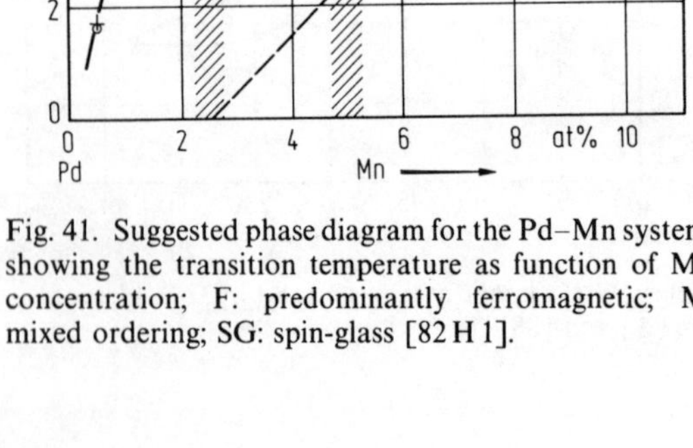

Fig. 41. Suggested phase diagram for the Pd–Mn system, showing the transition temperature as function of Mn concentration; F: predominantly ferromagnetic; M: mixed ordering; SG: spin-glass [82 H 1].

Table 6. Isolated-impurity Kondo (spin fluctuation) temperatures for Cr, Mn, Fe, and Co impurities in Rh, Pd, Pt [80 M 2].

	Cr	Mn	Fe	Co
Rh	–	10 K	50 K	1000 K
Pd	100 K	10 mK	20 mK	0.1 K
Pt	200 K	0.1 K	0.3 K	1 K

Table 7. Spin-glass and giant moment combinations of transition metals [80 M 2]. SG and GM represent favorable combinations for spin-glass or giant-moment behavior, respectively; XT means that a too high Kondo temperature limits the appearance of both the spin-glass or giant-moment states.

	Cr	Mn	Fe	Co
Rh	XT	SG	XT	XT
Pd	XT	SG+GM	GM	GM
Pt	XT	SG	GM	SG

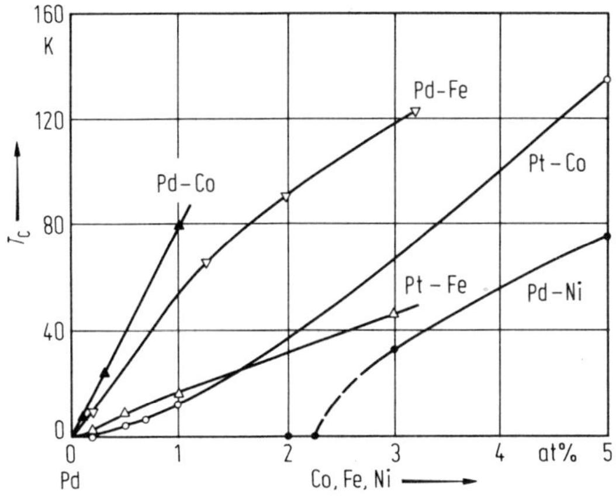

Fig. 42. Curie temperature as a function of concentration for Pd, Pt alloys with Fe, Co, Ni [62 B 1].

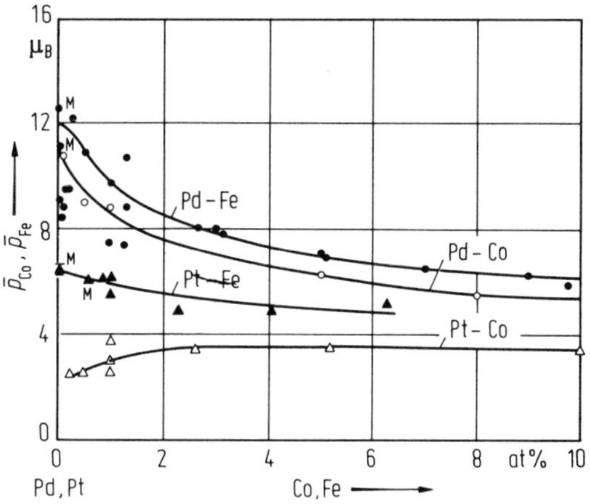

Fig. 43. Magnetic moment per solute atom as a function of concentration in alloys of Pd, Pt with Fe, Co [80 M 2].

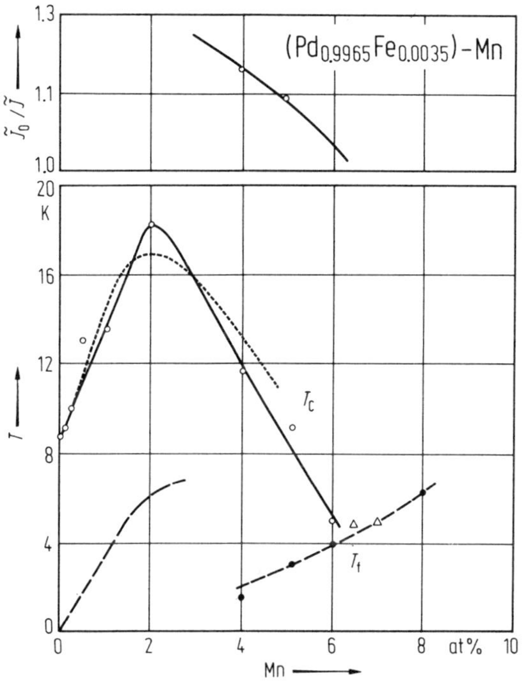

Fig. 44. Magnetic phase diagram for $(Pd_{0.9965}Fe_{0.0035})$–Mn showing the ferromagnetic transition (open circles) and the spin-glass transition (solid circles) temperatures; open triangles denote intermediate transitions; the dashed line represents the ferromagnetic transition temperature in binary Pd–Mn alloys [78 V 1]. The dotted line for T_C represents the result of a calculation. The upper figure shows the concentration dependence determined from experiment of the relative strength $\tilde{J}_0/\tilde{J}$ of ferromagnetic to spin-glass interaction parameter.

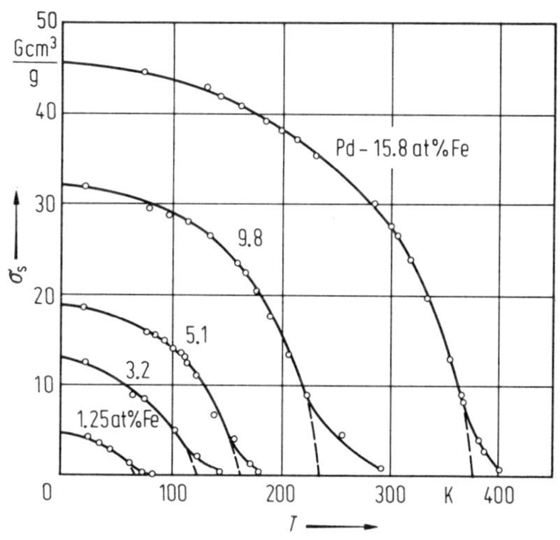

Fig. 45. Variation of spontaneous magnetization σ_s with temperature for Pd–Fe alloys; the iron content is given in the figure [60 C 2].

Table 8. Alloy compositions and magnetic parameters for dilute Pd–(Co, Ni, Fe) and Pt–(Co, Fe) alloys [74 B 1]. The Curie temperatures T_C have an accuracy $\pm 5\%$, the g values an accuracy of $\pm 2\%$, the magnetization extrapolated to 0 K, $M(0)$, is accurate to $\pm 10\%$, and the extrapolated anisotropy term $K_1(0)/M(0)$ is also accurate to $\pm 10\%$. The second order anisotropy terms $K_2(0)/M(0)$ are only reliable for the order of magnitude.

Alloy (at% solute)	T_C K	g	$M(0)$ G	$K_1(0)/M(0)$ G	$K_2(0)/M(0)$ G	Easy axis
Pd–Co(1.0)	70	2.30	60	−950	−500	[111]
Pd–Co(1.5)	85	2.33	69	−950	−500	[111]
Pt–Co(3.9)	67	2.29	62	200	0	[100]
Pd–Ni(5.0)	160	2.45	120	1200	550	[100]
Pd–Fe(0.25)	20	2.17	20	0	0	isotropic
Pd–Fe(1.7)	89	2.16	112	−65	0	[111]
Pt–Fe(4.5)	85	2.15	140	700	0	[100]

Table 9. Curie temperature, saturation magnetization and magnetic moments for Pd–Fe, Pt–Fe, Pt–Co, and Pd–Ni alloys [65 C 2].

	Pd–Fe			Pt–Fe				
Solute content [at%]	0.15	0.28	0.53	0.45	0.87	2.30	4.07	6.30
T_C [K]	4.3	9.5	23	3.3	5.4	21.5	59	104
σ_s [G cm^3 g^{-1}]	0.76	1.82	3.06		1.56	3.25	5.87	9.85
$\bar{p}_{at}$ [μ_B]	0.0145	0.0346	0.0581		0.0541	0.112	0.199	0.329
$\bar{p}_{Fe}$ [μ_B]	9.5	12.2	10.9		6.2	4.9	4.9	5.2

	Pt–Co						
Solute content [at%]	0.51	0.99	2.60	5.17	10.2	15.2	20.1
T_C [K]	2.7	6.8	39	104	218	315	
σ_s [G cm^3 g^{-1}]	0.36	0.84	2.56	5.30	10.52	15.37	19.6
$\bar{p}_{at}$ [μ_B]	0.013	0.029	0.087	0.18	0.34	0.48	0.59
$\bar{p}_{Co}$ [μ_B]	2.5	2.9	3.4	3.5	3.4	3.2	3.0

	Pd–Ni				
Solute content [at%]	2.5	3.33	4.29	5.55	11.78
T_C [K]	12	27	59	88	177
σ_s [G cm^3 g^{-1}]		2.88	5.40	7.37	14.50
$\bar{p}_{at}$ [μ_B]		0.054	0.10	0.14	0.26
$\bar{p}_{Ni}$ [μ_B]		1.6	2.4	2.5	2.2

4.2.4 References for 4.2

30 C 1 Constant, F.W.: Phys. Rev. **36** (1930) 1654.
32 S 1 Sadron, C.: Ann. Phys. Paris **17** (1932) 371.
36 F 1 Fallot, M.: Ann. Phys. **6** (1936) 305; **7** (1937) 420; **10** (1938) 29.
37 M 1 Marian, V.: Ann. Phys. Paris **7** (1937) 459.
40 G 1 Gebhardt, E., Köster, W.: Z. Metallkd. **32** (1940) 252.
50 A 1 Auwärter, M., Kussmann, A.: Ann. Phys. **7** (1950) 169.
50 K 1 Kussmann, A., Rittberg, G. v.: Ann. Phys. **7** (1950) 173.
52 K 1 Köster, W., Horn, E.: Z. Metallkd. **43** (1952) 444.
59 C 1 Crangle, J.: J. Phys. Paris **20** (1959) 435.

60 C 1 Crangle, J., Parsons, D.: Proc. R. Soc. **255** (1960) 509.
60 C 2 Crangle, J.: Philos. Mag. **5** (1960) 335.
62 B 1 Bozorth, R.M., Davis, D.D., Wernick, J.H.: J. Phys. Soc. Jpn. **17**, B–I (1962) 112.
62 C 1 Cable, J.W., Wollan, E.D., Koehler, W.C., Child, H.R.: Phys. Rev. **128** (1962) 2118.
62 P 1 Pickart, S.J., Nathans, R.: J. Appl. Phys. **33** (1962) 1336.
63 B 1 Bacon, G.E., Crangle, J.: Proc. R. Soc. London A **272** (1963) 387.
63 K 1 Kouvel, J.S., Hartelius, C.C., Osika, L.M.: J. Appl. Phys. **34** (1963) 1095.
63 P 1 Pickart, S.J., Nathans, R.: J. Appl. Phys. **34** (1963) 1203.
64 S 1 Shirane, G., Nathans, R., Chen, C.W.: Phys. Rev. **134** (1964) A 1547.
65 C 1 Cable, J.W., Wollan, E.O., Koehler, W.C.: Phys. Rev. **138** (1965) A 755.
65 C 2 Crangle, J., Scott, W.R.: J. Appl. Phys. **38** (1965) 921.
66 G 1 Geballe, T.H., Matthias, B.T., Clogston, A.M., Williams, H.J., Sherwood, R.C., Maita, J.P.: J. Appl. Phys. **37** (1966) 1181.
66 K 1 Krén, E., Kádár, G., Pál, L., Sólyom, J., Szabó, P.: Phys. Lett. **20** (1966) 331.
68 H 1 Hicks, T.J., Pepper, A.R., Smith, J.H.: J. Phys. C **1** (1968) 1683.
68 K 1 Kussmann, A., Müller, K., Raub, E.: Z. Metallkd. **59** (1968) 859.
68 K 2 Krén, E., Kádár, G., Pál, L., Sólyom, J., Szabó, P., Tarnóczi, T.: Phys. Rev. **171** (1968) 574.
68 P 1 Pál, L., Krén, E., Kádár, G., Szabó, P., Tarnóczi, T.: J. Appl. Phys. **39** (1968) 538.
68 P 2 Ponyatovskii, E.G., Kutsar, A.R., Dubovka, G.T.: Kristallografiya **12** (1967) 79 (Sov. Phys. Crystallogr. **12** (1968)).
69 A 1 Antonini, B., Lucari, F., Menzinger, F., Paoletti, A.: Phys. Rev. **187** (1969) 611.
69 K 1 Krén, E., Kádár, G.: Phys. Lett. **29** A (1969) 340.
69 K 2 Kouvel, J.S., Forsyth, J.B.: J. Appl. Phys. **40** (1969) 1359.
70 J 1 Jayaraman, A., Rice, T.M., Bucher, E.: J. Appl. Phys. **41** (1970) 869.
71 C 1 Crangle, J., Goodman, G.M.: Proc. R. Soc. (London) A **321** (1971) 477.
71 C 2 Chakravorty, S., Panigrahy, P., Beck, P.A.: J. Appl. Phys. **42** (1971) 1698.
71 C 3 Chouteau, G., Tournier, R.: J. Phys. Paris **32** (1971) C1–1002.
71 M 1 Moon, R.M.: Int. J. Magnetism **1** (1971) 219.
71 Y 1 Yamaoka, T., Mekata, M., Takaki, H.: J. Phys. Soc. Jpn. **31** (1971) 301.
72 F 1 Ferrando, W.A., Segnan, R., Schindler, A.I.: Phys. Rev. B **5** (1972) 4657.
73 A 1 Abdul-Noor, S.S., Booth, J.G.: Phys. Lett. **43** A (1973) 381.
73 A 2 Arajs, S., Rao, K.V., Åström, H.U., DeYoung, T.F.: Physica Scripta **8** (1973) 109.
73 B 1 Besnus, M.J., Meyer, A.J.P.: Phys. Status Solidi (b) **58** (1973) 533.
73 B 2 Besnus, M.J., Meyer, A.J.P.: Phys. Status Solidi (b) **55** (1973) 521.
73 K 1 Kelarev, V.V., Vokhmyanin, A.P., Dorofeyev, Yu.A., Sidorov, S.K.: Phys. Met. Metallogr. (USSR) **35**, **6** (1973) 1302.
74 A 1 Alberts, H.L., Beille, J., Bloch, D., Wohlfarth, E.P.: Phys. Rev. B **9** (1974) 2233.
74 B 1 Baggurley, D.M.S., Robertson, J.A.: J. Phys. F **4** (1974) 2282.
74 D 1 Dubovka, G.T.: Sov. Phys. JETP **38** (1974) 1140.
74 M 1 Men'shikov, A.Z., Dorofeyev, Yu.A., Kazantsev, V.A., Sidorov, S.K.: Phys. Met. Metallogr. (USSR) **38**, **3** (1974) 47.
74 Y 1 Yamaoka, T.: J. Phys. Soc. Jpn. **36** (1974) 445.
74 Y 2 Yamaoka, T., Mekata, M., Takaki, H.: J. Phys. Soc. Jpn. **36** (1974) 438.
75 A 1 Arajs, S., Rao, K.V., Anderson, E.E.: Solid State Commun. **16** (1975) 331.
75 F 1 Fukamichi, K., Saito, H.: J. Less-Common Met. **40** (1975) 357.
75 J 1 Jamieson, H.C.: J. Phys. F **5** (1975) 1021.
76 K 1 Kortekaas, T.F.M., Franse, J.J.M.: J. Phys. F **6** (1976) 1161.
76 V 1 Vinokurova, L., Vlasov, A. V., Pardavi-Horváth, M.: Phys. Status Solidi (b) **78** (1976) 353.
77 F 1 Franse, J.J.M.: Physica **86–88** B (1977) 283.
78 P 1 Pardavi-Horvath, M., Vinokurova, L.I., Vlasov, A.V.: Inst. Phys. Conf. Proc. **39** (1978) 603.
78 V 1 Verbeek, B.H., Nieuwenhuys, G.J., Stocker, H., Mydosh, J.A.: Phys. Rev. Lett. **40** (1978) 586.
79 P 1 Parette, G., Kajzar, F.: J. Phys. F **9** (1979) 1867.
80 M 1 Moyer, C.A., Arajs, S., Eroglu, A.: Phys. Rev. B **22** (1977) 3277.
80 M 2 Mydosh, J.A., Nieuwenhuys, G.J., in: Ferromagnetic Materials I (Wohlfarth, E.P., ed.), Amsterdam: North-Holland Publishing Company **1980**, p. 71.
81 C 1 Cheung, T.D., Kouvel, J.S., Garland, J.W.: Phys. Rev. B **23** (1981) 1245.
81 K 1 Kuentzler, R.: Inst. Phys. Conf. Ser. **55** (1981) 397.

81 V 1 Vinokurova, L.I., Vlasov, A.V., Kulikov, N.I., Pardavi-Horvath, M.: Hungarian Acad. Sci. Budapest **1981**.

82 H 1 Ho, S.C., Maartense, I., Williams, G.: J. Appl. Phys. **53** (1982) 2235.

82 T 1 Takahashi, Y., Jacobs, R.L.: J. Phys. F **12** (1982) 517.

83 S 1 Sumiyama, K., Shiga, M., Nakamura, Y.: J. Mag. Magn. Mater. **31–34** (1983) 111.

83 T 1 Takahashi, Y., Jacobs, R.L.: J. Mag. Magn. Mater. **31–34** (1983) 49.

5 Alloys and compounds of 3d elements with main group elements

5.1 3d elements and Cu, Ag or Au

Survey on tables and figures containing data on alloys and compounds between 3d-elements and Cu, Ag and Au, respectively.

Property	3d-Cu		3d-Ag		3d-Au	
Magnetic and	Mn-Cu	Fig. 8			VAu_4	Figs. 17, 18
crystalline		Tab. 4			$3d\,Au_4$	Fig. 4
structures					$MnAu_3$	Fig. 22
					Mn_2Au_5	Fig. 23
					$MnAu_2$	Fig. 24
					MnAu	Fig. 26
Magnetic phase	Mn-Cu	Fig. 1			Cr-Au	Fig. 19
diagram					Mn-Au	Fig. 1
					Fe-Au	Figs. 1, 28
Transition	Mn_xCu_{1-x}	Tab. 1			VAu_4	Fig. 16
temperatures		Figs. 1, 9, 10				
	Fe-Cu	Fig. 11, Tab. 1			Cr-Au	Fig. 19, Tab. 1
	Co-Cu	Tab. 3	Mn-Ag	Tab. 1	Mn-Au	Tab. 1
	Ni-Cu	Figs. 7, 12, 13		Fig. 14	$MnAu_4$	Fig. 21
					Fe-Au	Figs. 6, 27, 28, Tab. 1
					Co-Au	Figs. 6, 30, Tabs. 1, 3
					Ni-Au	Fig. 7
Paramagnetism	Mn-Cu	Fig. 2, Tabs. 1.5, 6	Mn-Ag	Tab. 1	VAu_4	Fig. 16
	Fe-Cu	Tabs. 1, 8		Figs. 14, 15	Cr-Au	Tab. 1
	Co-Cu	Tab. 3			Mn-Au	Fig. 2, Tab. 1
	Ni-Cu	Fig. 13			Fe-Au	Fig. 2, Tab. 1
					Co-Au	Fig. 2.9, Tabs. 1, 3
Spinglass,	Mn_xCu_{1-x}	Tabs. 1, 7	Mn_xAg_{1-x}	Tab. 1	Cr-Au	Tab. 1, Fig. 5
Mictomagnetism,						
Magnetic cluster	Fe_xCu_{1-x}	Tab. 1, Fig. 5	Mn-Ag	Fig. 5	Mn-Au	Tab. 1
					Fe-Au	Tab. 1
	Co-Cu	Tab. 3				
	Ni-Cu	Fig. 13			Co-Au	Tab. 1, Fig. 5
Antiferromagnetism,	Mn-Cu	Tab. 6, Fig. 8			VAu_4	Figs. 17, 18
Ferromagnetism	Ni-Cu	Figs. 7, 12			$MnAu_4$	Figs. 20, 21
					$MnAu_2$	Fig. 25
					Ni-Au	Fig. 7

5.1.1 Tables and figures containing data on various alloys and compounds of the type 3d–(Cu, Ag, Au)

Table 1. Properties obtained from low-field susceptibility measurements on some 3d-noble metal alloys (spin glasses [80 m 1]. For Θ of Mn–Au, Fe–Au and Mn–Cu. See also Fig. 2.

	x at%	T_f K	p_{eff} μ_B	$\chi(0)/\chi(T_f)$	Θ K
Mn_xCu_{1-x}	0.1	1.8	5.5(5)	0.58	≈ − 2
	0.2	3.0	–	0.58	–
	0.5	6.4	5.4(5)	≈0.5	≈ 0
	1.3	12.3	6.0(5)	0.53	≈ 0
	5.2	27.8	8.0(10)	0.50	≈ 2
	8.8	42.5	10.0(10)	0.29	≈ 9
	10.3	45.0	11.0(10)	0.28	≈ 6
Fe_xCu_{1-x}	0.3	2.3	–	0.58	–
	0.7	5.0	6.0(5)	0.50	≈ − 2
	0.8	5.2	7.5(10)	–	≈ − 1
	1.0	6.0	7.5(5)	0.30	≈ 0
Mn_xAg_{1-x}	0.25	–	4.5(5)	–	≈ 0
	0.5	2.7	4.2(5)	–	≈ 1
	1.0	5.1	5.5(5)	0.58	≈ 0
	2.0	9.0	6.5(5)	0.57	≈ 2
	5.0	17.6	6.8(5)	0.48	≈ 2
	10.0	30.8	7.4(5)	0.42	≈ 6
Cr_xAu_{1-x}	0.2	4.1	4.0(5)	0.67	≈ 0
	0.6	11.3	3.5(5)	0.68	≈ 0
	0.9	14.3	4.4(5)	0.68	≈ − 5
	1.8	25.0	4.1(5)	0.69	≈ −11
	3.5	40.0	–	–	–
	3.7	41.5	–	0.75	–
	6.1	60.0	–	≈0.75	–
	10.8	92.0	–	–	–
Mn_xAu_{1-x}	0.2	1.6	–	–	–
	0.5	3.5	7.0(10)	0.60	≈ 0
	1.1	5.7	7.3(5)	≈0.65	≈ 1
	2.0	8.3	10.6(10)	0.60	≈ − 1
	5.7	17.5	11.5(10)	0.25	≈ 7
	10.3	29.3	>20	0.12	–
Fe_xAu_{1-x}	0.1	2.1	≈ 3.3	0.60	≈ 0
	0.2	3.0	≈ 3.3	0.60	≈ 0
	0.5	5.4	≈ 3.3	0.60	≈ 0
	0.9	8.5	3.4	0.50	≈ − 2
	1.0	9.0	–	–	–
	2.0	13.9	3.6	0.50	≈ − 1
	5.0	22.2	4.9	0.39	≈ 1.5
	7.6	27.9	7.1	0.27	≈ 4
	12.0	36.0	11.4	0.11	≈12
	13.0	38.0	15.5	0.11	≈17
Co_xAu_{1-x}	2.0	–	1.3(2)	–	–
	3.0	1.7	–	–	–
	4.0	3.2	–	–	–
	5.0	4.1	3.0(5)	0.28	≈ − 1
	7.0	6.7	4.5(5)	0.34	≈ 2

Table 2. 3d-Cu, Ag, Au. Measured effective magnetic moments p_{eff} in μ_B [68 d 1].

Matrix	V	Cr	Mn	Fe	Co
Au	3.1	4.0	5.8	3.7	4.5
Ag			5.6		
Cu		3.9	4.9	3.7	3⋯5

Table 3. Co–Cu, Au. Magnetic properties of Co–Cu and Co–Au alloys derived from magnetization measurements at very low temperatures down to less than 0.1 K [62 L 1].
IRM (0 K): isothermal remanent magnetization extrapolated to 0 K; H_c (0 K): coercive force extrapolated to 0 K; Θ: paramagnetic Curie temperature; T_f: spin-glass freezing temperature (from χ); C_{ex}/C_{iso}: ratio of Curie constants determined experimentally (ex) and of C_{iso} obtained by assuming isolated Co atoms with $p_{eff} = 1.71\ \mu_B$:

	Co_xCu_{1-x}				Co_xAu_{1-x}		
x [at%]	0.6	1	2	3	2	3	4
IRM (0 K)·10^3 [G cm³ g⁻¹]	< 1	8	115	217	8	17.6	41
H_c (0 K) [Oe]	–	600	1300	1300	500	950	1900
Θ [K]	−38	− 8.5	− 19	− 32	− 40	− 36	− 24
T_f [K]	3.7	5.6	11.5	10	4.5	5	6.5
C_{ex}/C_{iso}	1	0.86	4.7	5.5	3.34	3.79	3.51

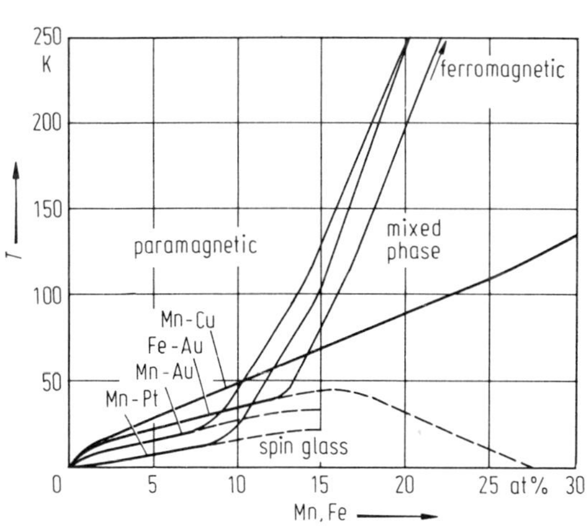

Fig. 1. Mn–Pt, Au, Cu; Fe–Au. Magnetic phase diagram for Mn–Pt [82 W 1], Mn–Au [81 V 1], Fe–Au [81 S 1] and Mn–Cu [78 b 1]. The term "mixed phase" is used to denote a spatially disordered ferromagnet [83 M 2].

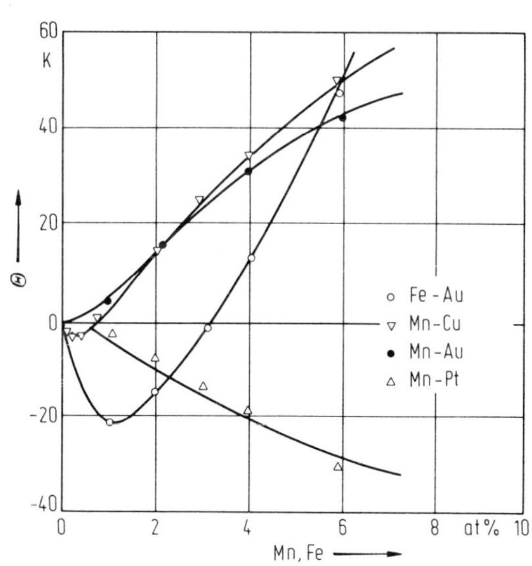

Fig. 2. Mn–Pt, Au, Cu; Fe–Au. Paramagnetic Curie temperature Θ vs. 3d-impurity concentration. The alloys are cold rolled. The values for Θ are extrapolated from high-temperature susceptibility data using Curie-Weiss analysis. The solid lines represent the configuration cluster calculation values [83 M 1].

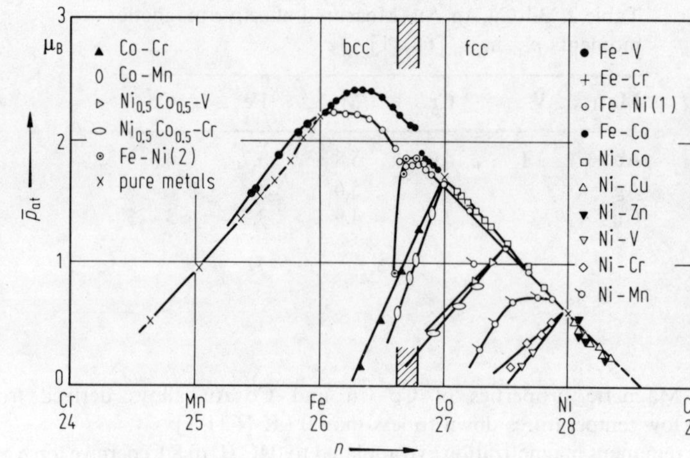

Fig. 3. Mean magnetic moment per atom, $\bar{p}_{at}$, as a function of the number of electrons per atom for ferromagnetic metals and alloys (Slater-Pauling curve) [64 c 1].

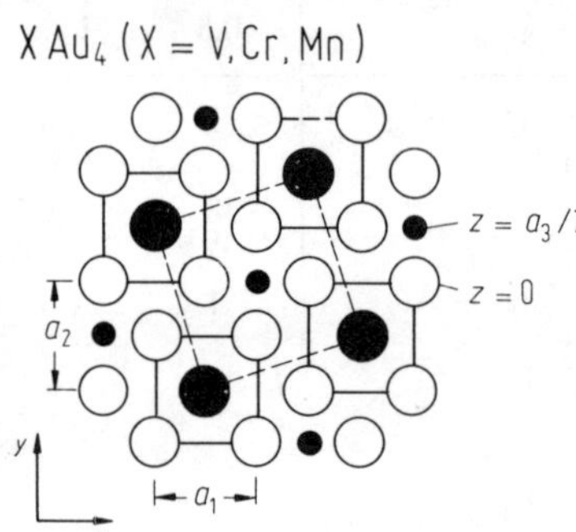

Fig. 4. 3dAu$_4$. Ordered structure of XAu$_4$ (X = V, Cr, Mn). $a_1 = a_2 = 4.08$ Å, $a_3 = 4.03$ Å. Full dots denote X atoms, empty dots denote Au atoms [75 v 1, 61 S 1].

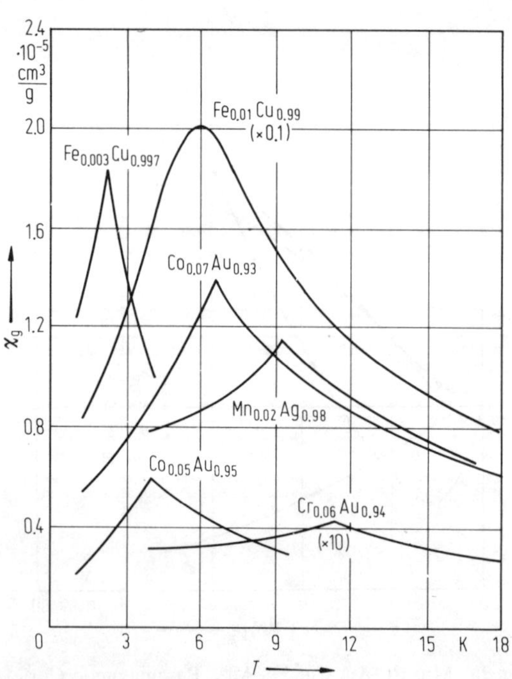

Fig. 5. Fe–Cu; Cr, Co–Au; Mn–Ag. Magnetic mass susceptibility χ_g vs. temperature for various spin glass alloys. Note that the magnitude of the data for $Fe_{0.01}Cu_{0.99}$ and $Cr_{0.06}Au_{0.94}$ have been multiplied by 0.1 and 10, respectively [74 C 1].

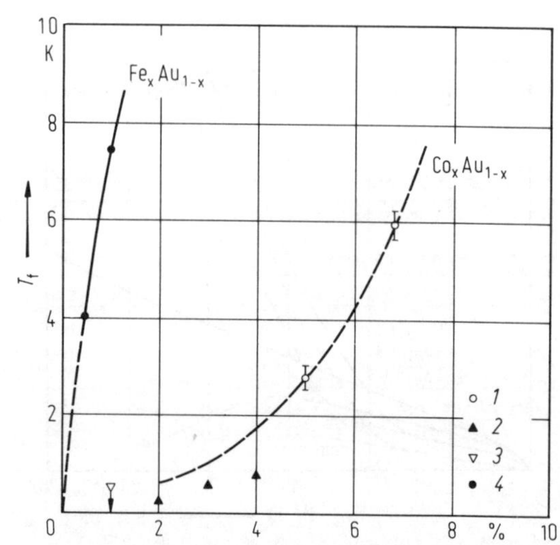

Fig. 6. Co, Fe–Au. Spin-glass freezing temperature T_f vs. x for bulk alloys [70 B 1]. *1* [70 B 1], *2* [62 L 1], *3* and *4* [64 L 1]. After [70 B 1].

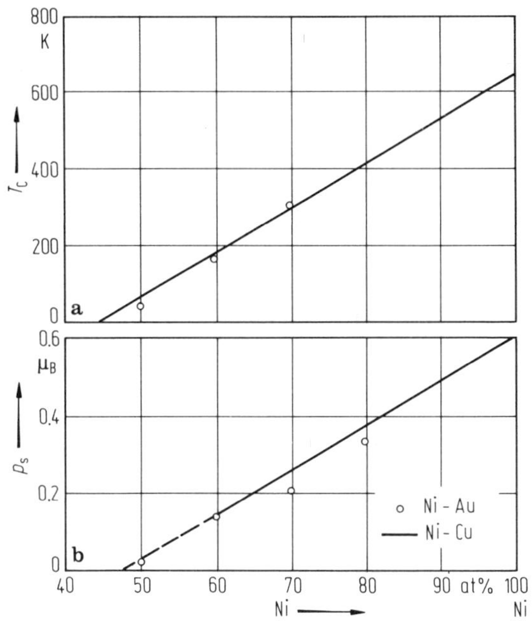

Fig. 7. Ni–Au, Cu. (a) Ferromagnetic Curie temperature T_C and (b) saturation magnetic moment per atom, p_s, vs. composition for Ni–Au [74 C 2] and Ni–Cu alloys [58 A 1]. After [74 C 2].

5.1.2 Data on individual alloys and compounds

5.1.2.1 3d–Cu

Ti–Cu

The solubility of Ti in Cu is 5.2 at% at 850 °C, and 0.26 at% at 400 °C [69 s 1]. Dilute alloys of Ti in Cu have been investigated magnetically by [67 W 1] and [76 F 1]. The compounds Ti_2Cu, TiCu, Ti_3Cu_3, Ti_2Cu_3 and $TiCu_4$ have been studied by [77 K 1] with respect to magnetic susceptibility above room temperature. The solubility of Cu in α-Ti was found to be 1.25 at% at 776 °C [58 h 1].

V–Cu

The solid solubility of V in Cu is 0.8 at% V at 1120 °C, and 0.1 at% V at room temperature [69 s 1]. Dilute alloys of V in Cu have been investigated by [67 C 1]. The solid solubility of Cu in V is about 6 at% Cu at room temperature [69 s 1]. The magnetic susceptibility is independent on temperature [74 K 1]. V-rich alloys are superconductors.

Cr–Cu

The solid solubility of Cr in Cu is about 0.1 at% at 1075 °C, and about 0.12 at% at 840 °C [65 e 1]. A miscibility gap exists, extending from 42 at% to about 94 at% Cr in Cu. Dilute alloys of Cr in Cu are paramagnetic below about 400 °C and a Curie-Weiss law is observed at low temperature. The solid solubility of Cu in Cr seems to be very small [58 h 1].

Mn–Cu

An uninterrupted series of solid solutions exists at high temperature [69 s 1]. This behaviour is typical of solid solutions of two monovalent components. As a result, dilute alloys of Mn in Cu show an effective magnetic moment $p_{eff} \approx 4.9 \mu_B$/Mn as for monovalent Mn^+ ions [75 v 1], corresponding to 4 magnetic electrons per solute ion. On the other side, Mn-rich alloys are antiferromagnetic like pure γ-Mn. In Mn-rich alloys the antiferromagnetic order is accompanied by a tetragonal cell distortion $c/a < 1$ up to the critical concentration $x_{cr} = 72$ at% Mn for disappearance of antiferromagnetic order. The magnetic structure of Mn-rich antiferromagnetic γ-Mn–Cu is AF1 (Fig. 8a), changing to AF3 as more Cu is added (Fig. 8b). Due to this change in magnetic structure, the magnetic moment per Mn atom decreases. At the Néel temperature a structural transformation fct–fcc of the lattice takes place accompanied by magnetic disordering [81 K 1]. In alloys containing more Mn than the critical Mn concentration, coexistence of antiferromagnetic long-range order and mictomagnetism (due to short-range order) are observed at low temperature (Fig. 9). Below the critical Mn concentration, 72 at% Mn mictomagnetism occurs (Figs. 9, 10, Table 6). At concentrations near 1 at% Mn typical spin-glass properties can be observed in small magnetic fields.

Table 4. γ-Mn–Cu. Values of the magnetic moment per Mn atom and the lattice parameters of the tetragonal unit cell for antiferromagnetic γ-Mn–Cu alloys at 4.2 K as derived from neutron and X-ray diffraction [81 C 1].

Mn at%	p_{Mn} μ_B	a Å	c Å
82.5	1.33(2)	3.765(2)	3.617(2)
80.0	1.06(4)	3.763(2)	3.634(2)
77.5	0.68(2)	3.759(2)	3.657(2)
75.0	0.41(3)	3.752(2)	3.681(2)
74.0	0.34(3)	3.749(2)	3.700(2)
72.5	0.21(3)	3.754(2)	–

Table 5. Mn–Cu. Slow-cooled samples: annealing 2 h at 950 K. Curie-Weiss analysis of the high-temperature, statically measured, paramagnetic susceptibility curves. T_d: temperature at which the magnetic susceptibility deviates more than the experimental scattering from the high-temperature Curie-Weiss fit.

Mn at%	p_{eff} μ_B/Mn	Θ K	T_d K	Ref.
0.0430	5.2(5)	− 0.8(5)		73 F 1
0.0848	5.0(5)	0.0(5)		73 F 1
0.131	5.0(5)	1.0(5)		73 F 1
0.343	4.9(5)	2.0(5)		73 F 1
1.03	5.16	3.8	51	81 M 1
2.08	5.06	14.1	80	81 M 1
2.08 [1])	5.06	17.6	96	81 M 1
2.96	5.19	25.2	106	81 M 1
4.04	5.04	34.2	130	81 M 1
5.94	4.90	55.0	160	81 M 1

[1]) Sample quenched from 900 °C into water.

Table 6. Mn–Cu. Curie constant per Mn atom, C_{Mn}, temperature-independent magnetic mass susceptibility χ_{g0}, paramagnetic Curie temperature Θ and concentration c_{AF} of antiferromagnetically coupled Mn atoms, in the paramagnetic state, as fraction of the total number of alloy atoms [76 T 1].
aged: aged for four weeks at 373 K after treatment qu.
qu.: annealed at 1123 K, quenched into ice water;
def.: more than 30% compression in three mutually perpendicular directions.

Mn at%	C_{Mn} 10^{-5} μ_B K Oe^{-1}			χ_{g0} 10^{-6} cm^3 g^{-1}			Θ K			c_{AF}		
	aged	qu.	def.	aged	qu.	def.	aged	qu.	def	aged	qu.	def.
25	–	39.5	29.6	–	0.2	1.70	–	97.5	0.78	–	0.066	0.112
50	2.48	4.0	3.20	10.2	8.98	9.01	19.3	14.4	7.69	0.472	0.463	0.470
60	1.46	1.70	1.56	9.91	9.98	9.82	21.8	15.2	11.1	0.584	0.581	0.583
66	1.02	1.08	0.90	9.83	9.86	10.0	15.2	6.25	15.4	0.648	0.647	0.649

Table 7. Mn–Cu. Total magnetic cluster moment per Mn atom, p_{cl}, and the fraction f of Mn magnetic moments in magnetic clusters, as derived from the low-temperature magnetization after aligning the magnetic moments by field-cooling and assuming $p_{Mn} = 4\,\mu_B$ [81 B 1].

Mn at%	p_{cl} μ_B/Mn	f %
9	0.027 [1])	0.68
12	0.26	6.5
25	0.135	3.4
66	0.00049	0.019

[1]) Calculated from paramagnetic data for 50 and 60 K.

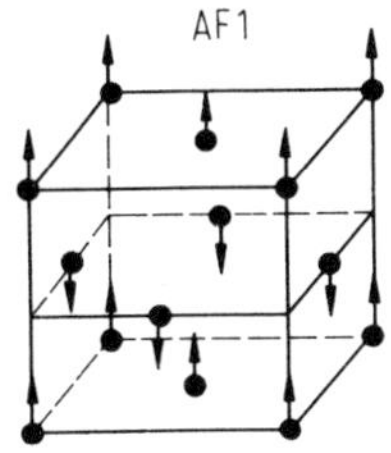

a

Fig. 8a. Mn–Cu. Type AF1 antiferromagnetic structure in Mn-rich Mn–Cu alloys, >72 at% Mn, as derived from neutron scattering [81 C 1]. For an indication that the direction of the spins is inclined from the *c* axis, see [80 T 1].

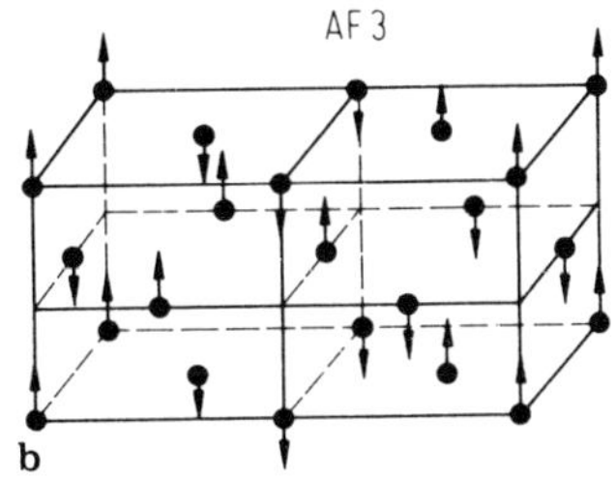

b

Fig. 8b. Mn–Cu. Type AF3 antiferromagnetic short-range order in γ-Mn–Cu alloys with a composition above the critical composition for antiferromagnetism, ≈60 at% Mn [81 C 1].

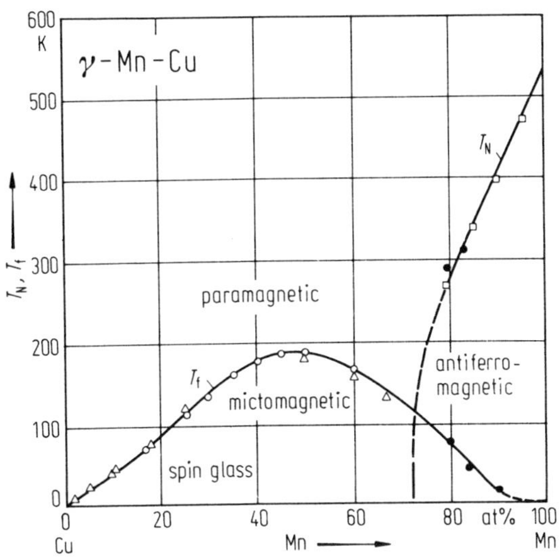

Fig. 9. γ-Mn–Cu. Variation of Néel temperature T_N and spin freezing temperature T_f with composition [80 G 1]. Squares [69 S 1], triangles [76 T 1].

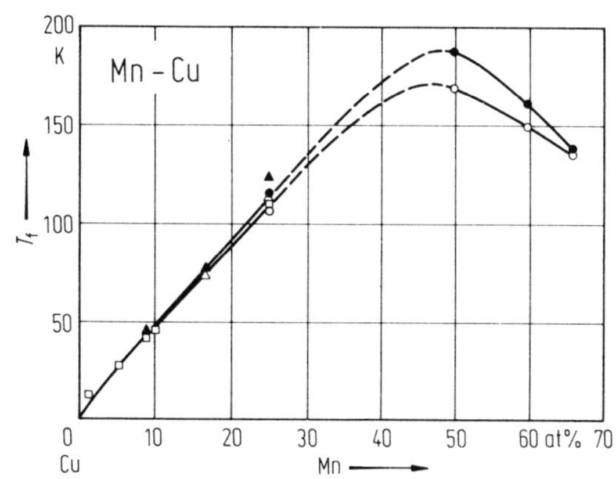

Fig. 10. Mn–Cu. Mictomagnetic freezing temperature T_f vs. Mn concentration. Solid circles: aged, open circles: quenched, plastically deformed [73 C 1], solid triangles: aged, open triangles: quenched [75 M 1].

Fe–Cu

The solubility of Fe in Cu is low, about 3 at% Fe at 1000 °C, and 0.35 at% Fe at 780 °C [58 h 1, 65 e 1, 78 m 1]. In quenched alloys, annealing at a few hundred °C causes the formation of γ-Fe clusters [54 K 1, 61 B 1]. Precipitated γ-Fe clusters have a spin structure consisting of ferromagnetic (001) sheets with antiparallel spin directions in neighbour sheets, the spin vector being inclined by about 19° from the (001) axis [62 A 1]. At lower aging temperatures superparamagnetic clusters are formed [61 B 1]. The tendency to superparamagnetism manifests in high values for the effective magnetic moment $p_{eff} \approx 5 \cdots 7 \mu_B$/Fe for about 1 at% Fe in Cu. For dilute Fe in Cu $p_{eff} \approx 3.4 \mu_B$/Fe are known corresponding to nearly 3 magnetic electrons per solute ion. The paramagnetic Curie temperature Θ depends on Fe concentration and on the method of alloy preparation. Susceptibility maxima are observed at low temperature in small magnetic fields. The high-field susceptibility of bulk alloys is not sensitive to annealing. Freezing temperatures deduced from magnetic and resistivity measurements coincide near 2 at% Fe in Cu (Fig. 11). Also Mössbauer studies reveal the freezing-in of the spins (Fig. 11b); in addition they show the precipitation of α-Fe in Cu. The solubility of Cu in γ-Fe is lower than 3.5 at% at 850 °C [58 h 1] and that of Cu in α-Fe is only about 1.2 at% Cu at 800 °C [65 e 1] (according to [68 A 1] it is 3 at% at 835 °C). The saturation magnetic moment per atom of bcc Fe–Cu alloys decreases linearly on alloying with some at% Cu and the solute atoms do not have magnetic moments [68 A 1].

Table 8. Fe–Cu. Curie-Weiss analysis of the high-temperature paramagnetic properties. Data for quench-condensed films can be found in [85 K 1]. Data from neutron scattering can be found in [79 D 1].

Fe at%	p_{eff} μ_B/Fe	Θ K	Measuring method	Ref.
0.0447	3.3(3)	− 10.4(5)	high dc field	73 F 1
0.0842	3.3(3)	− 6.8(5)		
0.137	3.5(3)	− 4.6(5)		
0.175	3.9(4)	− 3.6(5)		
0.230	4.0(4)	− 1.6(5)		
0.614	5.9(5)	+ 0.6(5)		
0.57	4.81	− 18	low dc field	59 S 1
1.19	4.92	67		
1.76	4.96	101		
2.16	4.89	112		
0.7	6.0(5)	≈ − 2	mutual induction bridge	80 m 1
0.8	7.5(10)	≈ − 1		
1	7.5(5)	≈ 0		

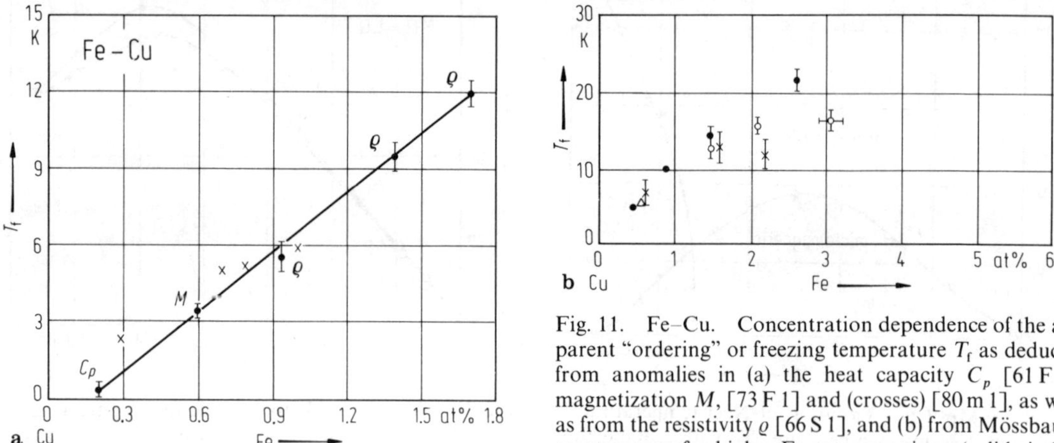

Fig. 11. Fe–Cu. Concentration dependence of the apparent "ordering" or freezing temperature T_f as deduced from anomalies in (a) the heat capacity C_p [61 F 1], magnetization M, [73 F 1] and (crosses) [80 m 1], as well as from the resistivity ϱ [66 S 1], and (b) from Mössbauer spectroscopy for higher Fe concentrations: (solid circles) as rolled: (open circles) fast quenched [76 C 1]; (crosses) [65 G 1]; (triangle) [70 B 2].

Co–Cu

The solubility of Co in Cu is about 5 at% at 1110 °C [58 h 1]. Thermodynamically stable solid solutions do not exist at any temperature in the concentration range between 12 at% and 95 at% Cu. Isothermal ageing produces precipitates of a Co-rich β-phase containing about 10 at% Cu of fcc structure, coherent with the Cu matrix. Also during quenching from high temperature small Co-rich precipitates can be formed. Concerning Co concentrations up to 2 at% Co in Cu, susceptibility and magnetization have been measured for the mictomagnetic and superparamagnetic alloys. For higher Co concentration, only metastable solid solutions can be produced, e.g., by quench-condensation, and have been studied with respect to susceptibility, magnetization and electrical resistivity. $Co_{0.08}Cu_{0.92}$ is found to be ferromagnetic [81 Z 1]. In Co-rich bulk alloys, mean magnetic moments have been determined [70 B 3].

Ni–Cu

A continuous series of solid solutions exists for Ni–Cu alloys. The Ni–Cu alloys have been investigated in a large number of publications covering nearly all compositions. Up to about 30 at% Ni, their susceptibility is almost temperature-independent [58 P 1], and up to about 10 at% Ni it is increasing almost proportional to the Ni content [58 P 1]. The alloyed Ni has no magnetic moment before Ni atoms with 10 to 11 nearest Ni neighbours are present. These nearest neighbour Ni atoms form large polarization clouds, causing superparamagnetic behaviour with giant magnetic moments of 10 μ_B to 12 μ_B above $\approx$30 at% Ni in Cu [70 K 1]. The critical concentration for ferromagnetism is 47.6 at% Ni. At concentrations higher than the critical concentration the ferromagnetic Curie temperature T_C and the average magnetic moment of Ni increase almost linearly with rising Ni concentration towards the values of pure Ni.

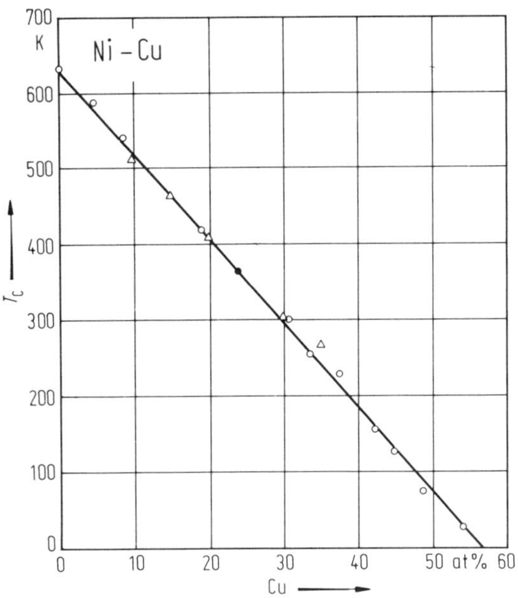

◀ Fig. 12. Ni–Cu. Variation of the ferromagnetic Curie temperature T_C with composition [58 A 1]. Open circles [58 A 1], solid circle [53 O 1], triangles [58 M 1].

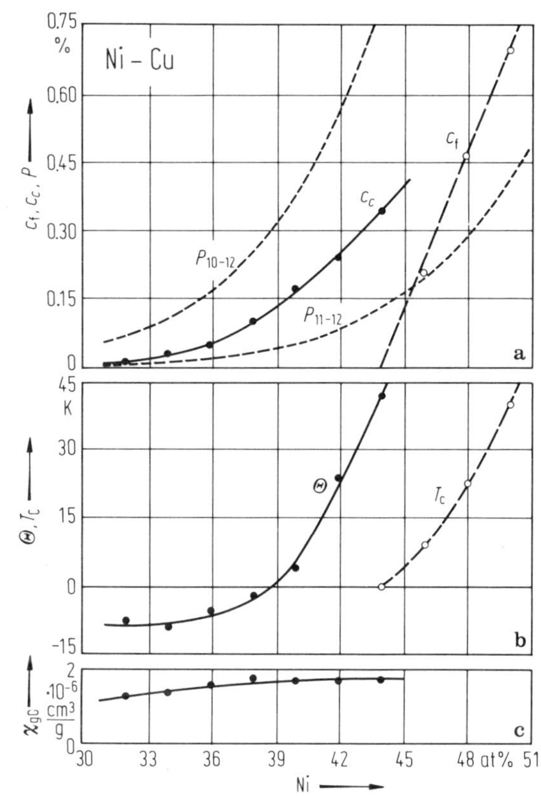

Fig. 13. Ni–Cu. Dependence on Ni concentration of (a) concentration c_c of magnetic clouds, probabilities P of Ni–Ni neighbour numbers and concentration c_f of ferromagnetic clouds, (b) ferromagnetic- and paramagnetic Curie temperatures T_C and Θ, respectively, and (c) temperature-independent part, χ_{g0}, of the initial mass susceptibility [70 K 1]. Data: solid circles [70 K 1], open circles [69 H 1].

5.1.2.2 3d–Ag

Mn–Ag

The solid solubility of Mn in Ag extends to about 40 at% Mn at 900 °C and to 14 at% Mn at 300 °C. A miscibility gap in the liquid state extends from 46.8 at% to 96.2 at% Mn at 1180 °C. The solubility of Ag in γ-Mn is 2.6 at% Ag at 990 °C. In β-Mn and in α-Mn the solubility of Ag is smaller than 0.2 at% Ag and practically nil, respectively [58 h 1]. The magnetic properties of dilute Mn in Ag are similar to those of dilute MnCu alloys, e.g., a Curie-Weiss law is observed for the susceptibility with an effective magnetic moment of $\approx$ 5.1 μ_B/Mn and paramagnetic Curie temperatures near 0 K for extreme dilution.

Quench-condensed alloy films show $\Theta \approx 0$ for Mn concentrations up to about 16 at% Mn and negative values of Θ for concentrations higher than 16 at% Mn, the critical concentration for percolation in the fcc lattice,

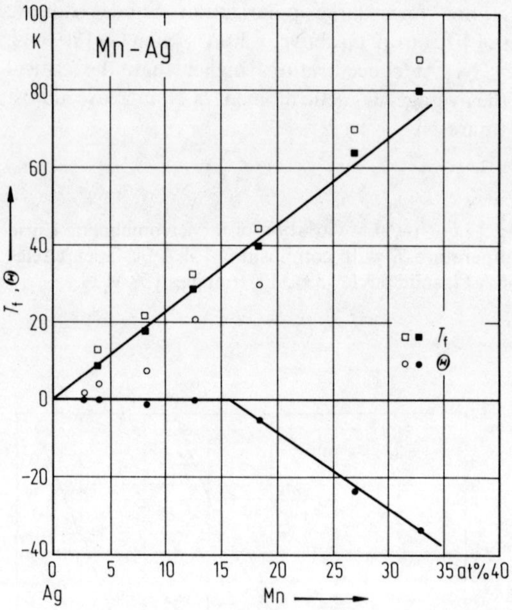

Fig. 14. Mn–Ag. Spin glass freezing temperature T_f and paramagnetic Curie temperature Θ vs. Mn concentration for films. Solid symbols apply for annealing at 140 K (Mn atoms distributed at random in the Ag host) and open symbols apply for annealing at 300 K [68 K 1].

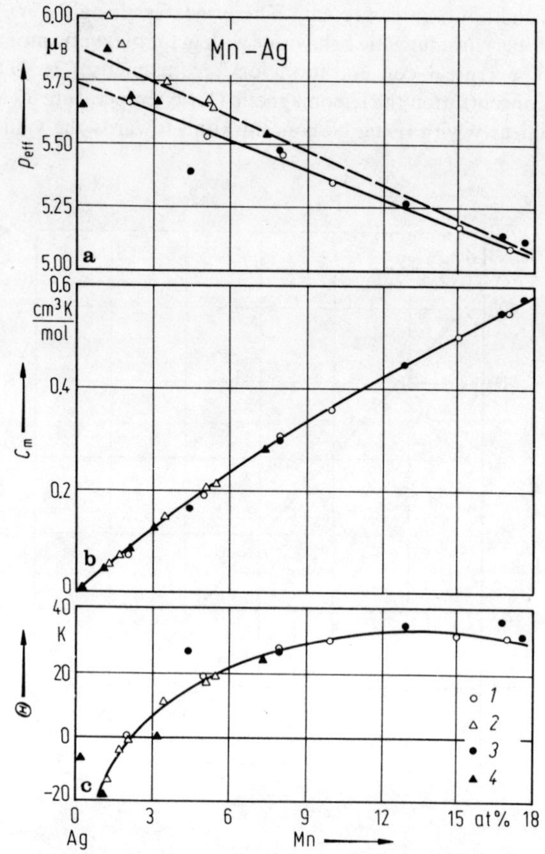

Fig. 15. Mn–Ag. Effective magnetic moment p_{eff}, Curie constant per mole C_m and paramagnetic Curie temperature Θ vs. Mn concentration [62 H 1]. *1* [62 H 1], *2* [60 V 1], *3* [59 M 1], *4* [36 G 1].

5.1.2.3 3d-Au

Ti–Au

The solid solubility of Ti in Au extends to about 11 at% Ti at 1090 °C and to 0.85 at% Ti at 500 °C. The following compounds have been identified $TiAu_4$, $TiAu_2$, TiAu, Ti_3Au having the structures: face-centered tetragonal, Ni_4Mo type, body-centered tetragonal, orthorhombic AuCd type, cubic β-W type, respectively. The solubility of Au in β-Ti is 6 at% Au at 1000 °C and in α-Ti is less than 2 at% Au at 700 °C [69 s 1]. The susceptibility of the alloys at low Ti concentration is temperature-independent and has to be explained with localized impurity d states without magnetic moment, a behaviour similar to dilute Ni–Au, Ni–Ag, and Ni–Cu [62 v 1, 75 v 1]. The impurity susceptibilities per mole 3d metal have the following values χ_m

Ti in Au: $2.0 \cdot 10^{-4}\,cm^3\,mol^{-1}$ [75 v 1]
Ni in Au: $3.8 \cdot 10^{-4}\,cm^3\,mol^{-1}$ [75 v 1]
Ni in Ag: $5.0 \cdot 10^{-4}\,cm^3\,mol^{-1}$ [80 H 1]
Ni in Cu: $1.3 \cdot 10^{-4}\,cm^3\,mol^{-1}$ [58 P 1].

The intermetallic compound $TiAu_4$ does not show anomalies in the resistivity vs. temperature curve.

V–Au

The solid solubility of V in Au extends to about 59 at% V at 1400 °C [65 e 1, 78 m 1]. The intermetallic compound VAu_4 has a tetragonal structure isotypic with Ni_4Mo. The compound VAu_2 exists, showing an

A-centered orthorhombic structure. In V_3Au quenched rapidly from the melt, a disordered W-type solid solution can be produced, instead of the stable β-W structure [69 s 1]. NMR data on V_3Au are given in [74 S 1]. The magnetic susceptibility is temperature-dependent up to 30 at% V. The magnetic properties depend on heat-treatment, especially the paramagnetic Curie temperature. The ordered structure shows the V ion to be surrounded by 12 Au ions. VAu_4 is ferromagnetic although it is composed of nonmagnetic metals. The ferromagnetic Curie temperature is reported to be 40⋯60 K depending on heat-treatment. The magnetic moment p_v per V ion is 0.3⋯0.8 μ_B at 0 K and it depends on mechanical treatment. Disordered VAu_4 shows a Pauli-type paramagnetism. Magnetization measurements on a VAu_4 single crystal yield $\bar{p}_v \approx 1\,\mu_B$ and $T_C = 61$ K, a large uniaxial anisotropy a remarkable field-cooling effect.

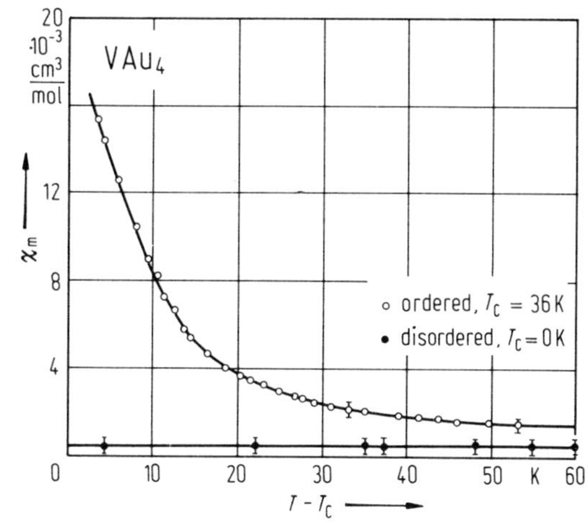

Fig. 16. VAu_4. Magnetic molar susceptibility χ_m vs. reduced temperature, $T - T_C$, of ordered and disordered VAu_4 [67 C 2].

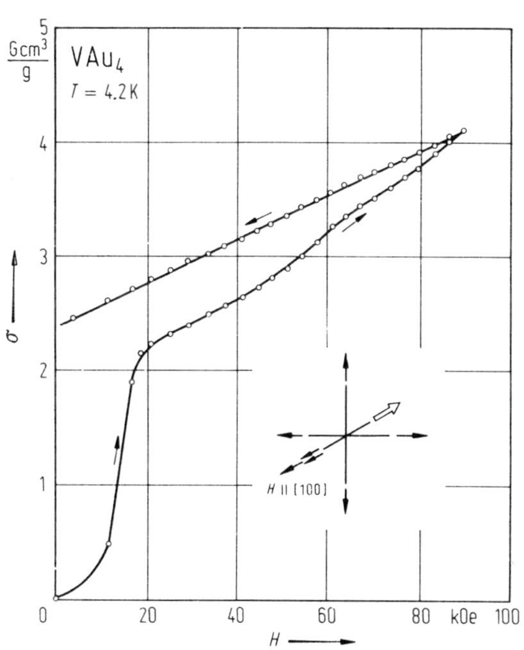

Fig. 17. Ordered VAu_4. High-field specific magnetization curve of a single crystal measured along [100] at 4.2 K in $H \leqq 90$ kOe after zero-field cooling. Inset shows spin configuration of magnetized or field-cooled state [80 A 1].

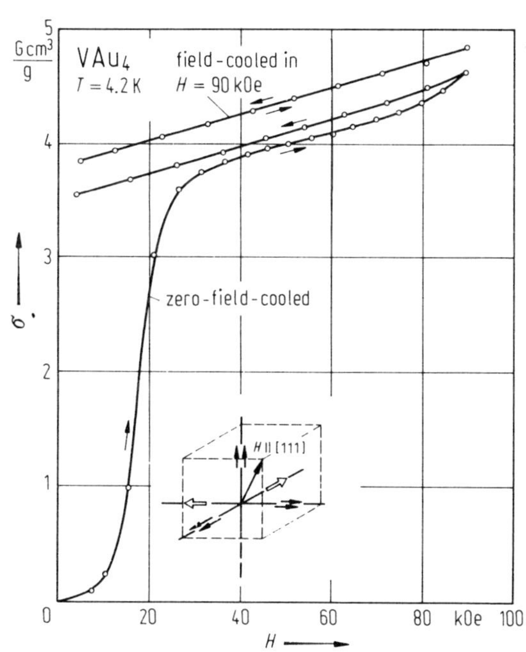

Fig. 18. Ordered VAu_4. High-field magnetization curve of a single crystal measured along [111] at 4.2 K in $H \leqq 90$ kOe after zero-field cooling and field-cooling in 90 kOe. Inset shows spin configuration of magnetized or field-cooled state [80 A 1].

Cr–Au

The solid solubility of Cr in Au is about 40 at% Cr at 1000 °C, and 25 at% Cr at 400 °C [65 e 1]. An ordered phase of Ni_4Mo type structure is known to be antiferromagnetic. Localized magnetic moments are parallel within the chains of nearest neighbours and antiparallel from chain to chain. The second nearest neighbour interaction decides the ferromagnetic or antiferromagnetic state of the system [69 T 1] as observed for the ordered solid solutions of $MnAu_4$ and $CrAu_4$. For small Cr concentration and low temperature spin-glass freezing is observed whereas for Cr concentrations higher than about 12 at% Cr antiferromagnetism exists. An antiferromagnetic to ferromagnetic transition is found in the series of compounds $Mn_{1-x}Cr_xAu_4$.

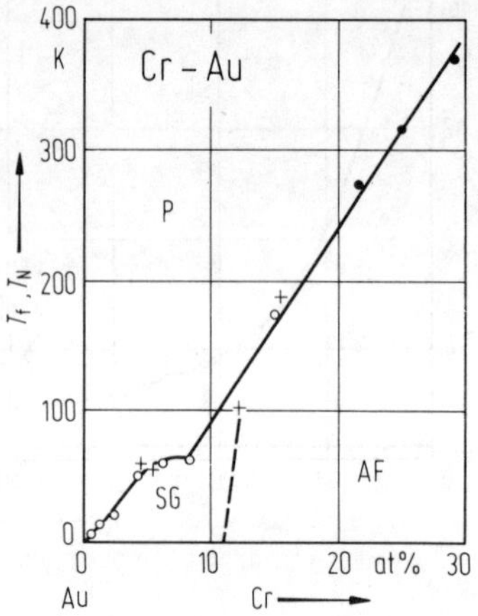

Fig. 19. Cr–Au. Transition temperatures T_f and T_N vs. Cr concentration from susceptibility measurements: solid circles [61 W 1], open circles [67 Y 1], and from diffuse scattering of neutrons: crosses [71 N 1]. $Cr_{0.121}Au_{0.879}$ is antiferromagnetic [71 N 1]. P: paramagnetic, AF: antiferromagnetic, SG: spin glass.

Mn–Au

At about 900 °C a series of solid solutions exists: from 0 to 30 at% Mn showing fcc structure, from 35 to 65 at% Mn showing bcc structure and from 70 to 93 at% Mn with fcc structure. The solid solubility of Au in β-Mn and in α-Mn is probably below 1 at% Au [58 h 1]. At small Mn-concentration spin-glass and mictomagnetic behaviour is observed in the magnetic susceptibility and magnetization at low temperature. Ferromagnetic interaction is predominant as indicated by positive paramagnetic Curie temperature and a decrease of the ac susceptibility peak in superposed dc magnetic fields. Mössbauer spectra show hyperfine splitting.

From the fcc solid solutions three phases arise: $MnAu_4$, $MnAu_3$ and Mn_2Au_5. $MnAu_4$ has a Ni_4Mo type structure as have the alloys of Au with Cr, V, and Ti. Depending on heat-treatment $MnAu_4$ transforms from the disordered to the ordered state. In an ordered alloy field-cooling can be of strong influence on the magnetization. At different temperature intervals such an alloy can show para-, superpara- or mictomagnetism. The antiferro- to ferromagnetic transition of $Mn_{1-x}Cr_xAu_4$ is $Mn_{1-x}V_xAu_4$, where only ferromagnetism is observed at low temperature.

The structure of $MnAu_3$ is shown in Fig. 22. Nearest neighbour Mn pairs are aligned antiferromagnetically and next nearest neighbour pairs ferromagnetically. The ordered $MnAu_3$ is an antiferromagnet showing a maximum in the magnetization at $T_N = 145$ K and an uprupt increase of the magnetization with magnetic field at about 47 kOe. Disordered $MnAu_3$ is mictomagnetic, characterized by a displaced hysteresis loop upon field-cooling [61 J 1] and remanence.

The ordered compound Mn_2Au_5 has a monoclinic unit cell containing 14 atoms. All Mn atoms are arranged in antiferromagnetic double chains nearly perpendicular to the plane of the figure. As compared to $MnAu_3$ there are more antiferromagnetic couplings in Mn_2Au_5 resulting in a higher Néel temperature $T_N = 355$ K. The transition at T_N manifests itself as a change in slope of the resistivity vs. temperature curve and as a kink in the Youngs modulus vs. temperature curve.

The ordered compound $MnAu_2$ has a body-centered tetragonal structure in which normal to the c axis, layers of Mn atoms are separated by two layers of Au atoms. In the layers, Mn atoms are coupled ferromagnetically with a constant angle from layer to layer. This helical arrangement causes $MnAu_2$ to be metamagnetic meaning ferromagnetic or antiferromagnetic depending on temperature and magnetic field. In the magnetization vs. temperature curve antiferromagnetism is present at low fields and ferromagnetic behaviour at high fields. At a critical field a spin flop turns the layer magnetic moments with components antiparallel to the magnetic field into orientations near the magnetic field direction.

The ordered compound MnAu has a p3 antiferromagnetic structure with ferromagnetic sheets lying normal to the short axis of its tetragonally distorted CsCl-type cell. The t_1 phase has $c/a < 1$ and the t_2 phase has $c/a > 1$. For each Mn atom $4.2\mu_B$ magnetic moment point in some unknown direction within the plane of the sheets and an additional component of $1.6\mu_B$ magnetic moment lies perpendicular to the magnetic sheets [81 P1]. The magnetic susceptibility shows a sharp peak at the Néel temperature $T_N = 500$ K.

Mn_3Au has a structure similar to that of $MnAu_2$. Its magnetic susceptibility is small and nearly independent of temperature [75 v 1].

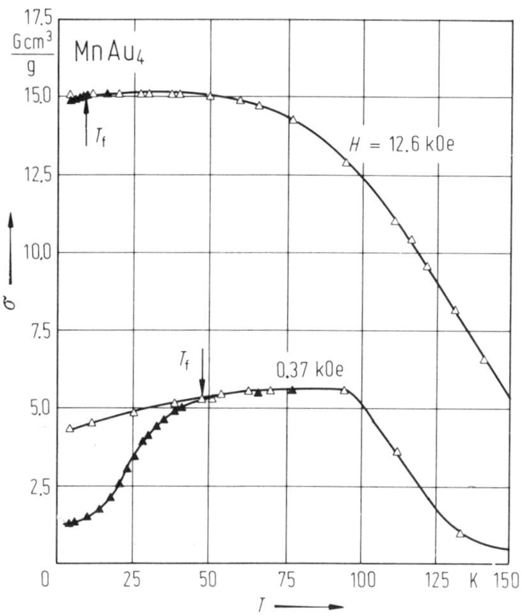

◄ Fig. 20. $MnAu_4$. Magnetic moment per gram, σ, vs. temperature after aging 2 days at 100 °C, measured in 0.37 and 12.6 kOe after field-cooling in 12.6 kOe (open triangles) and after zero-filed-cooling (solid triangles) [72 C 1].

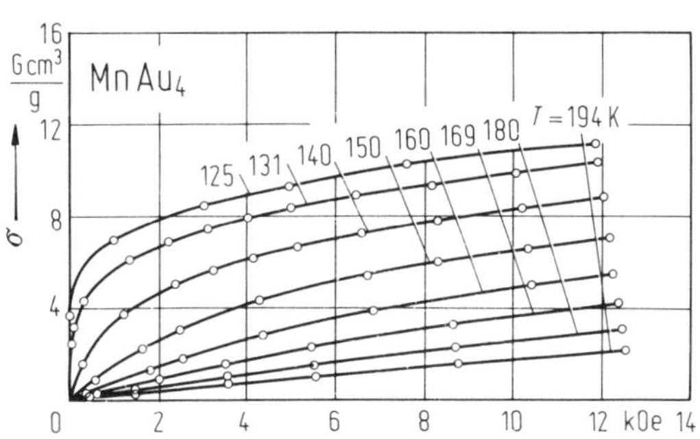

Fig. 21. $MnAu_4$. Magnetic moment per gram, σ, vs. magnetic field after aging 15 days at 100 °C. Parameter is the temperature. Different phases are present: at $T > 180$ K paramagnetic, $131\,K < T < 169$ K superparamagnetic, $T < 131$ K mictomagnetic [72 C 1].

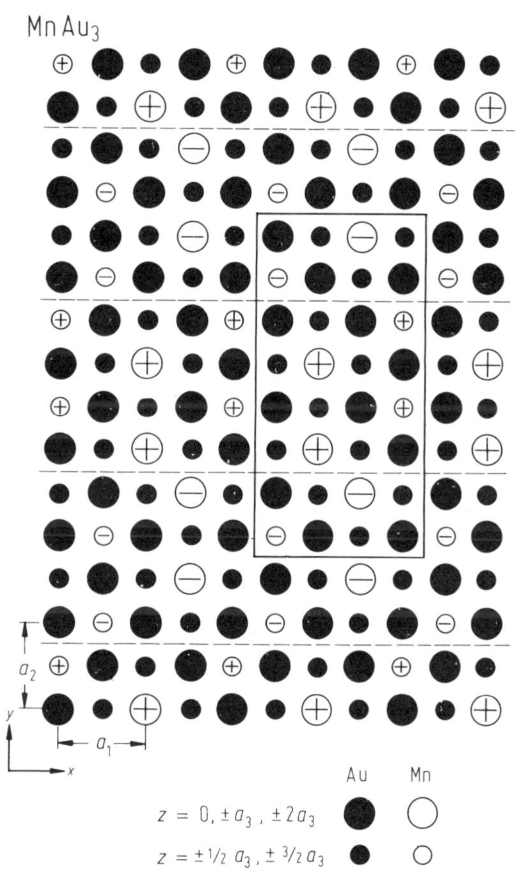

Fig. 22. $MnAu_3$. Atomic and momental order of the orthorhombic superlattice. Nearest neighbour Mn pairs are aligned antiferromagnetically as marked by + and − and next-nearest neighbour pairs ferromagnetically [61 J 1, 61 S 1].

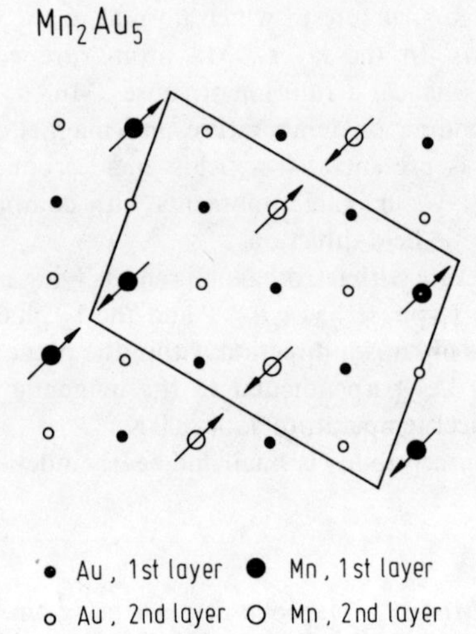

Fig. 23. Mn_2Au_5. Atomic and momental order. Projection on a plane perpendicular to the *b* direction. Solid and open circles represent atoms of the first and second layer, respectively [69 S 2].

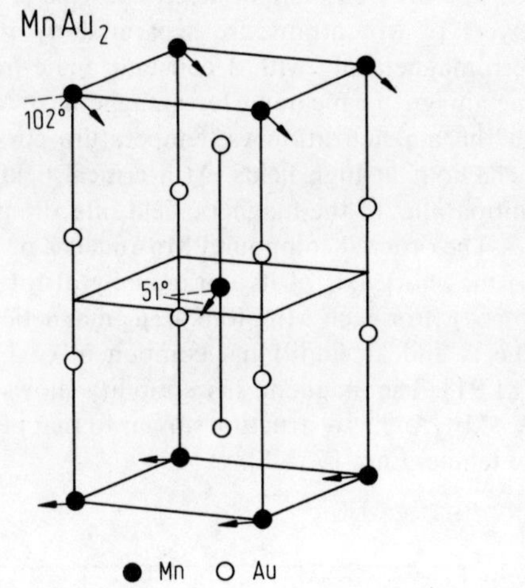

Fig. 24. $MnAu_2$. Atomic order and helical order of magnetic moments [61 H 1, 75 v 1].

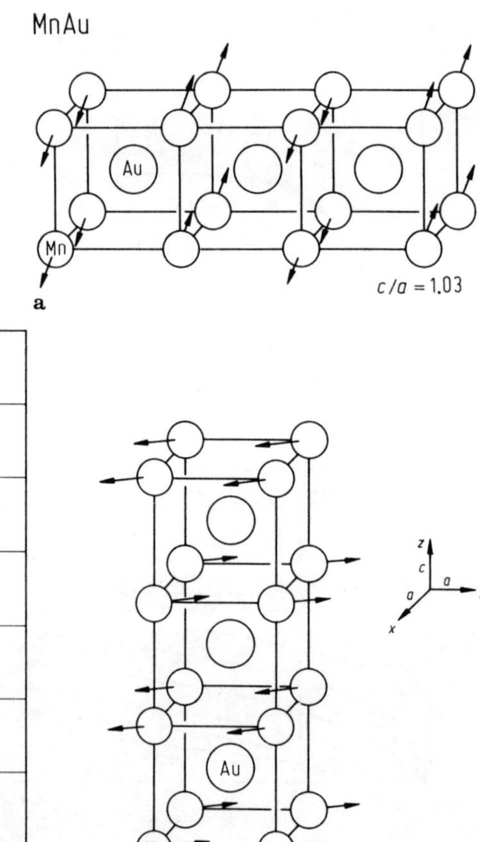

Fig. 26. MnAu. Atomic and momental structure. (a) t_2 phase and (b) t_1 phase [62 B 1].

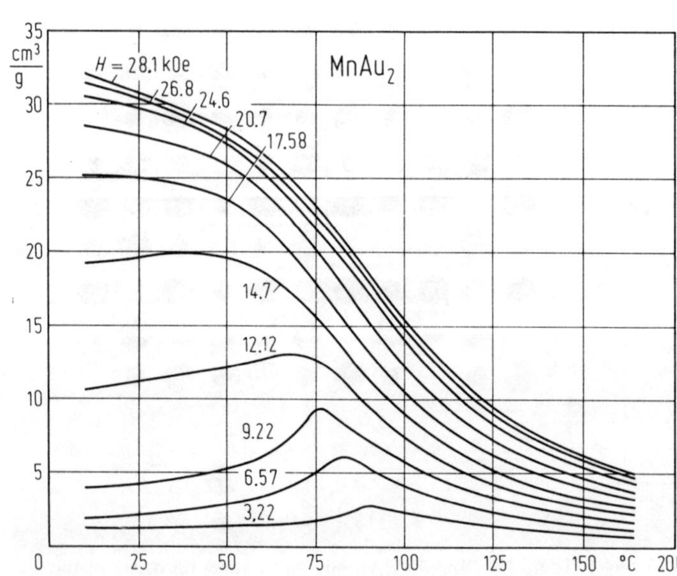

Fig. 25. $MnAu_2$. Magnetic moment per unit mass, σ, vs. temperature for the metamagnet $MnAu_2$ in different magnetic fields [56 M 1].

Fe–Au

The solid solubility of Fe in Au is about 60 at% Fe at 1000 °C, and 20 at% Fe at 450 °C. The solubility of Au in γ-Fe is 4.1 at% Au at 1171 °C [69 s 1, 78 m 1]. The solubility of Au in α-Fe at 850 °C is less than 1.5 at% Au. No intermetallic compounds are known [58 h 1]. Amorphous ferromagnetic Fe–Au alloys exist between 45 and 97 at% Fe at low temperature [70 F 1]. The magnetic phase diagram shows at low Fe-concentration and low temperature spin-glass and mictomagnetic behaviour: a sharp cusp in the magnetic susceptibility, dependence of T_f on annealing, Fe concentration, and the time constant of the measurement; remanent magnetization and magnetic viscosity. The tendency of Fe atoms to form atomic clusters coupled ferromagnetically manifests itself, e.g., in the broad concentration range of the magnetic transition region, 12 ⋯ 16 at% Fe (Fig. 27). Furthermore Fig. 28 shows a phase diagram with the inverse annealing temperature as a parameter comparable to the concentration or the degree of atomic clustering [81 C 2]. Annealing the alloys causes an increase of the magnetic susceptibility and of the magnetization. The effective magnetic moment increases with the Fe concentration from 3.7 μ_B/Fe [67 H 1] in very dilute alloys to about 5 μ_B/Fe at 10 at% Fe.

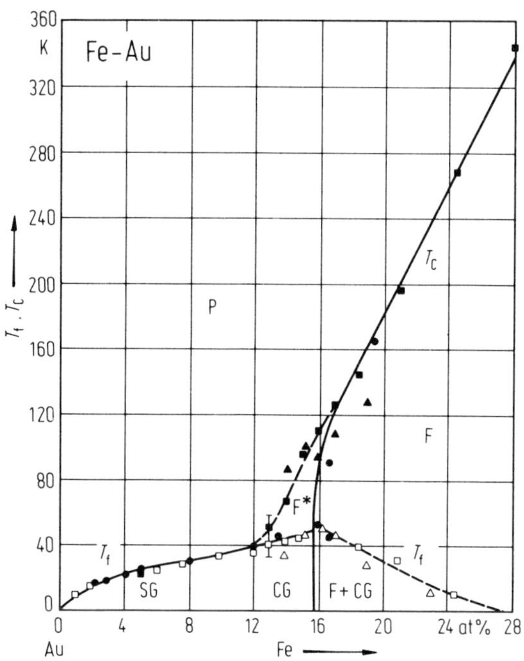

Fig. 27. Fe–Au. Variation of spin freezing temperature T_f and ferromagnetic Curie temperature T_C with composition, determined from (circles) Mössbauer effect anomalies [72 B1, 65G1], (squares) magnetic susceptibility [72 C 2, 81 S 1] and (triangles) NMR [79 S 1]. P: paramagnetic, F: ferromagnetic, F*: quasi-critical region, SG: spin glass and CG: cluster glass [81 S 1].

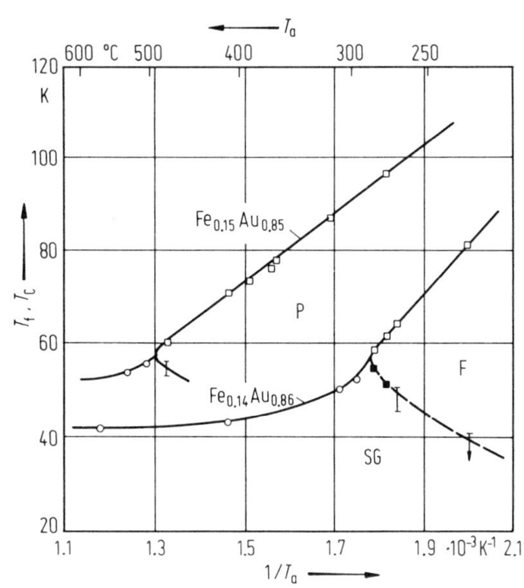

Fig. 28. Fe–Au. Variation of spin freezing temperature T_f and ferromagnetic Curie temperature T_C vs. reciprocal annealing temperature T_a^{-1}. Annealing varied between 1 day at T_a = 200 °C to 15 min at T_a = 750 °C, afterwards water quenching [80 C 1]. T_a^{-1} is proportional to the degree of atomic clustering and comparable to the Fe concentration. Open squares: paramagnetic (P) to ferromagnetic (F) transition, circles: paramagnetic to spin-glass (SG) transition and solid squares: ferromagnetic to spin-glass transition [81 C 2].

Co–Au

The solid solubility of Co in Au is 24 at% Co at about 1000 °C and falls to 0.2 at% Co at 400 °C. In α-Co only about 2 at% Au are soluble at 1000 °C, and about 0.3 at% at 500 °C [58 h 1, 65 e 1]. The strong decrease of the solubility of Co in Au with decreasing temperature makes it difficult to distribute Co atoms at random in the bulk Au matrix. Consequently, magnetic properties of bulk alloys produced by quenching from below the melting point are expected to be different from those of alloys with random distribution of the components, e.g., quench-condensed films or splat-cooled samples. Bulk alloys with Co concentrations smaller than 4 at% Co show freezing of the spins at low temperature; however, the magnetic susceptibility is Curie-Weiss type only at temperatures smaller than 10 K. These data, together with magnetization measurements support the conclusion that only groups of three or more nearest neighbour Co atoms are magnetic in Co–Au, a result confirmed for Co–Au films as well. Annealing of the more concentrated bulk alloys collects most of the Co into incoherent ferromagnetic precipitates, the sample shows magnetic hardness.

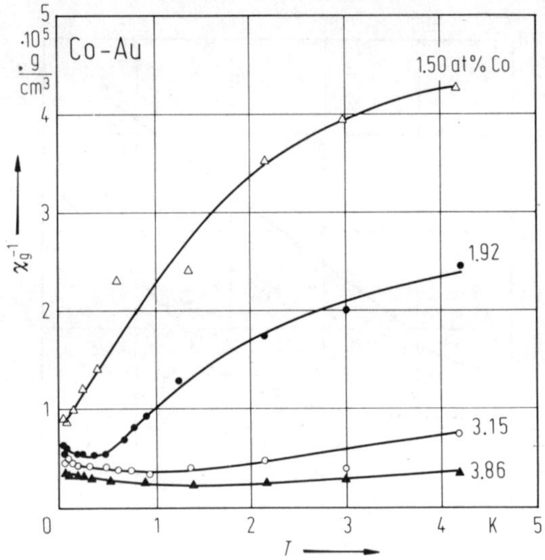

Fig. 29. Co–Au. Reciprocal of the initial magnetic mass susceptibility, χ_g^{-1}, vs. T. Samples annealed at 950 °C and quenched in a flow of hydrogen gas [71 B 1].

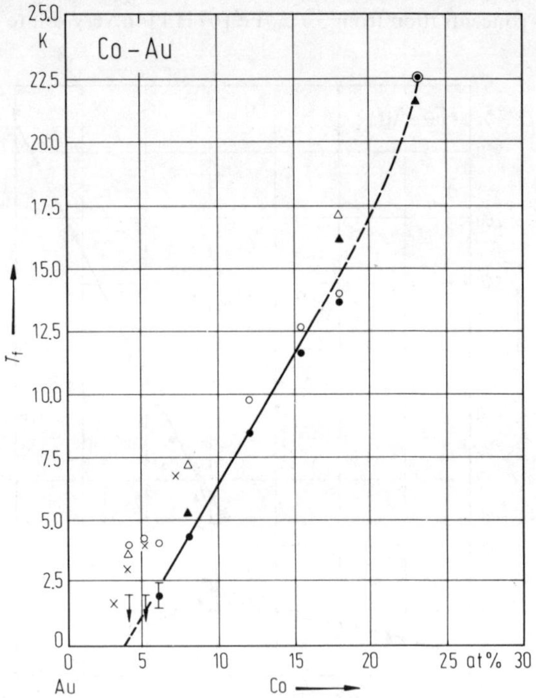

Fig. 30. Co–Au. Spin-glass freezing temperature T_f vs. Co concentration for (solid symbols) alloys annealed at 80 K (random alloys) and (open symbols) alloys annealed at 295 K, and (crosses) bulk alloys [74 C 1]. Arrows indicate upper bound on T_f values concerning random alloy films [79 K 1].

Ni–Au

Solid solubility at all concentrations exists at about 900 °C, whereas at 400 °C a miscibility gap extends from about 10 to 98 at% Ni [58 h 1, 69 s 1]. Concerning magnetic properties, a paramagnetic region starting at small Ni concentration and extending to about 35 at% Ni is followed by a critical region from 35 at% Ni to about 53 at% Ni and a ferromagnetic region above 55 at% Ni. In the concentration region from 20 to 53 at% Ni ferromagnetic Ni clusters are present in the quenched state of the alloys [81 H 1]. After annealing, a mixture of coherent precipitates exists, one Ni-rich and the other Au-rich.

5.1.3 References for 5.1

Review

58 h 1 Hansen, M., Anderko, K.: Constitution of binary alloys, London: McGraw Hill **1958**.
62 v 1 Vogt, E., Höhl, M., in: Landolt-Börnstein 6th ed., Vol. II/9, (Hellwege, K.H., Hellwege, A.M., eds.), Berlin: Springer, **1962**.
64 c 1 Chikazumi, S.: Physics of Magnetism, New York: John Wiley & Sons **1964**.
65 e 1 Elliot, R.P.: Constitution of binary alloys, first suppl., London: McGraw Hill **1965**.
68 d 1 Daybell, M.D., Steyert, W.A.: Rev. Mod. Phys. **40** (1968) 380.
69 s 1 Shunk, F.A.: Constitution of binary alloys, second suppl., London: McGraw Hill **1969**.
75 v 1 Vogt, E.: Phys. Status Solidi (a) **28** (1975) 11.
78 b 1 Beck, P.A.: Progr. Mater. Sci. **23** (1978) 1.
78 m 1 Moffat, W.G.: The handbook of binary phase diagrams, Schenectady: General Electric **1978**.
80 m 1 Mydosh, J.A., Nieuwenhuys, G.J., in: Ferromagnetic Materials Vol. 1, Chap. II (Wohlfarth, E.P., ed.), Amsterdam: North-Holland Publ. Co. **1980**, p. 71.

Bibliography

36 G 1 Gustafsson, G.: Ann. Phys. Leipzig **25** (1936) 545.
53 O 1 Oliver, D.J., Sucksmith, W.: Proc. R. Soc. London Ser. A **219** (1953) 7.
54 K 1 Knappvost, A.: Z. Metallkd. **45** (1954) 137.
56 M 1 Meyer, A.J.P., Taglang, P.: J. Phys. Paris **17** (1956) 457.
58 A 1 Ahern, S.A., Martin, M.J.C., Sucksmith, W.: Proc. R. Soc. London Ser. A **248** (1958) 145.
58 M 1 Meyer, A., Wolff, C.: C.R. Acad. Sci. **246** (1958) 576.
58 P 1 Pugh, E.W., Ryan, F.M.: Phys. Rev. **111** (1958) 1038.
59 M 1 Morris, D.P., Williams, J.: Proc. Phys. Soc. (London) **73** (1959) 422.
59 S 1 Scheil, E., Wachtel, E., Kalkuhl, A.: Ann. Phys. Leipzig **4** (1959) 58.
60 V 1 Van Itterbeek, A., Peelaers, W., Steffens, F.: Appl. Sci. Res. Sect. B **8** (1960) 337.
61 B 1 Berghout, C.: Z. Metallkd. **52** (1961) 179.
61 F 1 Franck J.P., Manchester, F.D., Martin, D.L.: Proc. R. Soc. London Ser. A **263** (1961) 499.
61 H 1 Herpin, A., Meriel, P.: J. Phys. Radium **22** (1961) 337.
61 J 1 Jacobs, I.S., Kouvel, J.S., Lawrence, P.E.: J. Phys. Soc. Jpn. (Suppl.) **17 B-1** (1961) 157.
61 S 1 Sato, H.: J. Phys. Chem. Solids **19** (1961) 54.
61 W 1 Wachtel, E., Vetter, U.: Z. Metallkd. **52** (1961) 525.
62 A 1 Abrahams, S.C., Guttmann, L., Kasper, J.S.: Phys. Rev. **127** (1962) 2052.
62 B 1 Bacon, G.E.: Proc. Phys. Soc. (London) **79** (1962) 938.
62 H 1 Henderson, B., Raynor, G.V.: J. Phys. Paris **23** (1962) 685.
62 L 1 Le Guillerm, J., Tournier, R., Weil, L.: Proc. LT 8 London, **1962** (Davies, R.O., ed.), London: Butterworths Publ. 1963, p. 236.
64 L 1 Lutes, O.S., Schmit, J.L.: Phys. Rev. **134** (1964) A 676.
65 G 1 Gonser, U., Grant, R.W., Meechan, C.J., Muir, A.H., Wiedersich, H.: J. Appl. Phys. **36** (1965) 2124.
66 S 1 Svensson, K.: LT 10 Proceedings, Moscow 1966 p. 267.
67 C 2 Claus, H., Sinha, A.K., Beck, P.A.: Phys. Lett. **26** (1967) 38.
67 C 1 Creveling, L., Luo, H.L., Knapp, G.S.: Phys. Rev. Lett. **18** (1967) 851.
67 H 1 Hurd, C.M.: J. Phys. Chem. Solids **28** (1967) 1345.
67 W 1 Weiss, W.D.: Z. Metallkd. **58** (1967) 909.
67 Y 1 Yasukochi, K., Kuwasawa, Y., Sekizawa, K.: J. Appl. Phys. **38** (1967) 1318.
68 A 1 Aldred, A.T.: J. Phys. C **1** (1968) 1103.
68 K 1 Korn, D.: Z. Physik **214** (1968) 136.
69 H 1 Hicks, T.J., Rainford, B., Kouvel, J.S., Low, G.G., Comly, J.B.: Phys. Rev. Lett. **22** (1969) 531.
69 S 1 Smith, J.H., Vance, E.R.: J. Appl. Phys. **40** (1969) 4853.
69 S 2 Smith, J.H., Wells, P.: J. Phys. C **2** (1969) 356.
69 T 1 Toth, R.S., Arrott, A., Shinozaki, S.S., Werner, S.A., Sato, H.: J. Appl. Phys. **40** (1969) 1373.
70 B 1 Bancroft, M.H.: Phys. Rev. B **2** (1970) 2597.
70 B 2 Bennett, L.H., Swartzendruber, L.J.: Acta Metall. **18** (1970) 485.
70 B 3 Besnus, M.J., Meyer, A.J.P., Berninger, R.: Phys. Lett. A **32** (1970) 192.

70 F 1 Felsch, W.: Z. Angew. Phys. **29** (1970) 217.
70 K 1 Kouvel, J.S., Comly, J.B.: Phys. Rev. Lett. **24** (1970) 598.
71 B 1 Boucai, E., Lecoanet, B., Pilon, J., Tholence, J.L., Tournier, R.: Phys. Rev. B **3** (1971) 3834.
71 N 1 Nakai, Y., Kunitomi, N., Endoh, Y., Ishikawa, Y.: Solid State Commun. **9** (1971) 921.
72 B 1 Borg, R.J., Lai, D.Y.F., Violet, C.E.: Phys. Rev. B **5** (1972) 1035.
72 C 1 Chakrabarti, D.J., Beck, P.A.: Int. J. Magn. **3** (1972) 319.
72 C 2 Cannella, V., Mydosh, J.A.: Phys. Rev. B **6** (1972) 4220.
73 C 1 Cannella, V., In: Amorphous Magnetism, New York: Plenum Press **1973**, p. 195.
73 F 1 Franz, J.M., Sellmyer, D.J.: Phys. Rev. B **8** (1973) 2083.
74 C 1 Cannella, V., Mydosh, J.A.: AIP Conf. Proc. **18** (1) (1974) 651.
74 C 2 Clinton, J.R., Tyler, E.H., Luo, H.L.: J. Phys. F **4** (1974) 1162.
74 K 1 Kumagai, K., Ohtsuka, T.: J. Phys. Soc. Jpn. **37** (1974) 384.
74 S 1 Schoep, G.K., Van de Valk, H.J., Frijters, G.A.M., Kok, H.B., Poulis, N.J.: Physica **77** (1974) 449.
75 M 1 Mukhopadhyay, A.K., Shull, R.D., Beck, P.A.: J. Less-Common Met. **43** (1975) 69.
76 C 1 Campbell, S.J., Clark, P.E., Hicks, T.J.: J. Phys. F **6** (1976) 249.
76 F 1 Follstaedt, D.M., Abbas, D., Stakelon, T.S., Slichter, C.P.: Phys. Rev. B **14** (1976) 47.
76 T 1 Tustison, R.W., Beck, P.A.: Solid State Commun. **20** (1976) 841.
77 K 1 Kuz'Menko, P.P., Kal'Naya, G.I., Suprunenko, P.A., Buyanov, Yu.I.: Metallofizika **70** (1977) 43.
79 D 1 Davis, J.R., Hicks, T.J.: J. Phys. F **9** (1979) L 7.
79 K 1 Korn, D., Schilling, D., Zibold, G.: J. Phys. F **9** (1979) 1111.
79 S 1 Sarkissian, B.V.B.: Philos. Mag. B **39** (1979) 413.
80 A 1 Adachi, K., Matsui, M., Fukuda, Y.: J. Phys. Soc. Jpn. **48** (1980) 6.
80 C 1 Crane, S., Claus, H.: Solid State Commun. **35** (1980) 461.
80 G 1 Gibbs, P., Smith, J.H.: J. Mag. Magn. Mater. **15–18** (1980) 155.
80 H 1 Hanson, M., Leonardsson, L.: J. Mag. Magn. Mater. **15–18** (1980) 75.
80 T 1 Tsunoda, Y., Nakai, Y.: Solid State Commun. **34** (1980) 413.
81 B 1 Beck, P.A.: Phys. Rev. B **23** (1981) 2290.
81 C 1 Cowlan, N., Shamah, A.M.: J. Phys. F **11** (1981) 27.
81 C 2 Crane, S., Claus, H.: Phys. Rev. Lett. **46** (1981) 1693.
81 H 1 Hurd, C.M., McAlister, S.P., Shiozaki, I.: J. Phys. F **11** (1981) 457.
81 K 1 Kondoh, H., Takeuchi, S.: J. Phys. Soc. Jpn. **50** (1981) 2135.
81 M 1 Morgownik, A.F.J., Mydosh, J.A.: Phys. Rev. B **24** (1981) 5277.
81 P 1 Plant, J.S.: J. Phys. F **11** (1981) 1901.
81 S 1 Sarkissian, B.V.B.: J. Phys. F **11** (1981) 2191.
81 V 1 Vaknin, D., Davidov, D., Nieuwenhuys, G.J., Hoekstra, F.R., Barberis, G.E., Mydosh, J.A.: Physica B **108** (1981) 765.
81 Z 1 Zibold, G., Korn, D.: Physica B, C **107** (1981) 99.
82 W 1 Wassermann, E.F.: Physica B **109–110** (1982) 1936.
83 M 1 Morgownik, A.F.J., Mydosh, J.A.: Solid State Commun. **47** (1983) 321.
83 M 2 Morgownik, A.F.J., Mydosh, J.A.: Solid State Commun. **47** (1983) 325.
85 K 1 Korn, D., Zibold, G.: J. Phys. F **15** (1985) 2497.

5.2 3d elements with Be, Mg or Zn

For properties of Kondo and spin fluctuation systems and spin glasses with Be, Mg or Zn as the host material and a 3d element as an impurity solvent, see K.H. Fischer in Landolt-Börnstein, New Series vol. III/15a, pp. 294ff. and 334ff.

5.2.1 Alloys and compounds of 3d elements with Be

Table 1. $TiBe_2$. Some physical parameters of three different samples. Parameters for fits to the lower-temperature data ($T < 9$ K), are given in parentheses. Specific heat constants are given by $C_p/T = \gamma + \varepsilon T^2 + \delta T^2 \ln T$, T in [K]. Θ_D is the Debye temperature and T_{sf} is the characteristic spin-fluctuation temperature [82 S 1].

	a Å	$\chi_m(0\,K)$ $cm^3\,mol^{-1}$	γ mJ mol^{-1}	ε mJ mol^{-1} K^{-2}	δ mJ mol^{-1} K^{-4}	T_{sf}[K] Θ_D: 400 K	500 K	600 K
$TiBe_{2.00}$	6.4538(4)	$0.802 \cdot 10^{-2}$	51.09	−0.2072	0.06563	89	45	33
			(52.67)	(−0.3692)	(0.1247)	(38)	(27)	(23)
$TiBe_{2.00}$	6.4510(5)	$0.947 \cdot 10^{-2}$	56.33	−0.2207	0.06724	97	50	38
			(56.86)	(−0.2556)	(0.07408)	(101)	(56)	(42)
$TiBe_{1.94}O_{0.06}$	6.4536(7)	$0.768 \cdot 10^{-2}$	37.88	−0.2208	0.0724	70	38	29
			(38.68)	(−0.3348)	(0.1189)	(34)	(24)	(20)

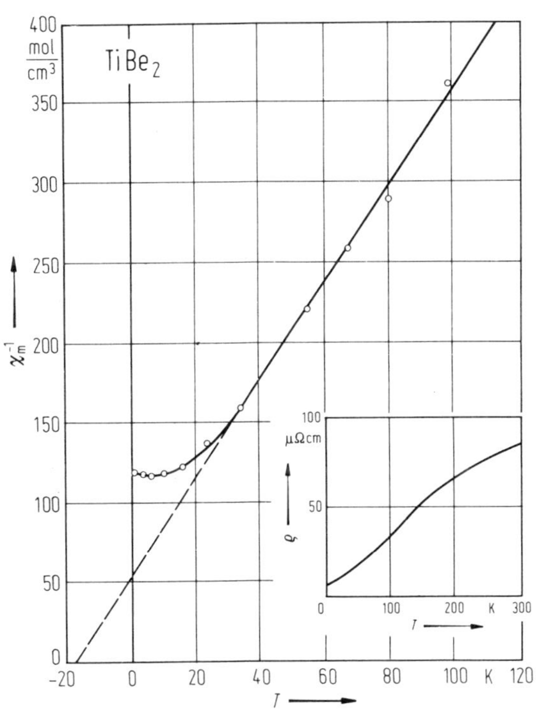

Fig. 1. $TiBe_2$. Inverse of the magnetic molar susceptibility, χ_m^{-1}, vs. temperature. The inset gives the electrical resistivity ϱ vs. temperature. The Néel temperature $T_N \approx 10$ K; the effective magnetic moment per molecule $p_{eff} = 1.64\,\mu_B$. The lattice parameter of the cubic laves phase at low T is $a = 6.4532(4)$ Å [78 M 1, 2].

Table 2. $TiBe_{2-x}Cu_x$. Magnetic properties [81 S 1]. The effective paramagnetic moment per formula unit, p_{eff}, derived from the Curie-Weiss curve (measured between $T = 35$ K and $T = 120$ K), as well as its extrapolated intercept with the T-axis, designated Θ. T_C is the ferromagnetic Curie temperature and $\bar{p}_{Ti}$ the average magnetic moment per Ti atom derived from the spontaneous magnetization.

x	p_{eff} μ_B	Θ K	T_C K	$\bar{p}_{Ti}$ μ_B
0	1.64	−20	–	–
0.012	1.69	−13.5	–	–
0.05	1.65	− 5.5	–	–
0.10	1.72	− 2.5	–	–
0.16	1.84	3.5	4.74	–
0.20	1.76	12.5	7.8	0.22
0.30	1.57	16	11.0	0.21
0.40	1.63	23	18.0	0.23
0.50	1.66	27	21.0	0.24

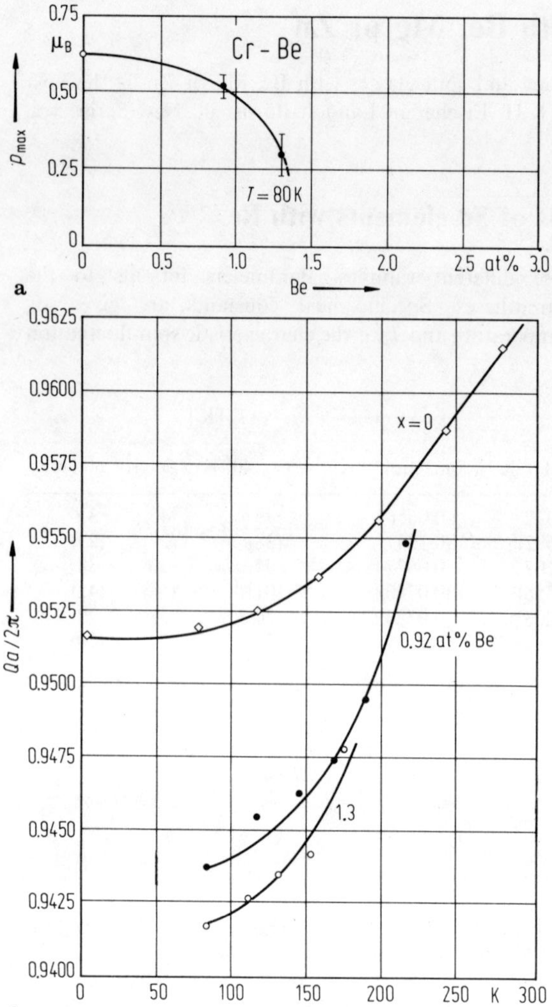

Fig. 2. Cr–Be. (a) Amplitude p_{max} of the spin modulation, i.e. the maximum value of the magnetic moment of the Cr atoms at 80 K [81 S 1]. Solid circles [81 S 2], open circle [66 K 1] (b) Spin density wave (SDW) vector Q expressed in reciprocal lattice constants, $2\pi/a$, vs. temperature [81 S 1]. Lozenges: [67 W 1].

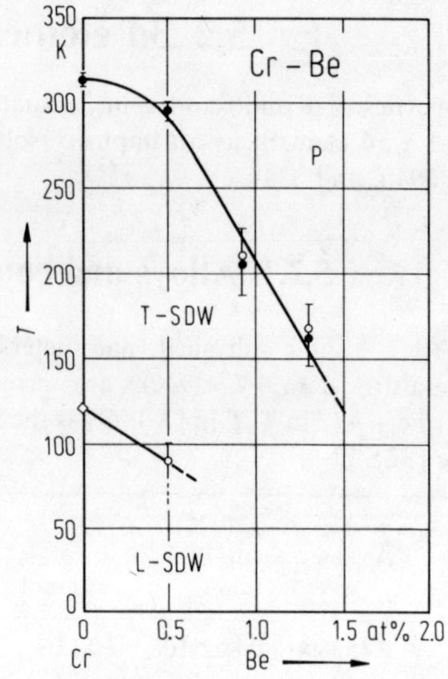

Fig. 3. Cr–Be. Magnetic phase diagram. The upper curve gives the Néel temperatures for the transition from a sinusoidal incommensurate antiferromagnetic to a paramagnetic structure. Above 1.3 at% Be up to 6 at% Be no antiferromagnetism was found. The regions designated T–SDW and L–SDW have transverse and longitudinal spin density waves, respectively. Open circles: neutron diffraction results, solid circles: resistivity results [81 S 1]. Lozenge: [67 W 1].

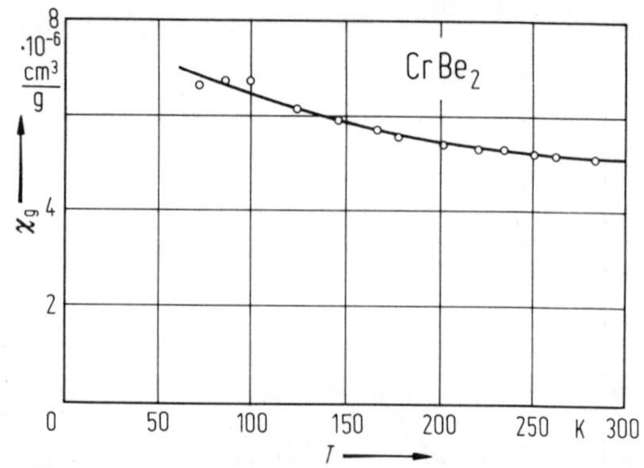

Fig. 4. $CrBe_2$. Low-field magnetic mass susceptibility χ_g vs. temperature [66 S 1].

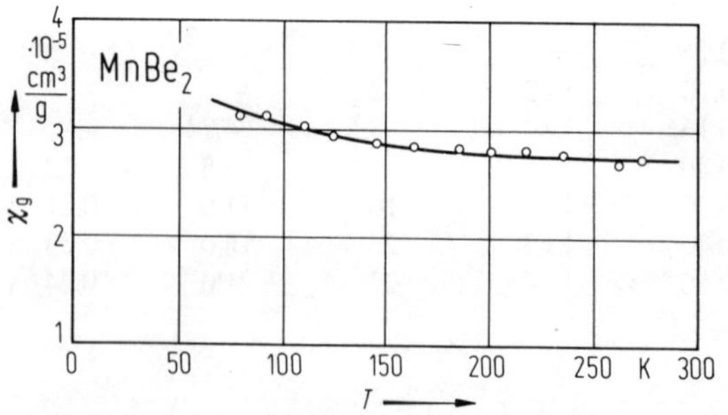

Fig. 5. $MnBe_2$. Magnetic mass susceptibility χ_g vs. temperature T [66 S 1].

Table 3. $CrBe_{12}$. Magnetic and related properties of the tetragonal body-centered $CrBe_{12}$ with $Mn_{12}Th$-structure. All Cr sites are equivalent [71 H 1].

	p_{Cr} μ_B	T_C K	p_{eff} μ_B	γ mJ mol^{-1} K^{-2}
$CrBe_{12}$	0.3	50	1.44	1.79

5.2.2 Alloys and compounds of 3d elements with Mg

Table 4. Crystallographic data of $MgCo_2$, $MgNi_2$ and Mg_2Ni [75 B 1].

	Crystal structure	a	c
		Å	
Mg_2Ni	hexagonal	5.212	13.247
$MgNi_2$	hexagonal (C36)	4.833	15.855
$MgCo_2$	hexagonal (C14)	4.859	7.954

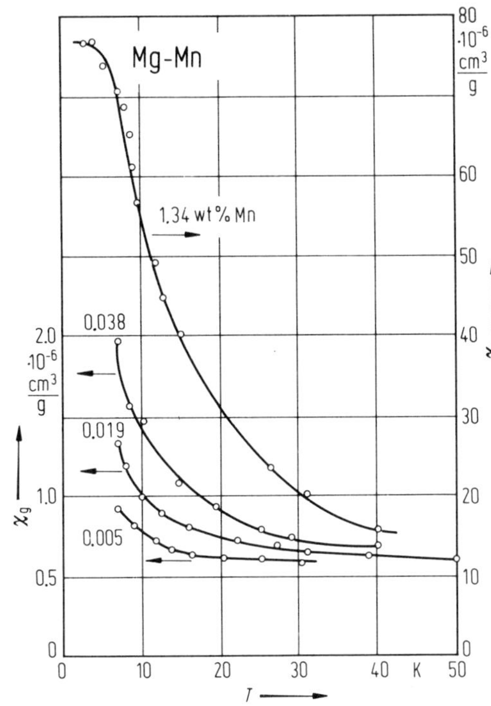

Fig. 6. Mg–Mn. Magnetic mass susceptibility χ_g vs. temperature [62 C 1]. The susceptibility is corrected for ferromagnetic impurities.

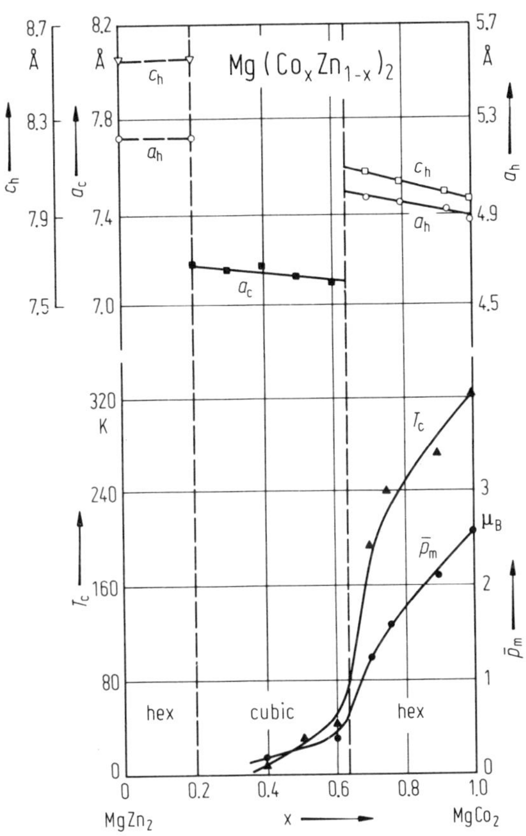

Fig. 7. $Mg(Co_xZn_{1-x})_2$. Concentration dependence of the lattice constants in the two hexagonal (C14) concentration ranges (a_h, c_h) and the cubic (C15) concentration range (a_c). The concentration dependence of the average magnetic moment per formula unit, $\bar{p}_m$, is indicated by solid circles. The concentration dependence of the ferromagnetic Curie temperature T_C is indicated by solid triangles [81 B 1].

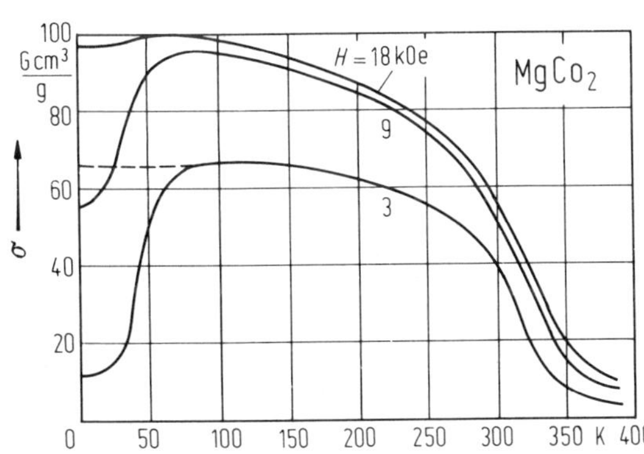

Fig. 8. $MgCo_2$ (2 at% excess of Mg). Temperature dependence of the magnetic moment per unit mass, σ, in various magnetic fields H. All measurements were made by heating the sample from 4.2 K to room temperature. The broken line indicates the results obtained with a slight excess of Co. An excess of Mg or Co does not influence the Curie point or the saturation magnetization. Cooling to 4.2 K in the presence of a magnetic field has no influence on the shape of the $\sigma(T)$ curves [75 B 1].

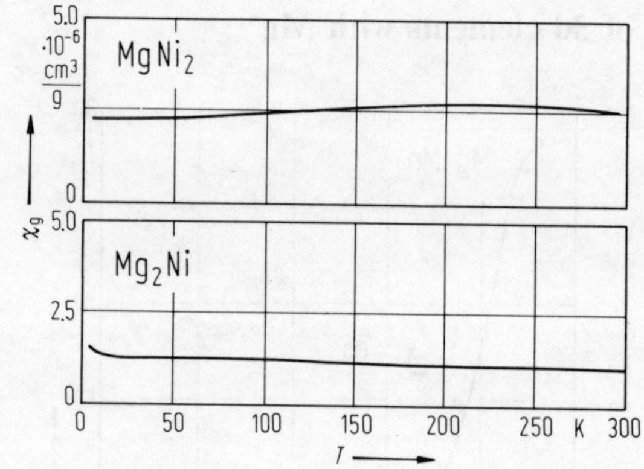

Fig. 9. Mg_2Ni; $MgNi_2$. Magnetic mass susceptibility χ_g vs. temperature T [75 B 1].

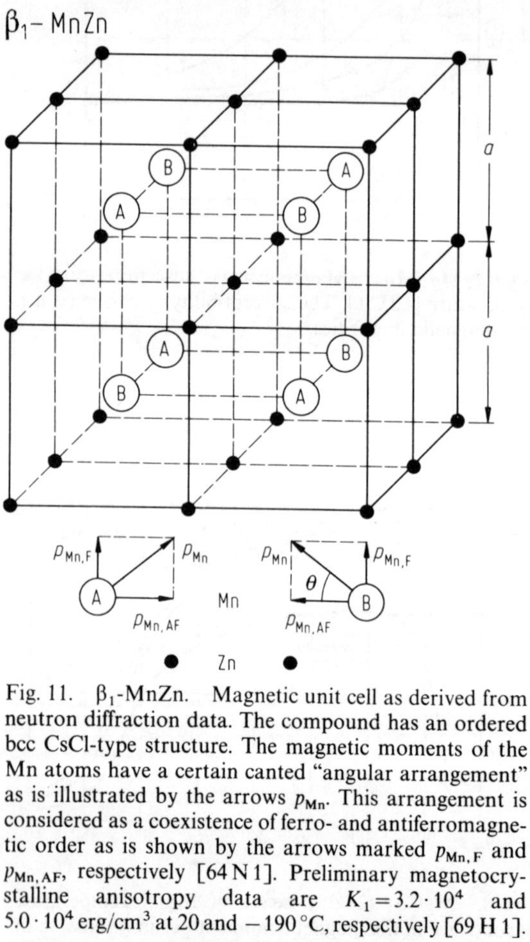

Fig. 11. β_1-MnZn. Magnetic unit cell as derived from neutron diffraction data. The compound has an ordered bcc CsCl-type structure. The magnetic moments of the Mn atoms have a certain canted "angular arrangement" as is illustrated by the arrows p_{Mn}. This arrangement is considered as a coexistence of ferro- and antiferromagnetic order as is shown by the arrows marked $p_{Mn,F}$ and $p_{Mn,AF}$, respectively [64 N 1]. Preliminary magnetocrystalline anisotropy data are $K_1 = 3.2 \cdot 10^4$ and $5.0 \cdot 10^4$ erg/cm³ at 20 and −190 °C, respectively [69 H 1].

	$p_{Mn,F}$	$p_{Mn,AF}$	p_{Mn}	Θ
	μ_B			°C
β_1-$Mn_{0.50}Zn_{0.50}$	1.72	2.96	3.4	30
β_1-$Mn_{0.56}Zn_{0.44}$	1.09	2.46	2.7	24

5.2.3 Alloys and compounds of 3d elements with Zn

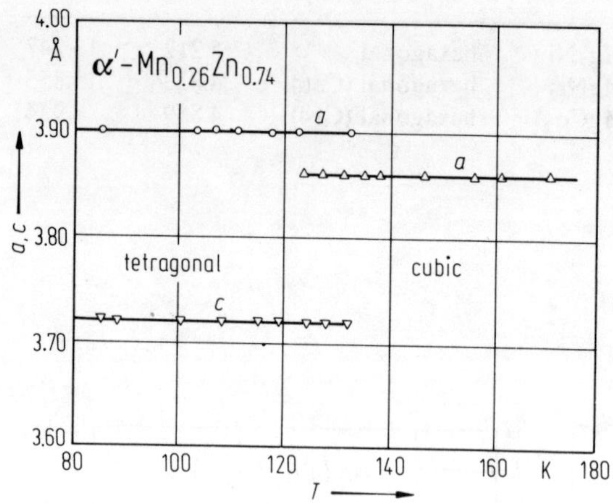

Fig 10. α'-$Mn_{0.26}Zn_{0.74}$. Lattice parameters a and c vs. temperature. The compound has an ordered fcc Cu_3Au-type structure at room temperature. Neutron diffraction experiments indicate a transition temperature of about 20 K lower. The compound becomes antiferromagnetic below 130 K [69 U 1].

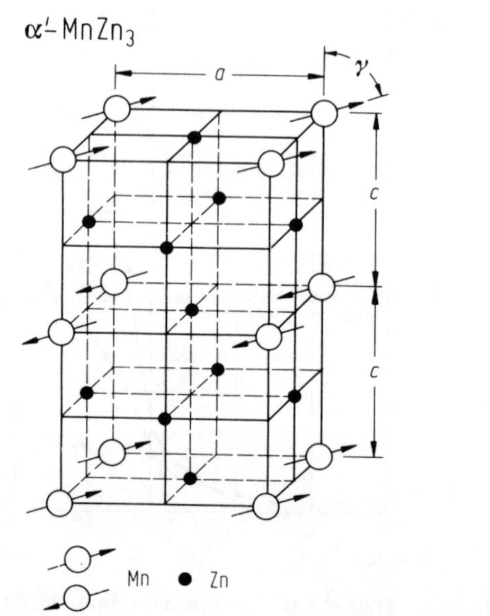

Fig. 12. α'-$MnZn_3$. Magnetic unit cell of tetragonally deformed α'-$MnZn_3$ at low temperatures as derived from neutron diffraction data. No conclusive results for the angle γ and the magnetic moments of the Mn atoms can be made. The neutron analysis gives for $\gamma = 70°$ and $90°$, values for $p_{Mn} = 2.7\,\mu_B$ and $2.5\,\mu_B$, respectively [64 N 1].

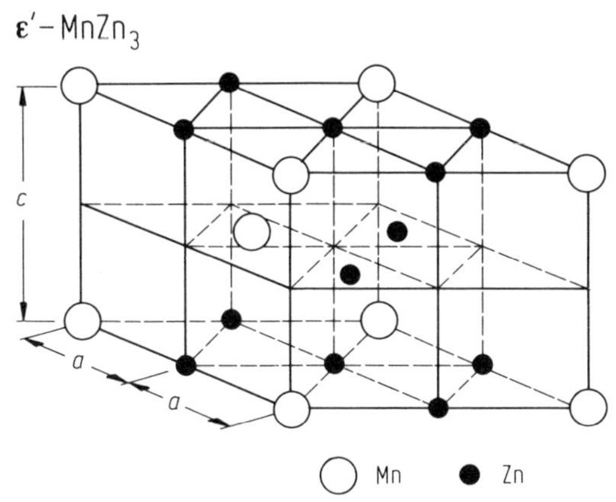

Fig. 13. ε'-$MnZn_3$. Unit cell of the metastable ε'-$MnZn_3$ phase, i.e., hcp structure with the long-range order of the Ni_3Sn-type. The phase is obtained by quenching a specimen in water after homogenization at 550 °C [64 N 1]. $a=2.75$ Å and $c/a=1.61$. The phase is strongly ferromagnetic with a Curie temperature above 400 K [60 T 1].

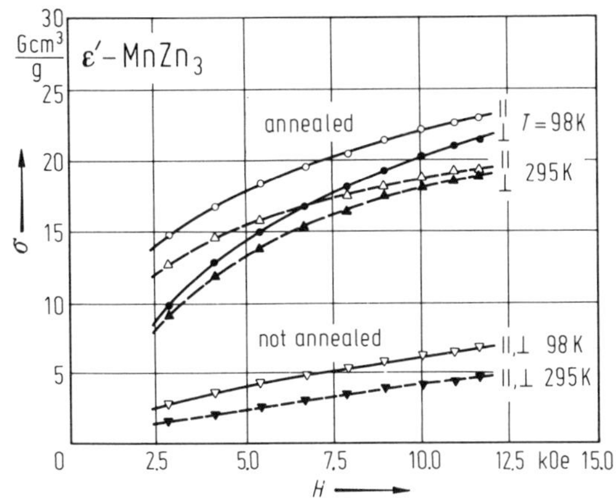

Fig. 14. ε'-$MnZn_3$, single crystal. Not annealed: quenched from 500 °C in water. Annealed: afterwards held for 50 h at 100 °C. Magnetic moment per gram, σ, vs. magnetic field H at two temperatures for measuring fields perpendicular or parallel to the c-axis, respectively [62 N 1].

Table 5. Ni–Zn. Average magnetic moment per atom, $\bar{p}_{at}$, spontaneous magnetic moment per unit mass, σ_s (RT) and σ_s (0 K), at room temperature and extrapolated to 0 K, respectively, as well as Curie temperature T_C [37 M 1]. For T_C and σ_s (325 K) see also [67 A 1].

Zn	σ_s (RT)	σ_s (0 K)	$\bar{p}_{at}$	T_C
at%	$G\,cm^3\,g^{-1}$		μ_B	K
4.12	45.34	49.83	0.52	573
5.57	41.46	45.11	0.48	530
10.81	25.38	35.18	0.37	430

5.2.4 References for 5.2

37 M 1 Marian, V.: Ann. Phys. Paris **7** (11) (1937) 459.
60 T 1 Tezuka, S., Sakai, S., Nakagawa, Y.: J. Phys. Soc. Jpn. **15** (1960) 931.
62 C 1 Collings, E.W., Hedgcock, F.T.: Phys. Rev. **126** (1962) 1654.
62 N 1 Nakagawa, Y., Sakai, S., Hori, T.: J. Phys. Soc. Jpn. **17**, Suppl. B.1 (1962) 168.
64 N 1 Nakagawa, Y., Hori, T.: J. Phys. Soc. Jpn. **19** (1964) 2082.
66 K 1 Koehler, W.C., Moon, R.M., Trego, A.L., Mackintosh, A.R.: Phys. Rev. **121** (1966) 405.
66 S 1 Saji, H., Yamadaya, T., Asanuma, M.: J. Phys. Soc. Jpn. **21** (1966) 255.
67 A 1 Annayer, R.G., Yusunov, T.M.: Phys. Met. Metallogr. **23** (1967) 210.
67 W 1 Werner, S.A., Arrott, A., Kendrick, H.: Phys. Rev. **122** (1967) 528.
69 H 1 Herr, A., Kuentzler, R., Meyer, A.J.-P.: J. Appl. Phys. **40** (1969) 1375.
69 U 1 Uchishiba, H., Hori, T., Nakagawa, Y.: J. Phys. Soc. Jpn. **27** (1969) 600.
71 H 1 Herr, A., Kuentzler, R.: Phys. Status Solidi b **45** (1971) K 1.
75 B 1 Buschow, K.H.J.: Solid State Commun. **17** (1975) 891.
78 M 1 Matthias, B.T., Giorgi, A.L., Struebing, V.O., Smith, J.L.: J. Phys. (Paris) Lett. **39** (1978) L441.
78 M 2 Mathias, B.T., Giorgi, A.L., Struebing, V.O., Smith, J.L.: Phys. Lett. A **69** (1978) 221.
81 B 1 Buschow, K.H.J., Kropp, H., Dormann, E.: J. Mag. Magn. Mater. **23** (1981) 257.
81 S 1 Sano, T., Kawarazaki, S., Iida, S., Kunitomi, N.: J. Phys. Soc. Jpn. **50** (1981) 1203.
81 S 2 Smith, J.L.: Physica B **107** (1981) 251.
82 S 1 Shaltiel, D., Monod, P., Felner, I.: J. Phys. F **12** (1982) 2703.

5.3 3d elements and B, Al or Ga

5.3.1 Survey of tables and figures

The alloys and compounds are listed according to the sequence of their constituting 3d elements in the periodic system of elements.

Alloy or compound	Basic		Data	
	Structure, Magnetic phase diagram	Paramagnetism	Magnetic Transition temperature	Ferromagnetism
		5.3.2 3d-B		
M_3B; M = 3d			—	—
Co_3B			Tab. 1, 2	Tab. 1, 2
Ni_3B		Fig. 5, Tab. 2	—	—
M_2B; M = 3d			Fig. 1	Figs. 4, 9
Cr_2B		Figs. 1, 2	Fig. 1	Fig. 1
Fe_2B		Fig. 1	Fig. 3, Tab. 1	Figs. 1, 9, Tab. 1
Co_2B		Fig. 1	Fig. 1, Tab. 1	Figs. 1, 4, Tabs. 1, 2
Ni_2B		Figs. 1, 5, Tab. 2	Tab. 1	Tab. 1
$(Fe_xCo_{1-x})_2B$	Fig. 8		Figs. 1, 8	Figs. 1, 7, 8
$(Ni_xCo_{1-x})_2B$			Fig. 1	Figs. 1, 7
M_5B_3; M = Cr		Fig. 2		
MB; M = 3d		Fig. 9		Fig. 9
CrB		Fig. 2		
FeB		Fig. 9		Figs. 3, 4, 9
CoB		Tab. 2		Tab. 1
NiB		Fig. 5, Tab. 2		
$Fe_xCo_{1-x}B$				Fig. 4
M_3B_4; M = Cr		Fig. 2		
MB_2; M = 3d		Fig. 7	Fig. 7	
CrB_2		Figs. 1, 2, 6	Figs. 6, 7	
		5.3.3 3d-Al		
Ti–Fe–Al	Tab. 3	Tab. 3		
Ti–Co–Al	Tab. 3	Tab. 3	Fig. 11	Fig. 11
Ti–Ni–Al	Tab. 3	Tab. 3		
V–Al alloys	Tab. 4	Tab. 4		
$V_{0.5-x}Mn_{0.5}Al_x$		Fig. 10	Fig. 10	
Cr–Al compounds		Tab. 5		
$Cr_{1-x}Al_x$			Fig. 12	
Cr_2Al_{11}		Fig. 13		
$CrAl_7$		Fig. 14		
Cr_xFeAl_{1-x}		Fig. 15		
$Cr_{2-x}Fe_xAl$	Fig. 16, Tab. 6		Fig. 16, Tab. 6	Fig. 15
Mn–Al compounds	Tab. 7		Tab. 7	
Mn–Fe–Al	Figs. 17, 18		Tab. 8	
Mn–Co–Al			Tab. 8	
Mn–Ni–Al	Fig. 19		Fig. 23, Tab. 8	Fig. 21
Mn–Cu–Al	Figs. 20, 22		Fig. 24, Tab. 8	

Alloy or compound	Basic		Data	
	Structure, Magnetic phase diagram	Paramagnetism	Magnetic Transition temperature	Ferromagnetism
Fe–Al	Figs. 25, 26	Figs. 28, 29, Tab. 9		Figs. 27, 30
$Fe_{0.7}Al_{0.3}$	Fig. 31			Fig. 31
Fe_3Al		Fig. 32		Figs 33, 35, Tab. 18
$Fe_{1-x}Co_xAl$		Fig. 34		
$R(Fe_xAl_{1-x})_2$ R = Rare Earth or Zr			Fig. 37	Figs. 36, 37
Co–Al		Figs. 28, 38, 39		
$Co_{1-x}Ni_xAl$		Fig. 42, Tab. 10		
NiAl	Fig. 41, Tab. 11	Fig. 40		
Ni_3Al		Figs. 43, 44		Figs. 43, 45, 46
		5.3.4 3d-Ga		
Ti–Fe–Ga	Tab. 3	Tab. 3		
Ti–Co–Ga	Tab. 3	Tab. 3	Fig. 11	Fig. 11
Ti–Ni–Ga	Tab. 3	Tab. 3		
V_3Ga		Fig. 47, Tab. 12		
V_xCoGa_{1-x}				Fig. 48
Cr–Ga		Tab. 13		
$(Cr_xFe_{1-x})_3Ga$	Fig. 49			Fig. 50
Cr_xCoGa_{1-x}		Tab. 14	Tab. 14	Fig. 48, Tab. 14
Mn–Ga	Figs. 54, 55, Tabs. 15, 16	Figs. 52, 56–58, Tab. 17	Fig. 51, Tab. 15	Figs. 53, 56, 57, 59, Tab. 15
$(Mn_xFe_{1-x})_3Ga$	Fig. 49			Fig. 50
Fe–Ga	Fig. 60	Fig. 61		Figs. 61, 62, 63
Fe_3Ga				Tab. 18
$(Fe_{1-x}Co_x)_3Ga$	Fig. 49			Fig. 50
$(Fe_{1-x}Ni_x)_3Ga$	Fig. 49			Fig. 50
Co–Ga	Figs. 64, 66	Figs. 28, 65, Tabs. 19, 20	Tab. 21	
Ni_3Ga		Fig. 67		Fig. 68
$Ni_3Fe_xGa_{1-x}$			Figs. 69, 70	Figs. 69, 70

5.3.2 3d-B alloys and compounds

Table 1. Fe, Co borides. Ferromagnetic properties [63 F 1]: saturation magnetic moment p_s and Curie temperature T_C.

	p_s μ_B/3d	T_C K
Co_3B	1.12	747
Co_2B	0.76	429
Fe_2B	1.90 [1]	1015
α-FeB	1.25	594
FeB	1.20	

[1] [29 W 1].

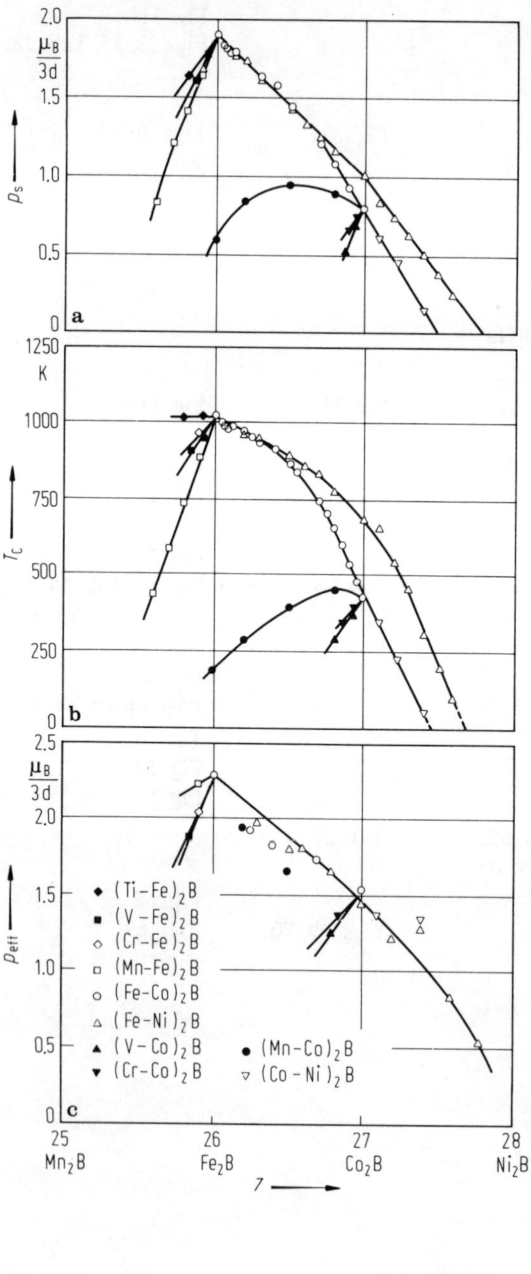

Fig. 1. 3d half-borides. Properties of the half-borides of 3d transition metals. (a) Saturation magnetic moment per metal atom, p_s, (b) Curie temperature T_C, and (c) p_{eff} per metal atom, plotted as a function of average atomic number Z of the transition metal atoms [66 C 2].

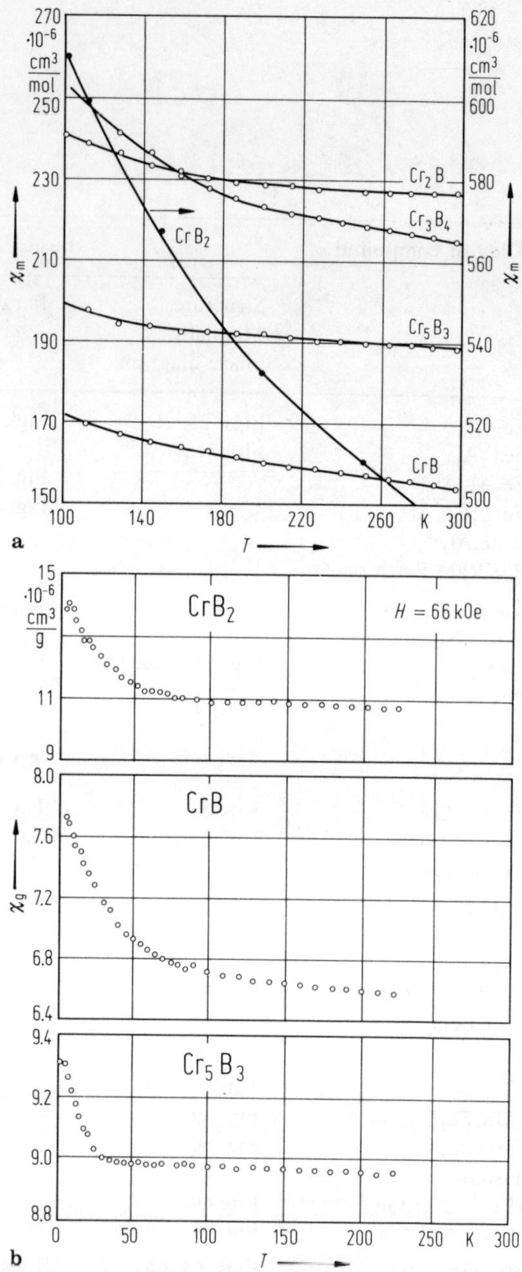

Fig. 2. Cr borides. (a) Temperature variation of magnetic molar susceptibility [76 G 1]. (b) Magnetic mass susceptibilities for $H = 66$ kOe [79 L 1].

Table 2. Co, Ni borides. Physical properties at 290 K [72 K 2].

	χ_g $10^{-6}\,cm^3\,g^{-1}$	χ_m $10^{-6}\,cm^3\,mol^{-1}$	R_H $cm^3\,C^{-1}$	ϱ $\mu\Omega\,cm$	Q $\mu V\,K^{-1}$	λ $W\,cm^{-1}\,K^{-1}$
Co_3B	ferromagnetic		–	50	−40	0.17
Co_2B	ferromagnetic		–	65	−27	0.14
CoB	−0.27	−18.8	−104	1080	−51	0.17
Ni_3B	+1.41	+88	0	21	−10.8	0.42
Ni_2B	+0.73	+47	+ 0.53	14	− 1.6	0.55
NiB	−0.08	− 5.5	+ 0.63	50	+ 2.3	0.22

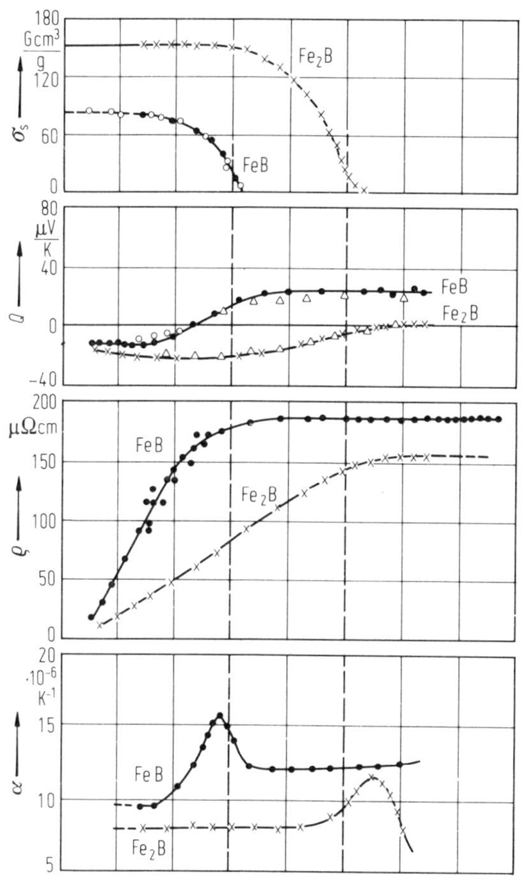

Fig. 3. Fe borides. Temperature dependence of Sturation magnetization, thermopower, electrical resistivity, and thermal expansion [71 K 1]. Solid circles and crosses: experimental points for specimen of FeB and Fe_2B, obtained by synthesis followed by hot pressing in vacuum; triangles: experimental points for thermo-emf, obtained for specimens synthesized and sintered in argon; open circles: experimental values taken from [61 L 1, 62 L 1].

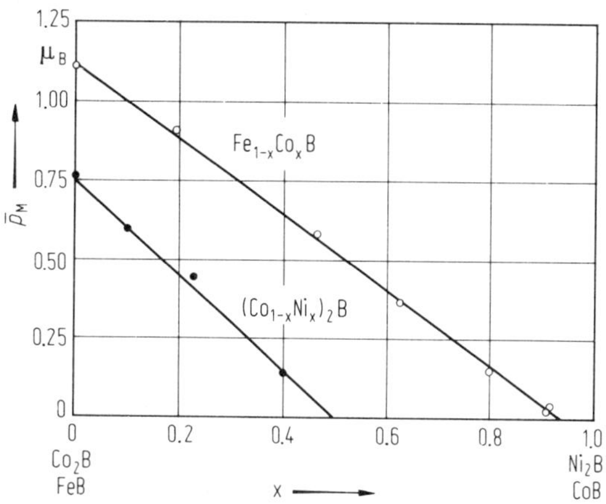

Fig. 4. $(Co_{1-x}Ni_x)_2B$, $Fe_{1-x}Co_xB$. Magnetic moment per metal atom vs. composition [70 K 1].

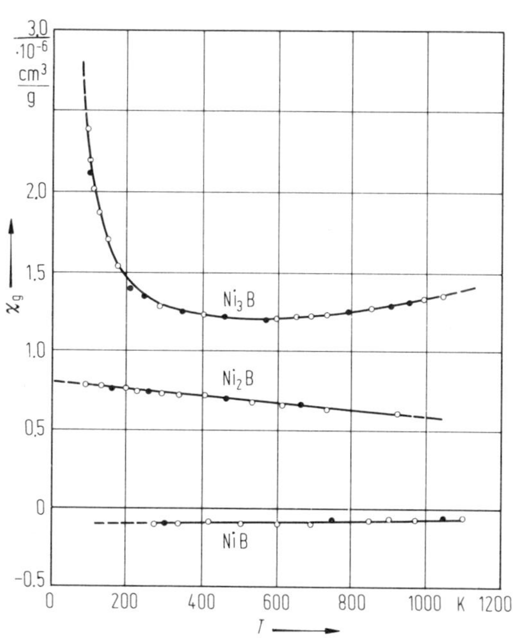

Fig. 5. Ni borides. Temperature dependence of the magnetic susceptibility during heating (open circles) and cooling (solid circles) [72 K 3].

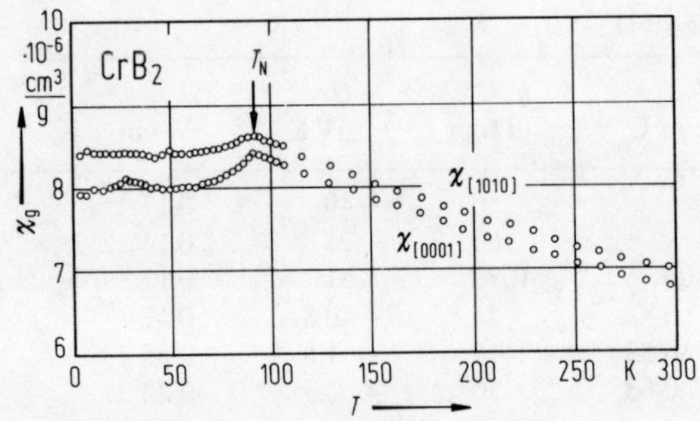

Fig. 6. CrB_2. Magnetic susceptibilities vs. temperature curves. T_N is 88 K and H is in the $[10\bar{1}0]$ and in the [0001] directions, respectively. $\chi_{[11\bar{2}0]}$ is so close to $\chi_{[10\bar{1}0]}$ that it is omitted from this figure [76 T 1].

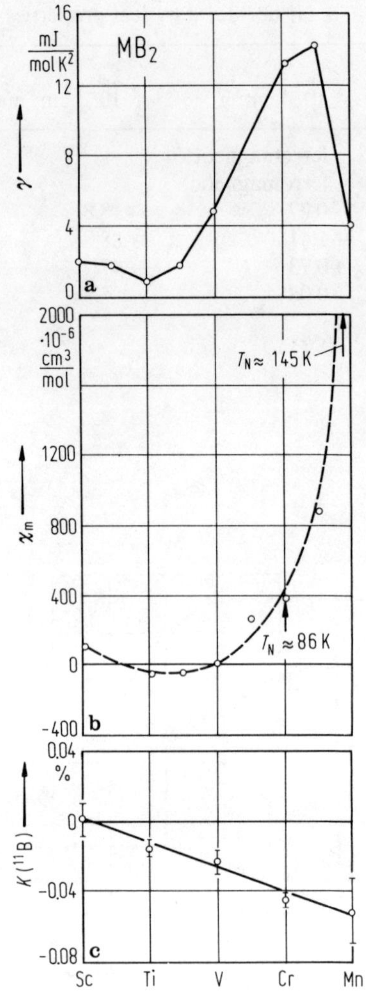

Fig. 7. 3d diborides. Properties of 3d diborides as a function of position of the transition metal in the periodic table. (a) Electronic specific heat coefficient γ [69 C 1]. The points midway between (ScB_2, TiB_2), (TiB_2, VB_2) and (CrB_2, MnB_2) are for the compounds $Sc_{0.5}Ti_{0.5}B_2$, $Ti_{0.5}V_{0.5}B_2$ and $Cr_{0.5}Mn_{0.5}B_2$. (b) Room-temperature magnetic susceptibility χ_m: ScB_2 [70 P 1], TiB_2 up to $Cr_{0.5}Mn_{0.5}B_2$ [69 C 1]. MnB_2 data from [69 B 1], [66 C 1] and [66 A 1]. (c) Knight shifts of the ^{11}B resonance; ScB_2, MnB_2 [71 C 1], TiB_2, VB_2 and CrB_2 ▶ [63 S 1].

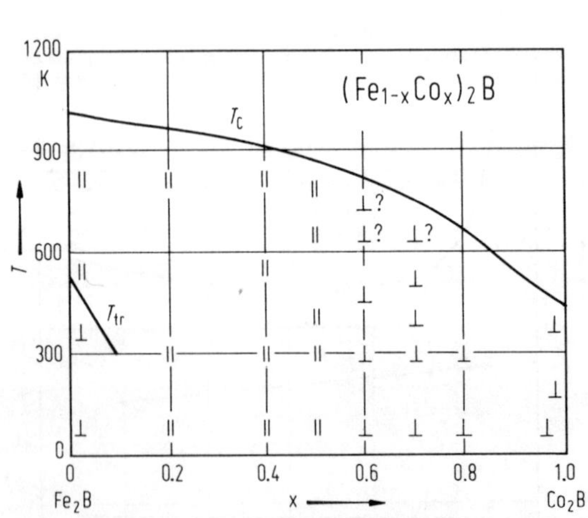

Fig. 8. $(Fe_{1-x}Co_x)_2B$. Direction of easy magnetization axis: the axis lies in the c plane ($\perp$), along the c axis ($\parallel$) and between them ($\perp$?). The direction of the easy magnetization changes at the temperature T_{tr} [75 T 1]. The figure includes the transition temperatures T_{tr} and Curie temperatures T_C measured by [65 C 1].

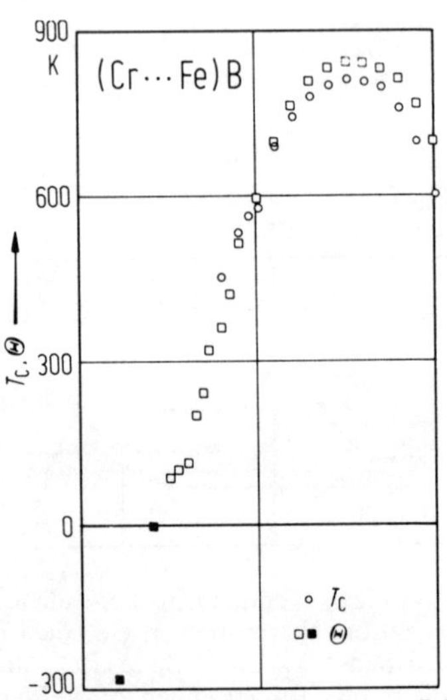

Fig. 9. 3d monoborides. Variation of the ferromagnetic, T_C, and paramagnetic, Θ, Curie temperatures. Solid squares are those of the CrB-type structure [81 K 1].

5.3.3 3d-Al alloys and compounds

For Heusler-type alloys, see also chapter 7.

Table 3. Ti-(Fe, Co, Ni)–(Al, Ga). Lattice parameters and magnetic characteristics [79 S 1].

Ti	Fe/Co/Ni	Al/Ga	a (RT)	p_{eff}	Θ	χ_0
at%			Å	μ_B	K	$10^{-4}\,cm^3\,mol^{-1}$
			Ti–Fe–Al			
25.8	24.2	50.0	11.99	1.66	4	1.80
27.8	24.2	48.0	12.02	1.80	−6	1.90
31.8	24.2	44.0	12.04	1.90	−10	2.14
33.3	24.2	42.5	12.08	1.98	−20	2.50
			Ti–Co–Al			
25.8	24.2	50.0	11.93	1.40	−7	2.60
			Ti–Ni–Al			
25.8	24.2	50.0	11.90	0.84	−8	1.60
			Ti–Fe–Ga			
32.5	24.2	43.3	12.01	1.80	30	3.20
			Ti–Co–Ga			
22.0	24.2	53.8	11.86	1.04	32	0.20
24.0	24.2	51.8	11.89	1.14	20	0.50
26.0	24.2	49.8	11.92	1.20	10	0.80
			Ti–Ni–Ga			
25.0	25.0	50.0	11.90	0.80	30	0.10
23.0	27.0	50.0	11.87	0.84	20	0.20
21.0	29.0	50.0	11.84	0.94	10	0.50

Table 4. V–A1. Room-temperature Knight shifts for ^{51}V ($H_0 = 6.7$ kOe) and ^{27}Al ($H_0 = 13.1$ kOe), NMR linewidth for ^{51}V, magnetic susceptibility and lattice parameter [64 O 1].

Al at%	Structure	K (^{51}V) %	ΔH (^{51}V) Oe	K (^{27}Al) %	χ_m $10^{-6}\,cm^3\,mol^{-1}$	a Å
V	bcc	0.564(6)			287	3.029
0.5	bcc	0.572(6)	10.8(6)			
1	bcc	0.568(6)	11.0(6)	−0.038(11)	284	
2	bcc	0.567(6)	10.6(6)	−0.036(9)		
4	bcc	0.570(10)	10.8(7)	−0.041(9)	272	3.033
8	bcc	0.573(6)	11.6(6)	−0.034(9)	253	3.037
15	bcc	0.573(6)	12.1(6)	−0.031(9)	231	3.044
20	bcc	0.584(6)	14.2(9)	−0.026(9)	207	
25	bcc	0.576(6)	13.4(7)	−0.028(9)		3.054
30	bcc	0.583(7)	14.1(9)	−0.024(9)	174	
35	bcc	0.598(10)	15.9(12)	−0.026(9)	160	
40	bcc	0.586(10)	16.7(12)	−0.024(9)		
61.5	γ-brass type	0.461(10)		−0.031(11)		
Al	fcc			0.161		

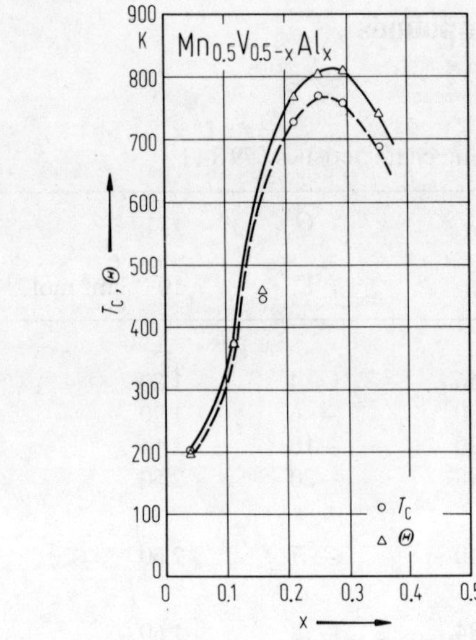

Fig. 10. $Mn_{0.5}V_{0.5-x}Al_x$. Curie temperature T_C together with paramagnetic Curie temperature Θ as a function of Al content [81 Y 1].

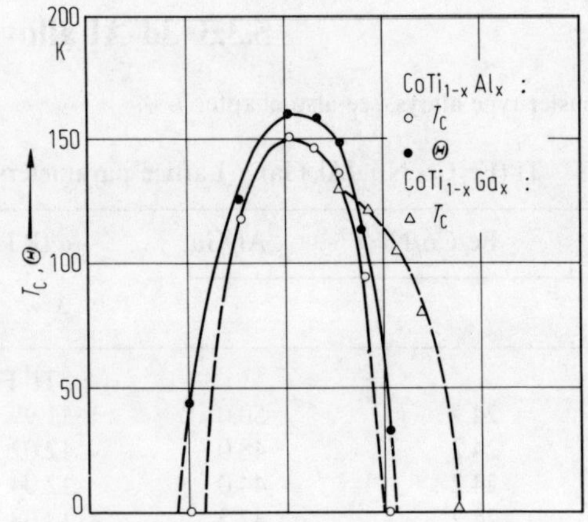

Fig. 11. $CoTi_{1-x}Al_x$. $CoTi_{1-x}Ga_x$ Ferromagnetic (open circles) and paramagnetic (full circles) Curie temperatures, T_C and Θ, respectively, for $CoTi_{1-x}Al_x$. The values take a maximum at x=0.4. Triangles are the ferromagnetic Curie temperatures of $CoTi_{1-x}Ga_x$ by [75 B 3].

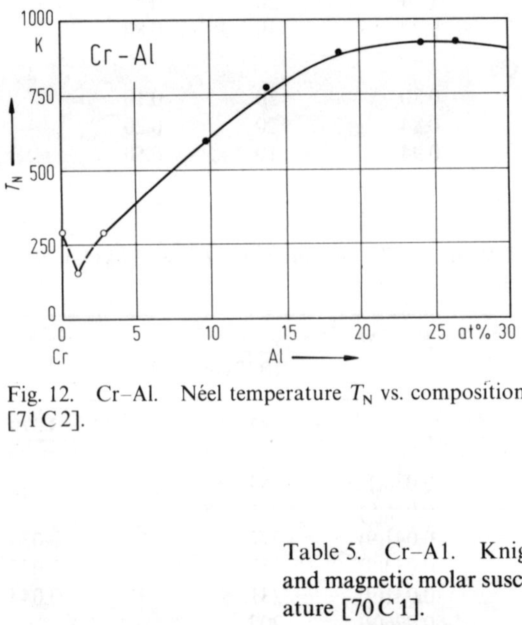

Fig. 12. Cr–Al. Néel temperature T_N vs. composition [71 C 2].

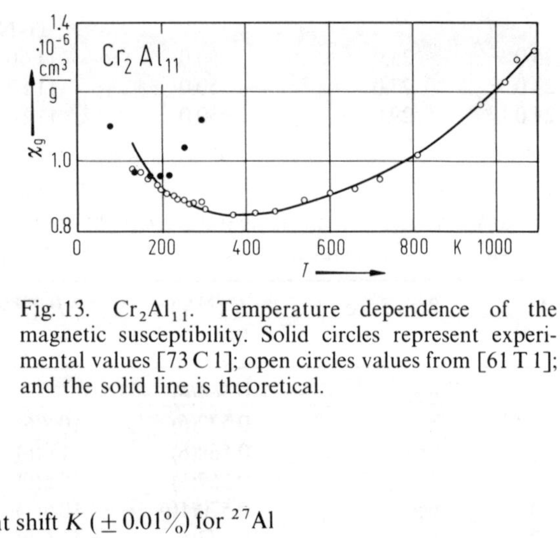

Fig. 13. Cr_2Al_{11}. Temperature dependence of the magnetic susceptibility. Solid circles represent experimental values [73 C 1]; open circles values from [61 T 1]; and the solid line is theoretical.

Table 5. Cr–Al. Knight shift K ($\pm 0.01\%$) for ^{27}Al and magnetic molar susceptibility χ_m at room temperature [70 C 1].

	K %	χ_m $10^{-6}\,cm^3\,mol^{-1}$
$CrAl_7$	0.16	17
Cr_2Al_{11}	0.16	31
$CrAl_4$	0.14	62
Cr_4Al_9	0.12	130
Cr_5Al_8	0.12	115
Cr_2Al	0.11	110
Al	0.16	18

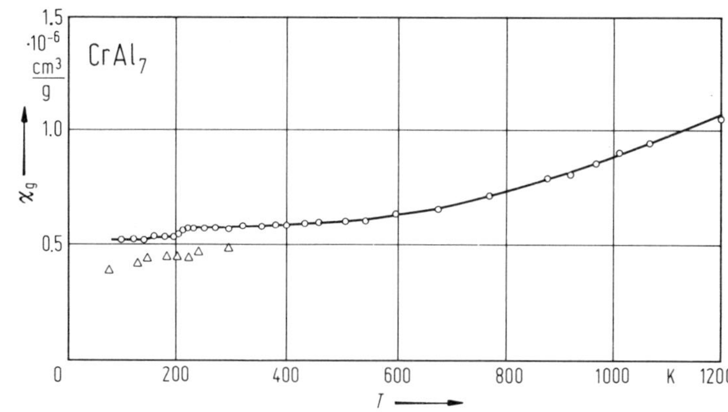

Fig. 14. $CrAl_7$. Temperature dependence of the magnetic susceptibility [74 C 1]. Triangles [61 T 1].

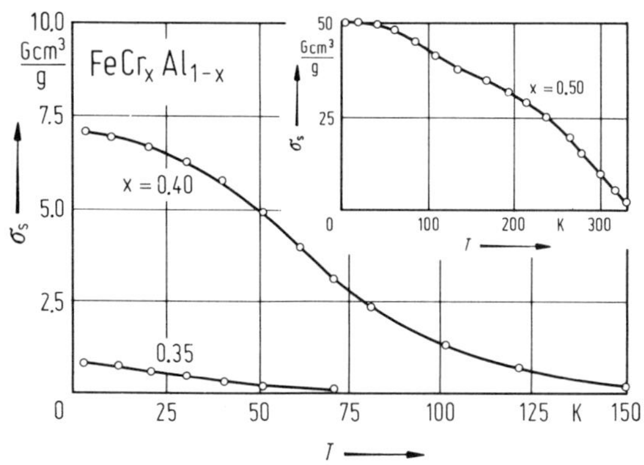

Fig. 15. $FeCr_xAl_{1-x}$. Thermal variation of spontaneous magnetization σ_s for x = 0.35, 0.40. Inset: x = 0.50. [85 O 1].

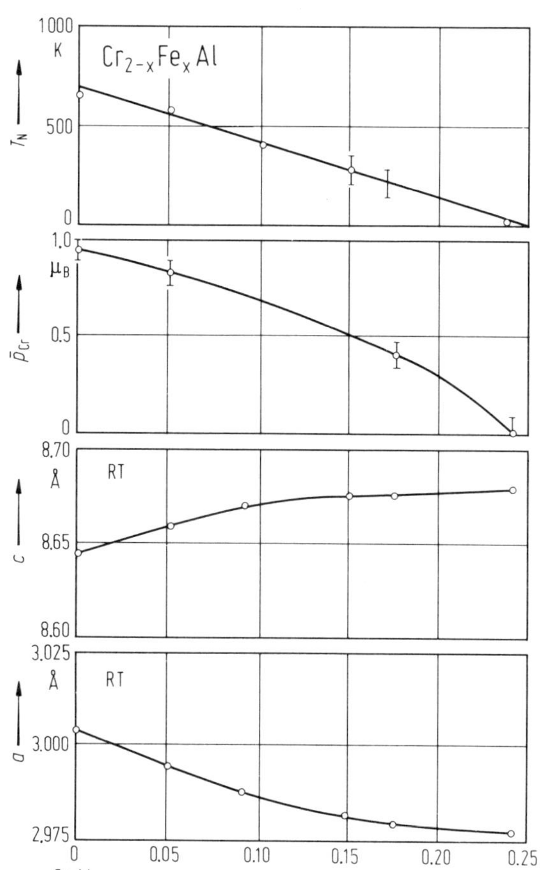

Fig. 16. $Cr_{2-x}Fe_xAl$. Composition dependence of lattice parameters at room temperature, *a* and *c*, magnetic moment per Cr atom, $\bar{p}_{Cr}$, in the ordered phase, and Néel temperature T_N [69 K 1].

Table 6. $Cr_{2-x}Fe_xAl$. Crystallographic and magnetic parameters [69 K 1]. *d*: distance between planes of Cr coupled antiferromagnetically.

x	*a* Å	*c* Å	$\bar{p}_{Cr}$ μ_B	T_N K	*d* Å
0	3.004	8.646	0.95(5)	663	1.193
0.05	2.995	8.658	0.84(5)	603	1.195
0.09	2.988	8.671	–	413	–
0.15	2.984	8.675	–	310	–
0.17	2.980	8.676	0.42(5)	[1])	1.232
0.24	2.977	8.680	–	0	1.233

[1]) T_N in the range of 100···300 K.

Table 7. Mn–Al. Crystal structure, lattice parameters and type of magnetic order of equilibrium and metastable phases of Mn–Al near stoichiometric composition [81 V 1].

Phase	Lattice	Lattice parameters	Magnetic order	Ref.
ε	hcp	$a=2.705$ Å $c=4.364$ Å $c/a=1.613$	Antiferromagnetic, $T_N=-183$ °C	67 M 1, 73 M 1
η	fcc	–	Antiferromagnetic $T_N=400$ °C	69 R 1
η	Face-centred rhombohedral	$a=8.96$ Å $\alpha=89°1'$		71 Y 1
β	B-19 Complex cubic	–	Paramagnetic	60 K 2
ε′	β-Mn Orthorhombic	$a=4.44$ Å $b=2.81$ Å $c=4.61$ Å	–	74 V 1
τ_{disord}	fcc	$a=3.8$ Å	Antiferromagnetic $T_N=260$ °C	72 B 1, 73 M 2, 74 V 1
τ_{order}	fct	$a=3.93$ Å $c=3.56$ Å	Ferromagnetic $T_C=380$ °C	60 K 1, 67 M 1

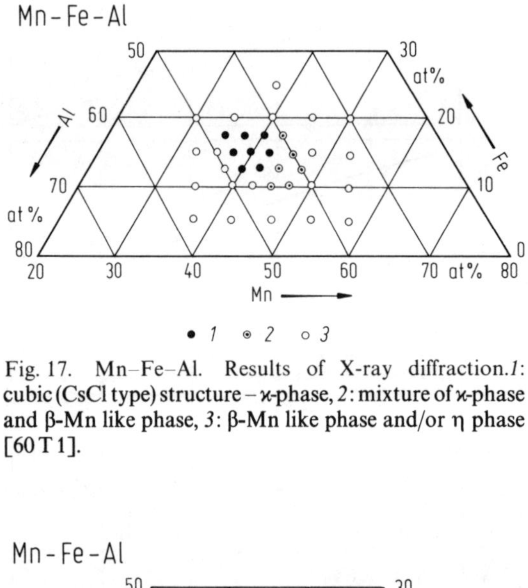

Fig. 17. Mn–Fe–Al. Results of X-ray diffraction. *1*: cubic (CsCl type) structure – ϰ-phase, *2*: mixture of ϰ-phase and β-Mn like phase, *3*: β-Mn like phase and/or η phase [60 T 1].

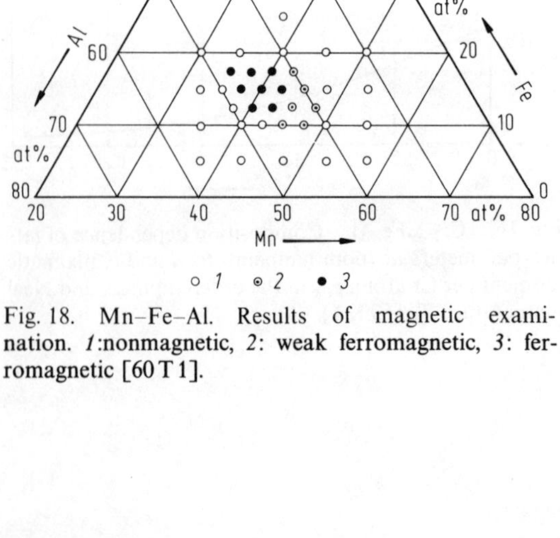

Fig. 18. Mn–Fe–Al. Results of magnetic examination. *1*: nonmagnetic, *2*: weak ferromagnetic, *3*: ferromagnetic [60 T 1].

Table 8. Mn–M–Al. Residual electrical resistivities ϱ_0 and Curie temperatures T_C of the ϰ-phase alloys [65 D 1].

	Al	Mn	ϱ_0	T_C
	at%		10^{-4} Ω cm	°C
Mn–Fe–Al	45	42.5	1.54	90
	45	40	1.68	30
	47.5	37.5	1.44	87
Mn–Co–Al	50	35	1.79	142
	50	30	1.91	110
	50	25	1.91	121
	45	45	2.02	195
	45	35	2.06	161
Mn–Ni–Al	50	40	1.38	175
	50	30	1.32	146
	50	25	1.55	151
	45	45	1.95	197
	45	35	1.57	205
Mn–Cu–Al	50	40	2.06	126
	50	35	1.82	82
	50	30	1.79	107
	45	45	1.95	133
	45	40	1.84	96

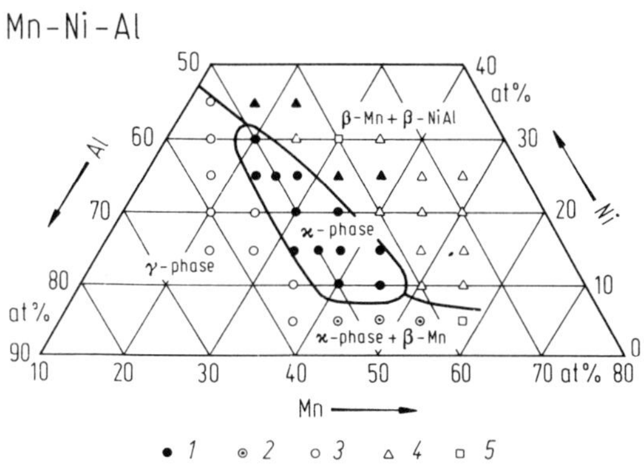

Fig. 19. Mn–Ni–Al. Results of magnetic examination and X-ray analysis. *1*: ϰ-phase; magnetic, *2*: ϰ-phase and β-Mn phase; weakly magnetic, *3*: orthorhombic γ phase; nonmagnetic, *4*: β-Mn phase and β-NiAl phase; nonmagnetic, *5*: β-Mn phase; nonmagnetic [61 T 3].

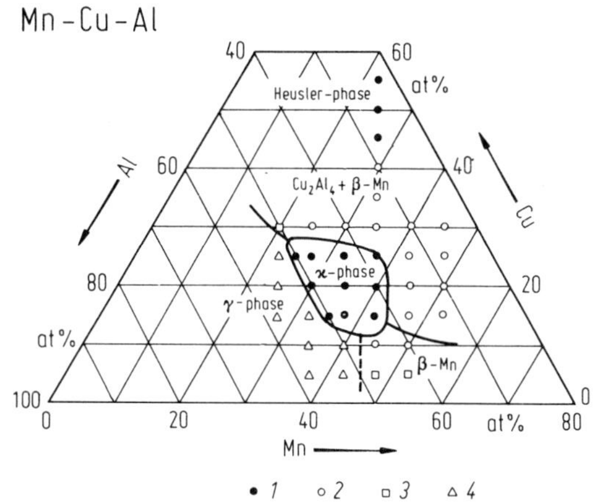

Fig. 20. Mn–Cu–Al. Results of X-ray analysis, *1*: Heusler phase in Cu-rich region and the ϰ-phase in Cu region, *2*: mixed structure of β-Mn and Cu_9Al_4 phase, *3*: β-Mn phase, *4*: orthorhombic γ phase. The Heusler and ϰ-phases are ferromagnetic and others are nonmagnetic [61 T 2].

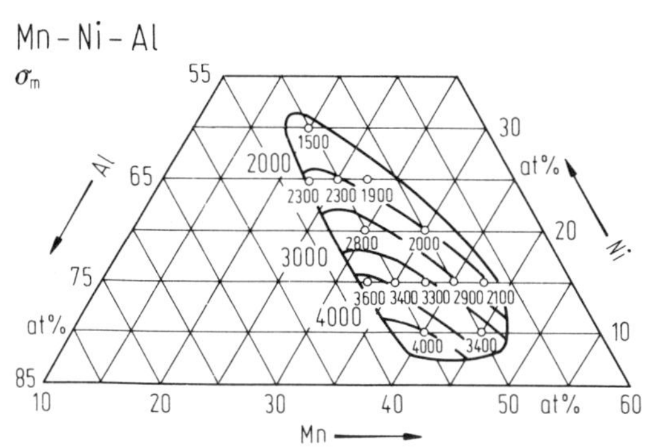

Fig. 21. Mn–Ni–Al. Composition dependence of the molar saturation magnetization σ_m, in $G\,cm^3\,mol^{-1}$ and the phase boundary of the Ni ϰ-phase [61 T 4].

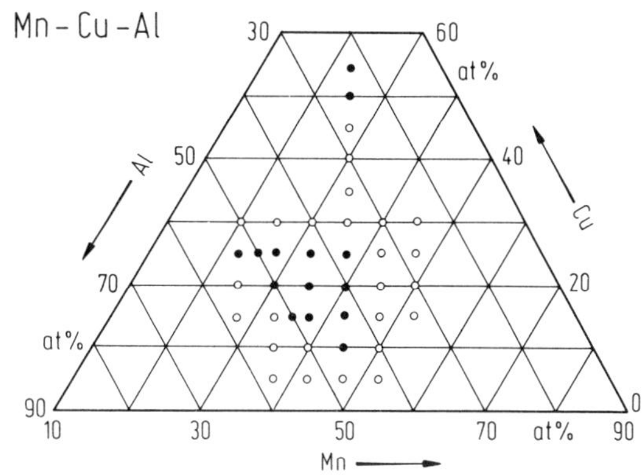

Fig. 22. Mn–Cu–Al. Composition range of specimens and results of magnetic testing. Solid circles: ferromagnetic specimens, open circles: nonmagnetic specimens [61 T 2].

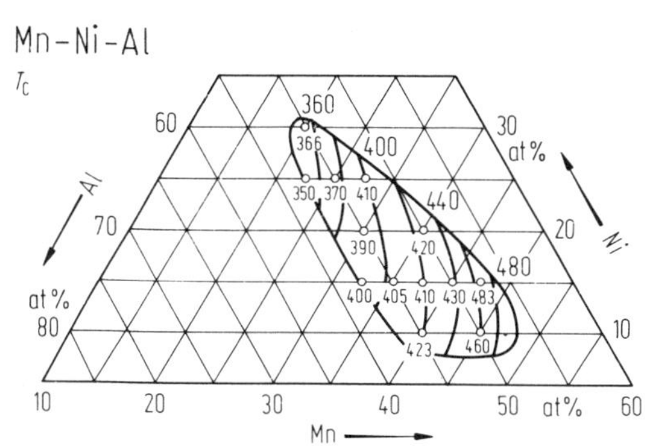

Fig. 23. Mn–Ni–Al. Composition dependence of the Curie temperature T_C, in K, [61 T 4].

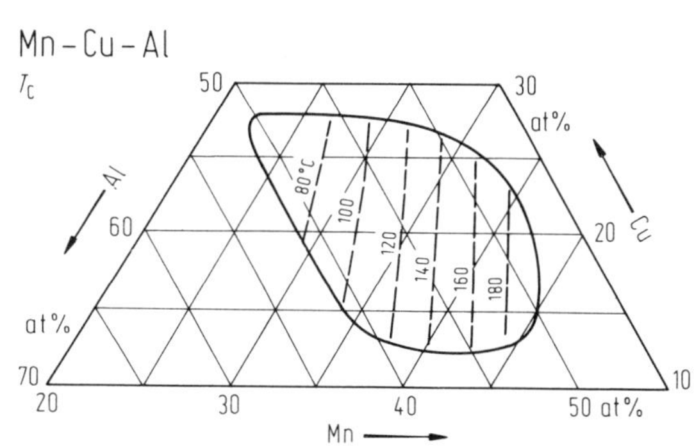

Fig. 24. Mn–Cu–Al. Curie temperature T_C, in °C, of the ϰ-phase [61 T 2].

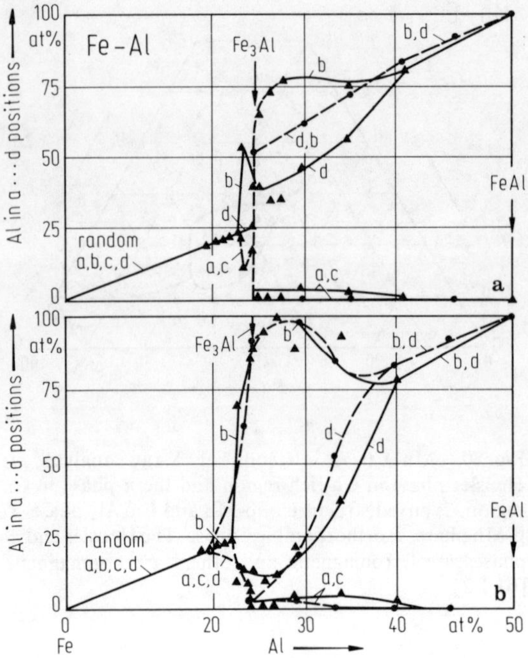

Fig. 25. Fe–Al. Distribution of Fe and Al atoms. (a) oil-quenched from 800 °C, (b) slow-cooled condition. The positions a, b, c, d refer to the (000), $(\frac{1}{4}\frac{1}{4}\frac{1}{4})$, $(\frac{1}{2}\frac{1}{2}\frac{1}{2})$ and $(\frac{3}{4}\frac{3}{4}\frac{3}{4})$ cell positions. The structure can be viewed as four interpenetrating fcc cells with origins at these positions [61 L 1]. Dashed lines and circles: [37 B 1].

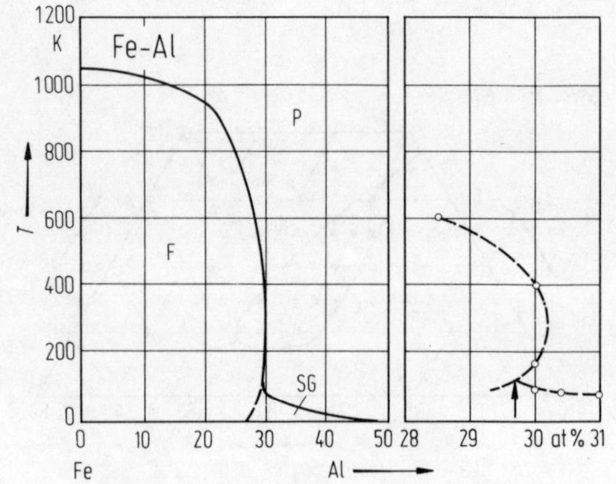

Fig. 26. Fe–Al. Magnetic phase diagram. Data include samples with both Fe_3Al and FeAl structures. Shown at right is a detail of the region near the multicritical point where the paramagnetic, ferromagnetic, and spin-glass phases meet. Data near the multicritical point are not sufficiently good to determine the multicritical temperature with precision [80 S 1].

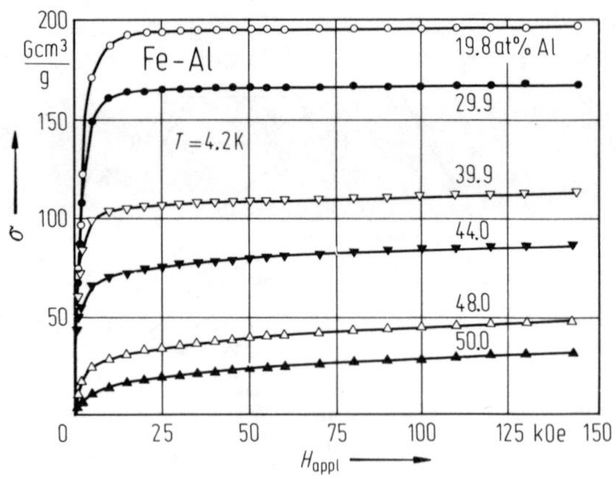

Fig. 27. Fe–Al. Magnetization σ against applied magnetic field for some cold-worked (disordered) alloys at 4.2 K [75 B 2].

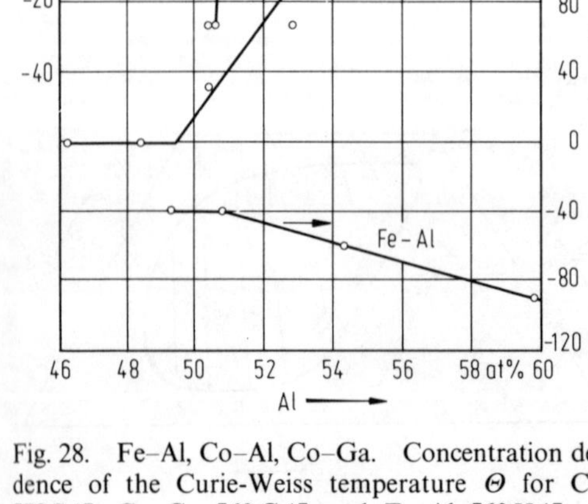

Fig. 28. Fe–Al, Co–Al, Co–Ga. Concentration dependence of the Curie-Weiss temperature Θ for Co–Al [72 S 1], Co–Ga [69 G 1] and Fe–Al [60 H 1] alloys [73 C 2].

Table 9. Fe–Al. Magnetic parameters between 1350 and 1500 K [78 A 1].

Al at%	C $10^{-2}\,cm^3\,K\,g^{-1}$	Θ K	p_{eff} μ_B/Fe	p_{para}
0			3.05	2.21
17.0	2.21	1059	3.29	2.44
18.7	2.24	1044	3.33	2.48
20.4	2.26	1031	3.37	2.52
22.0	2.27	1014	3.39	2.53
23.6	2.28	995	3.42	2.56
25.2	2.29	976	3.45	2.59
26.8	2.30	951	3.48	2.62

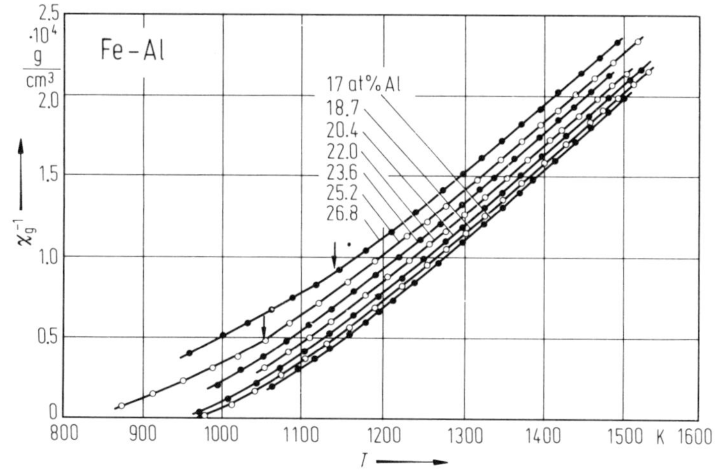

Fig. 29. Fe–Al. Inverse magnetic susceptibility of disordered alloys as a function of temperature [78 A 1].

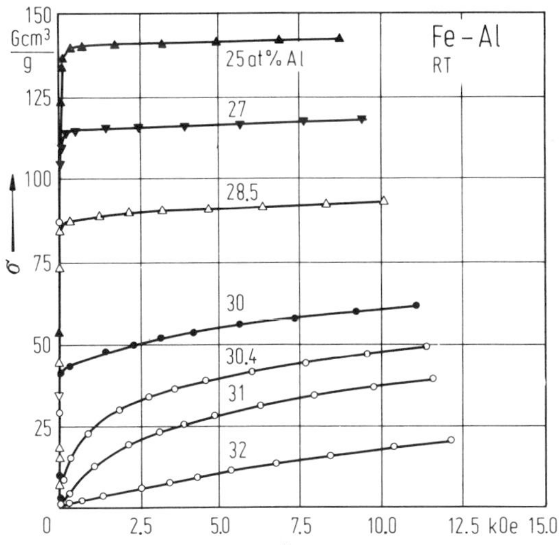

Fig. 30. Fe–Al. Magnetization σ vs. magnetic field H at room temperature for slow-cooled alloys with Fe_3Al-structures [76 S 1].

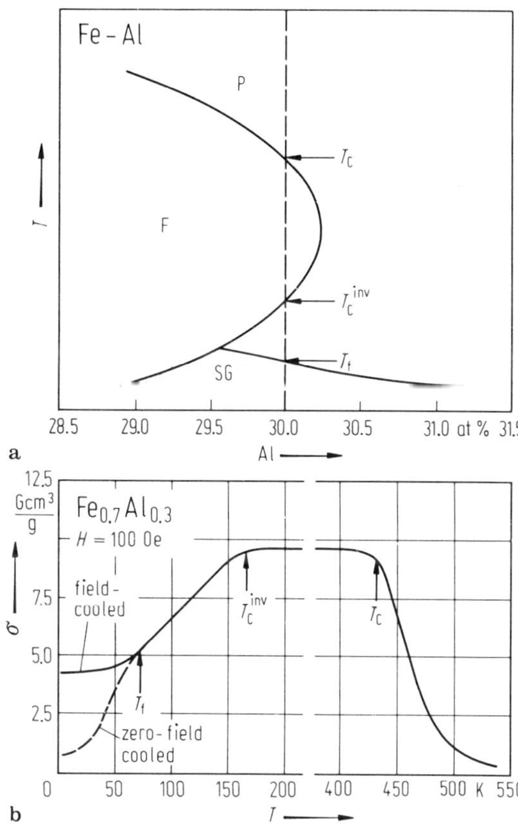

Fig. 31. Fe–Al. (a) Schematic magnetic phase diagram of Fe–Al alloy near 30 at% Al. (b) Temperature dependence of magnetization σ of $Fe_{0.7}Al_{0.3}$ with FeAl-type order measured in a magnetic field of 100 Oe [83 M 1].

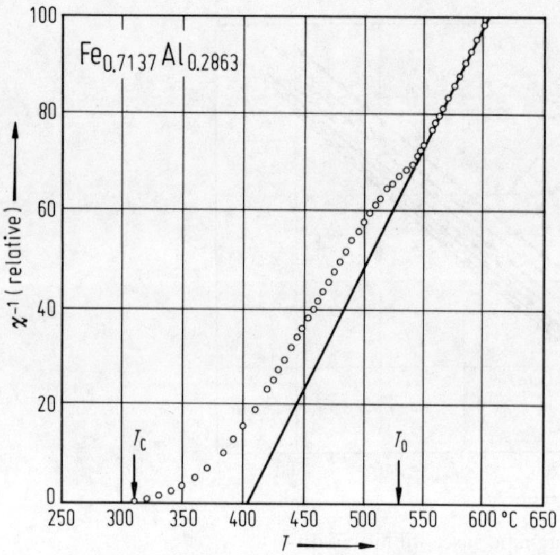

Fig. 32. $Fe_{0.7137}Al_{0.2863}$. Inverse magnetic susceptibility vs. temperature. Solid line indicates Curie-Weiss susceptibility in the disordered state, determined by least-squares fit for $T \gg T_0$ (the disordering temperature $= 530\,°C$) [73 H 1].

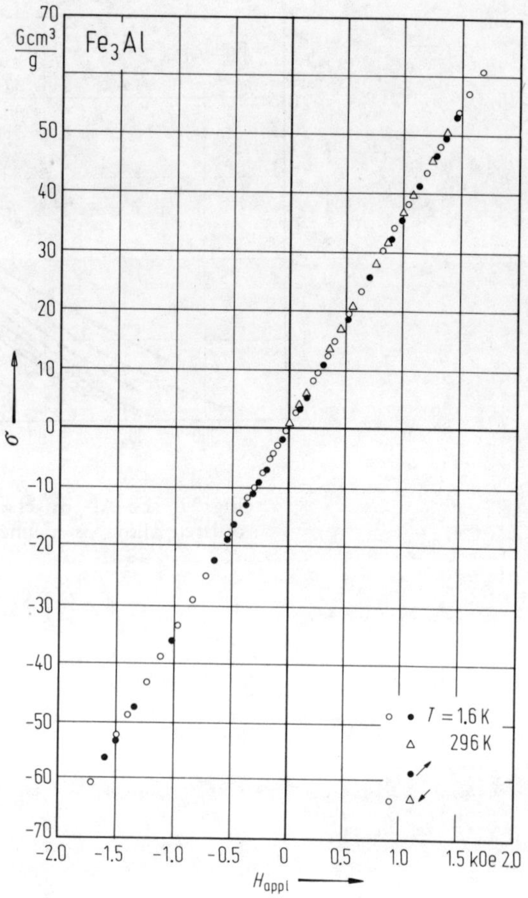

Fig. 33. Fe_3Al. Low-field magnetization σ vs. applied field H_{appl} at 1.6 and 300 K. The sequence of the measurements was (open symbols) descending from 12.6 kOe in the initial direction to zero at room temperature and to a reversed field of 12.6 kOe at 1.6 K, and (solid symbols) ascending in the opposite direction at 1.6 K [76 S 1].

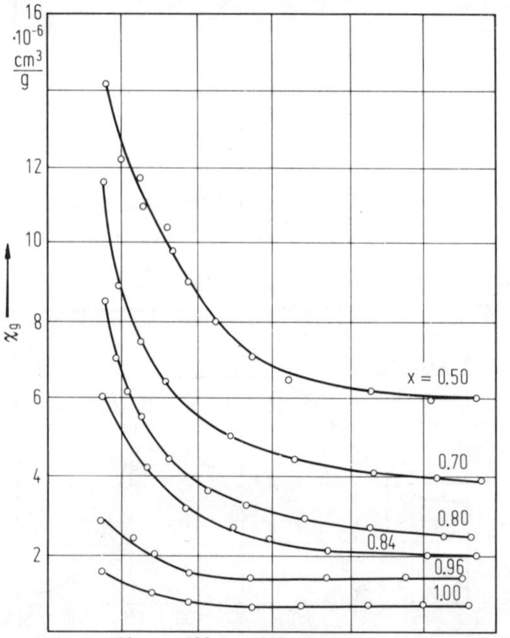

Fig. 34. $Fe_{1-x}Co_xAl$. Influence of temperature and composition on the magnetic susceptibility $0.5 \leqq x \leqq 1.0$ [73 S 1].

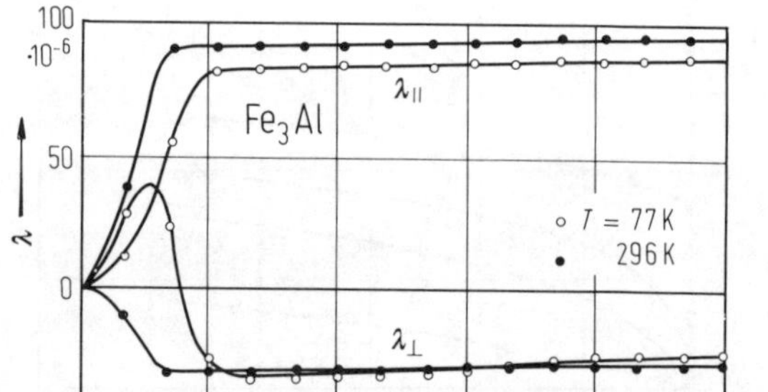

Fig. 35. Fe_3Al. Magnetostriction along the [100] direction as a function of the applied magnetic field H_{appl} parallel and perpendicular to [100] in the (001) plane at 296 and 77 K in ordered (slow-cooled) Fe_3Al [79 C 1].

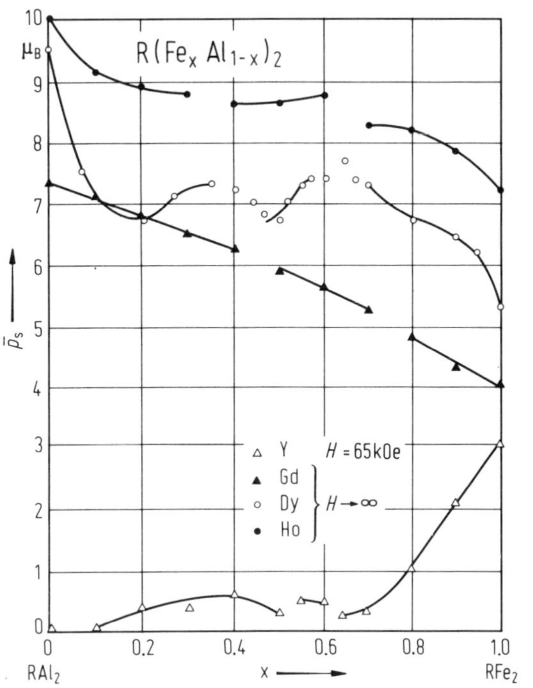

Fig. 36. R(Fe_xAl_{1-x})$_2$. Saturation magnetic moment with R = Y (H = 65 kOe), Gd, Dy, and Ho ($H \to \infty$) [76 G 1].

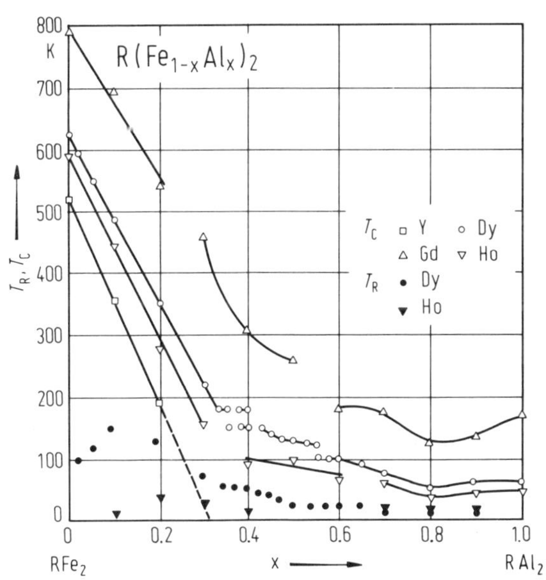

Fig. 37. R($Fe_{1-x}Al_x$)$_2$. Curie temperatures T_C (open symbols) of R($Fe_{1-x}Al_x$)$_2$ with R = Y, Gd, Dy, and Ho and the temperature T_R (solid symbols) where the remanent magnetization vanishes for Dy($Fe_{1-x}Al_x$)$_2$ and Ho($Fe_{1-x}Al_x$)$_2$ [76 G 1].

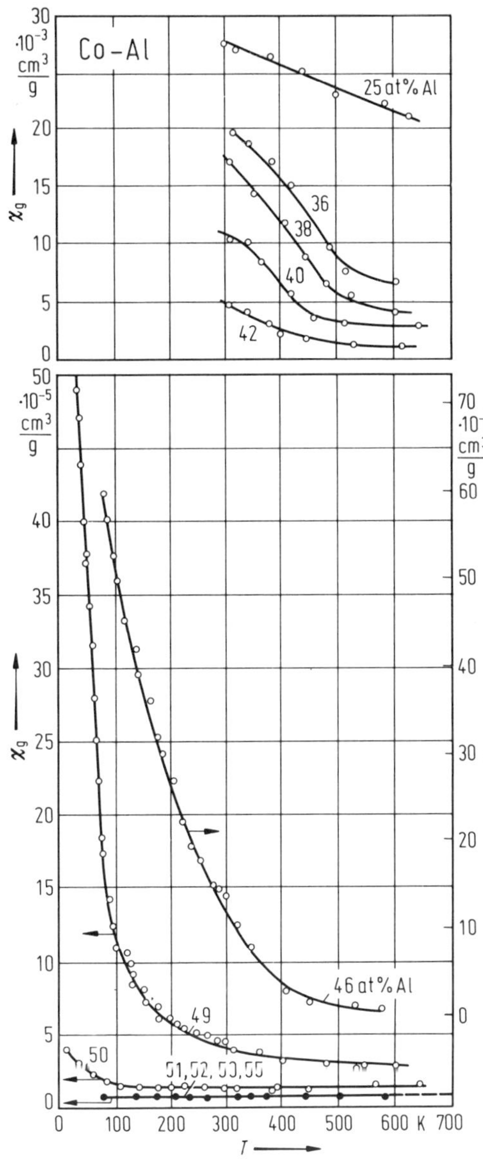

Fig. 38. Co–Al. Temperature-concentration curves of the magnetic susceptibility of the β-phase [71 B 1].

Table 10. $Co_{1-x}Ni_xAl$. Values of the Curie-Weiss parameters [74 P 1].

x	χ_0 $10^{-6}\,cm^3\,g^{-1}$	C_g $10^{-6}\,cm^3\,K\,g^{-1}$	Θ K
0.16	1.41	241	− 79
0.20	1.49	301	−113
0.30	1.34	532	−180
0.50	1.00	706	−266
0.70	0.88	233	−152

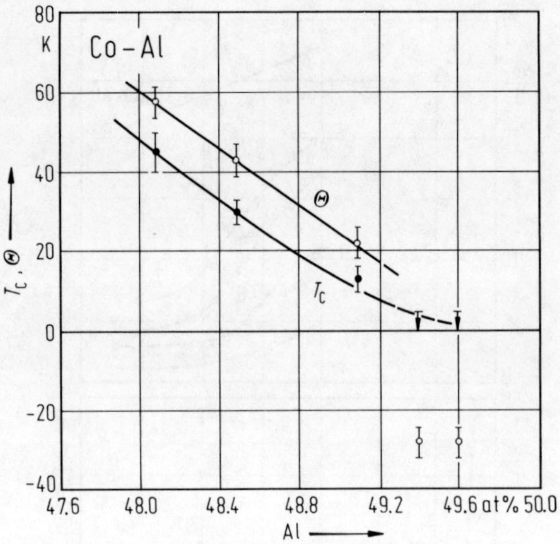

Fig. 39. Co–Al. Composition dependence of the Curie-Weiss temperature Θ and the ordering temperature T_C for samples with < 50 at% Al [72 S 1].

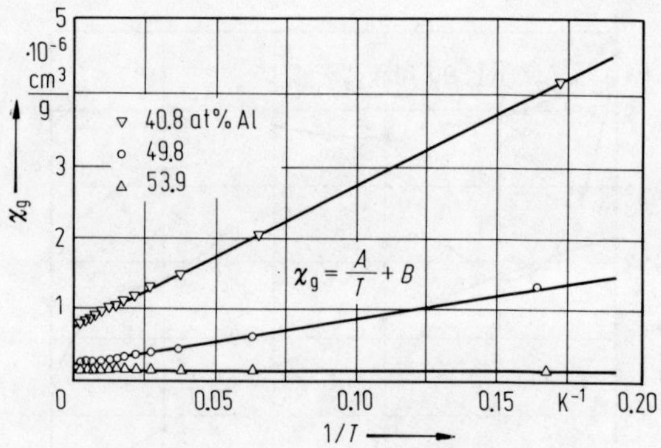

Fig. 40. Ni–Al. Temperature dependence of the magnetic susceptibility for several composition [69 B 2].

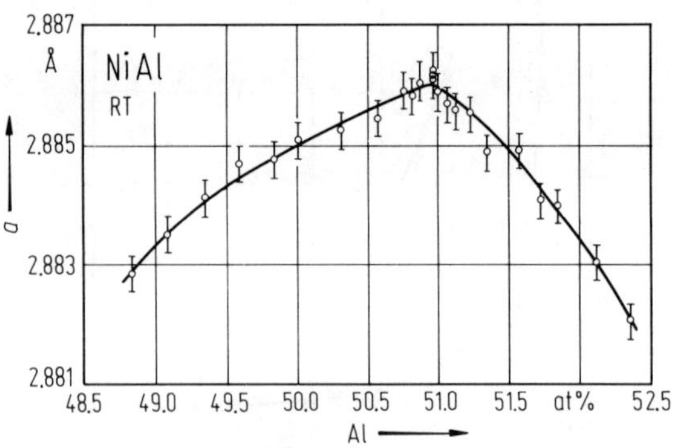

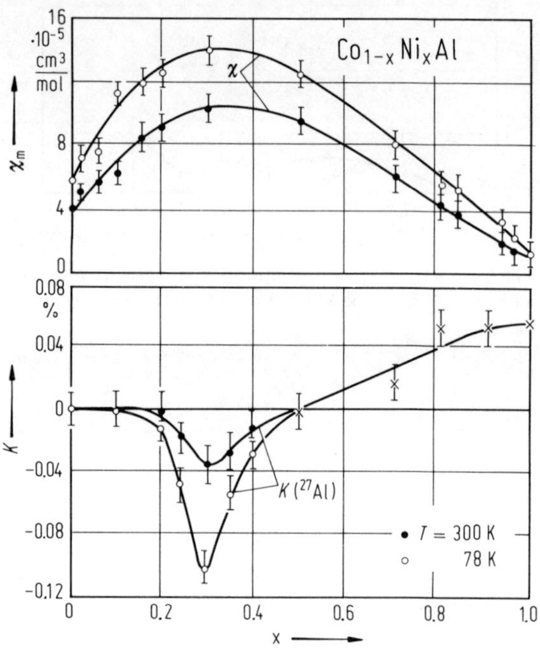

Fig. 41. Ni–Al. Observed variation of lattice parameter with initial composition. Error bars represent the estimated standard deviations of the experimental values [63 C 1].

Fig. 42. $Co_{1-x}Ni_xAl$. Magnetic susceptibility χ and Knight shift K at nuclei ^{27}Al vs. composition at 300 and 78 K [74 P 1].

Table 11. Ni_3Al. Number of neighbouring atoms around the host atoms before and after plastic deformation. There are two kinds of host Ni atoms with a different number of nearest neighbour (NN) atoms up to the antiphase boundary (APB) [83 T 1].

NN	1st		2nd		3rd		4th		5th		6th	
Host atoms	Ni	Al	Ni	Al	Ni	Al	Ni	Al	Ni	Al	Ni	Al
Normal state around Ni atom	8	4	6	0	16	8	12	0	16	8	8	0
around Al atom	12	0	0	6	24	0	0	12	24	0	0	8
APB around Ni atom												
at A site	8	4	6	0	16	8	12	0	16	8	8	0
at B site	9	3	3	3	19	5	9	3	20	4	3	5
around Al atom	11	1	3	3	21	3	3	9	20	4	4	4

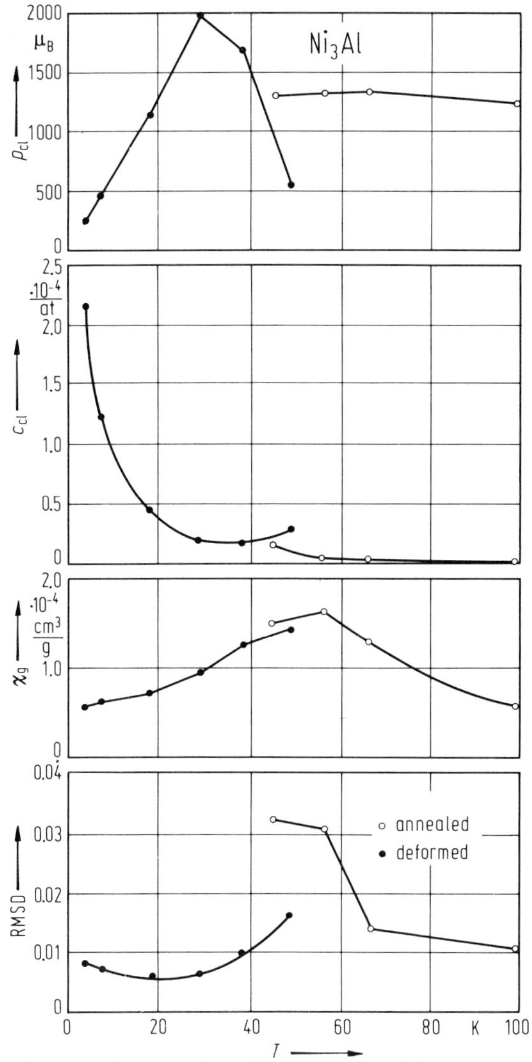

Fig. 43. Ni_3Al. Magnetic mass susceptibility χ_g, cluster concentration c_{cl} and cluster magnetic moment p_{cl}, vs. temperature for annealed and deformed Ni_3Al [72 R 1].

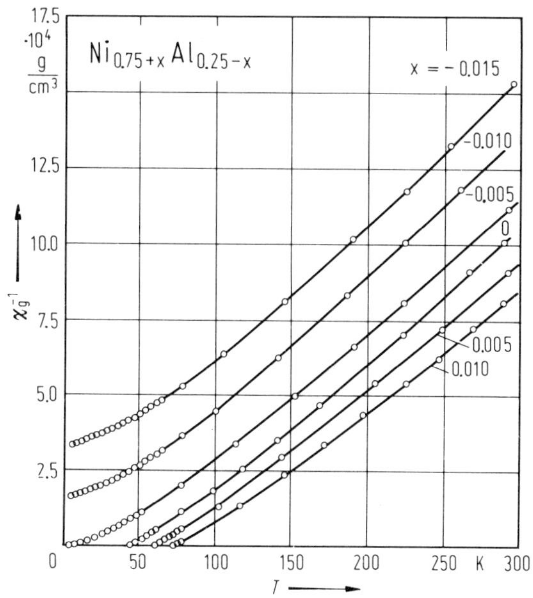

Fig. 44. Ni–Al. Reciprocal magnetic mass susceptibility vs. temperature [69 B 3].

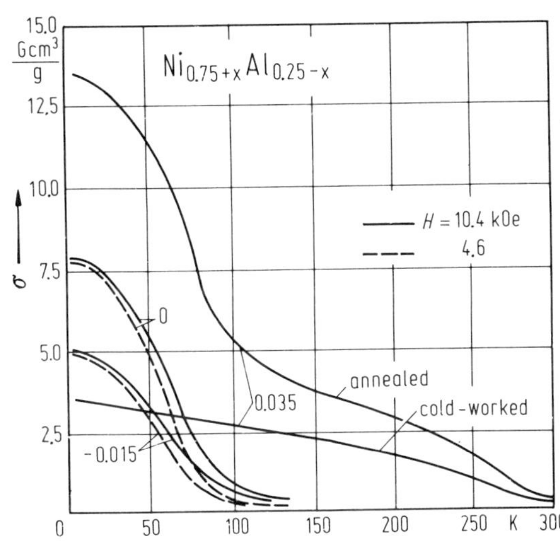

Fig. 45. $Ni_{0.75+x}Al_{0.25-x}$. Magnetization vs. temperature [67 B 1].

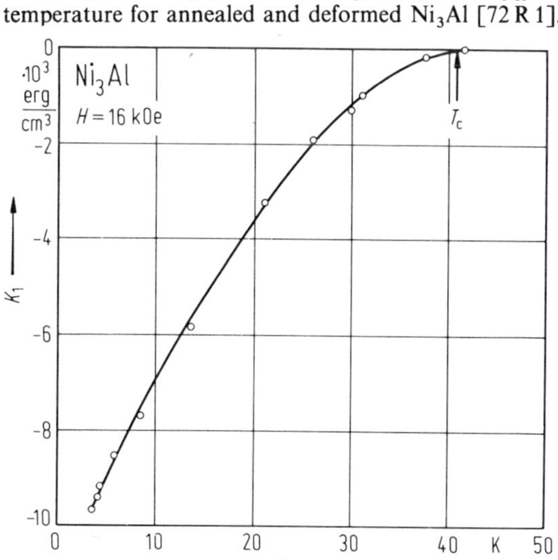

Fig. 46. Ni_3Al. Fitted first anisotropy constant as a function of temperature at $H = 16$ kOe. (The solid curve serves only to connect the data points). [82 S 1].

5.3.4 3d-Ga alloys and compounds

For Heusler-type alloys see also chapter 7.

Table 12. V_3Ga. Calorimetric and magnetic data, λ is the electron-phonon coupling parameter [76 J 1].

		Argon-quenched from 1250 °C		Slowly cooled 610 °C	
T_C (magnetic)	[K]	13.8		14.9	
T_C (calorimetric)	[K]	13.5		14.7 [1]	
γ	[mJ/K^2 mol]	19.9		22.9 [1]	
Θ_D (0 K)	[K]	310		300 [1]	
χ_m (0 K)	[10^{-6} cm^3/mol]	413		449	
λ		0.85	1.16	0.90	1.23
$N(E)$	[states/eV spin]	2.28	1.96	2.55	2.18
χ_P (0 K)	[10^{-6} cm^3/mol]	206	193	242	230
$\chi - \chi_P$	[10^{-6} cm^3/mol]	207	219	207	219
χ_{orb} [2]	[10^{-6} cm^3/mol]			130	190

[1]) Interpolated. [2]) [62 C 1].

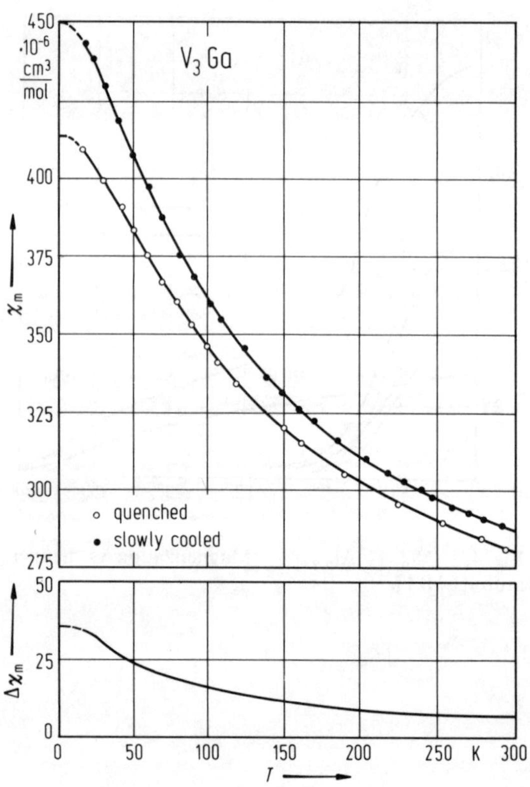

Fig. 47. V_3Ga. Magnetic susceptibility vs. temperature for annealed and quenched samples and difference curve [76 J 1].

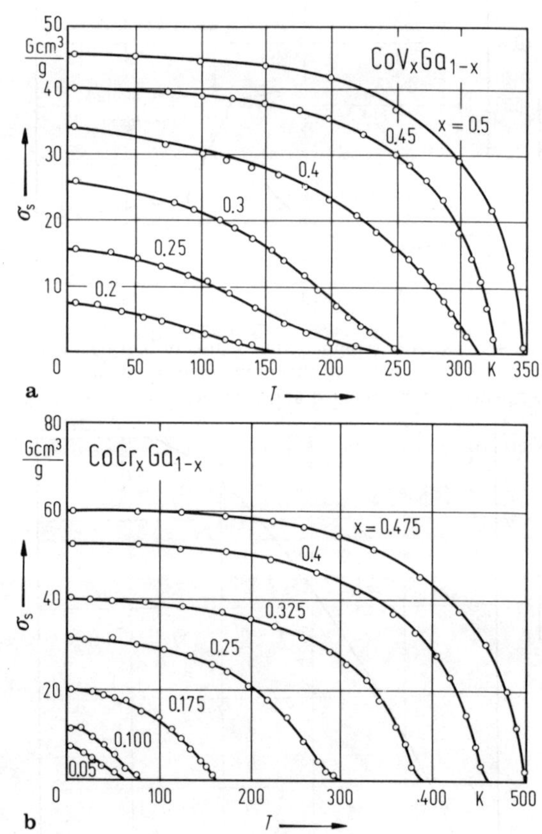

Fig. 48. CoM_xGa_{1-x}, M = V, Cr. Variation of spontaneous magnetization with temperature for quenched samples [77 C 1]. $CoV_{0.5}Ga_{0.5}$ [73 W 2].

Table 13. Cr–Ga. Mass susceptibilities χ_g at room temperature [85 B 1].

	χ_g (300 K) 10^{-6} cm^3 g^{-1}
Cr_3Ga	2.38 [1])
Cr_5Ga_6(CrGa)	2.70 [1])
$CrGa_4$	−0.13 [2])

[1]) Pauli paramagnetic 4···300 K.

[2]) Diamagnet plus very weak ferromagnetic impurity.

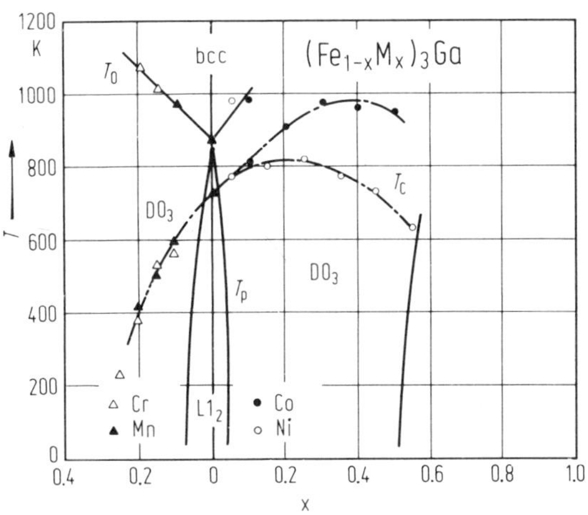

Fig. 49. $(Fe_{1-x}M_x)_3Ga$. Structural and magnetic phase diagram for M = Cr, Mn, Co and Ni, where T_0 stands for bcc⇄DO_3 transition temperature, T_p for the DO_3-phase limit, and T_C (broken lines) for the Curie temperature [83 K 1].

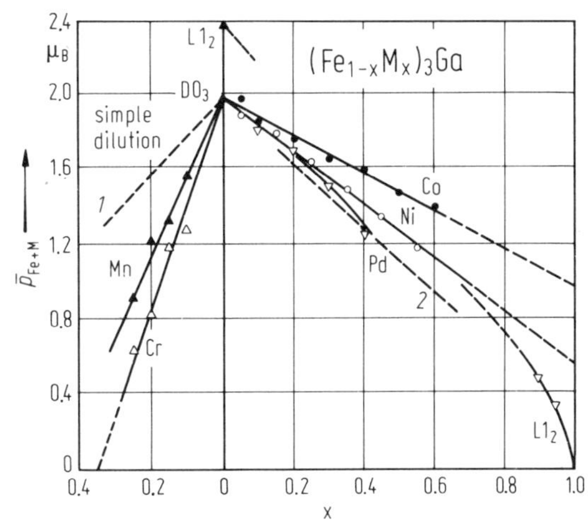

Fig. 50. $(Fe_{1-x}M_x)_3Ga$. Magnetic moment per magnetic atom (Fe and M). The dotted line *1* represents a simple dilution by M replacing Fe I (D site) alone, and the line *2*, replacing Fe II (A site) alone [83 K 1].

Table 14. $CoCr_xGa_{1-x}$. Bulk magnetic measurements on quenched samples [77 C 1].

x	T_C K	σ_s (0 K) G cm^3 g^{-1}	$\bar{p}_s$ (0 K) μ_B/f.u.	Θ K	C_g 10^{-6} cm^3 K g^{-1}	p_{eff} μ_B/f.u.	p_{eff}/p_s (0 K)
0.025	–	–	–	8(5)	1215(23)	1.12(2)	–
0.05	7(3)	3.4(3)	0.08(1)	46(5)	1260(24)	1.13(2)	14.13
0.075	27(4)	6.6(3)	0.15(1)	61(5)	1580(35)	1.41(3)	9.40
0.10	52(4)	11.0(4)	0:25(2)	78(5)	2170(41)	1.48(3)	5.92
0.175	148(4)	20.0(6)	0.45(2)	200(5)	2490(45)	1.58(3)	3.51
0.25	266(4)	31.5(10)	0.70(2)	295(5)	3150(63)	1.77(4)	2.53
0.325	374(4)	40.5(12)	0.89(3)	395(5)	3590(68)	1.87(4)	2.10
0.40	442(4)	53.0(12)	1.16(3)	445(5)	4640(77)	2.11(4)	1.82
0.475	495(4)	60.5(12)	1.30(3)				

Table 15. Mn–Ga. Lattice parameters and magnetic properties of the tetragonal δ-phase [65 B 1].

Mn	a	c	σ (77 K, 10 kOe)	H_c (77 K)	$\bar{p}_{Mn}$	T_C (4 kOe)
at%	Å		$G\,cm^3\,g^{-1}$	kOe	μ_B	K
70	2.77	3.58	Two-phase region			
66.7	2.76	3.60	31	4.9	0.502	748
65	2.753	3.619	39	2.6	0.646	718
60	2.750	3.642	48	1.15	0.877	658
55	2.74	3.69	Two-phase region			

Table 16. Mn–Ga. Chemical and phase composition at room temperature of samples quenched from 850 °C [79 G 1].

Mn	Phase composition	Lattice parameters		
		a	c	c/a
at%		Å	Å	
72.5	$\gamma_{fcc}+\varepsilon_{hcp}$ (D019)	–	–	–
72.8	$\gamma_{fcc}+\varepsilon_{hcp}$ (D019)	3.76	–	–
73.8	fcc	3.78	–	–
74.8	fcc	3.78	–	–
75.6	fcc	3.75	–	–
76.5	fcc	3.74	–	–
77.4	fcc	3.75	–	–
78.3	fct	3.76	3.70	0.98
79.2	fct	3.77	3.80	0.97
83.5	fct	–	–	–

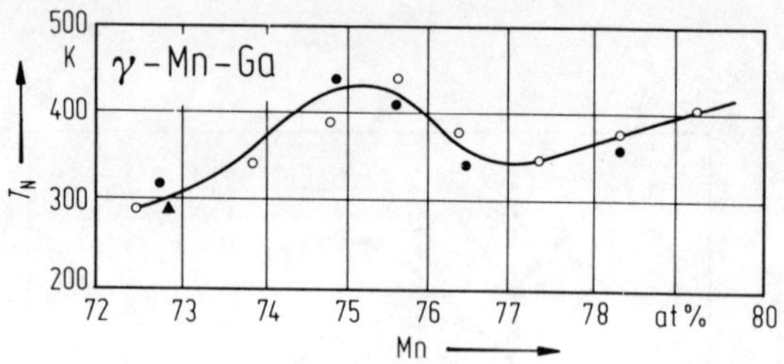

Fig. 51. γ-Mn–Ga. Concentration dependence of T_N of the γ-phase alloys quenched from 805 °C: open circles – dilatation; solid circles – neutron diffraction; triangle – metallographic analysis [79 G 1].

Table 17. Mn–Ga. Range over which the Curie-Weiss law is obeyed and values of Θ and p_{eff} for solid and liquid alloys [65 W 1].

Phase	Solid			Liquid		
	Curie-Weiss range °C	Θ K	p_{eff} μ_B/Mn	Curie-Weiss range °C	Θ K	p_{eff} μ_B/Mn
$MnGa_6$		420	0.63	750⋯940	– 700	5.47
$MnGa_4$	210⋯405	453	1.35	750⋯950	– 739	5.44
Mn_2Ga_5	190⋯450	451	2.45	730⋯950	– 742	5.36
Mn_3Ga_5	210⋯590	63	3.44	750⋯1000	– 737	5.12
Mn_5Ga_7	0⋯415	– 127	3.69	770⋯970	– 747	5.03
Mn_5Ga_7	415⋯620	– 273	4	–	–	–
Mn_5Ga_6	0⋯490	– 136	3.79	806⋯970	– 737	4.96
Mn_5Ga_6	490⋯700	– 252	4.03	–	–	–
MnGa						
$Mn_{0.50}Ga_{0.50}$	400⋯685	– 377	4.09	846⋯1135	– 766	4.85
$Mn_{0.49}Ga_{0.51}$	0⋯425	– 332	4.03	850⋯980	– 737	4.87
$Mn_{0.49}Ga_{0.51}$	425⋯710	– 487	4.33	–	–	–
$Mn_{0.47}Ga_{0.53}$	0⋯450	– 232	3.95	820⋯980	– 747	4.93
$Mn_{0.47}Ga_{0.53}$	450⋯735	– 335	4.17	–	–	–
Mn_7Ga_6	770⋯860	– 480	4.27	900⋯980	– 877	4.78
$(Mn_2Ga)_h$	840⋯890	– 953	4.43	–	–	–

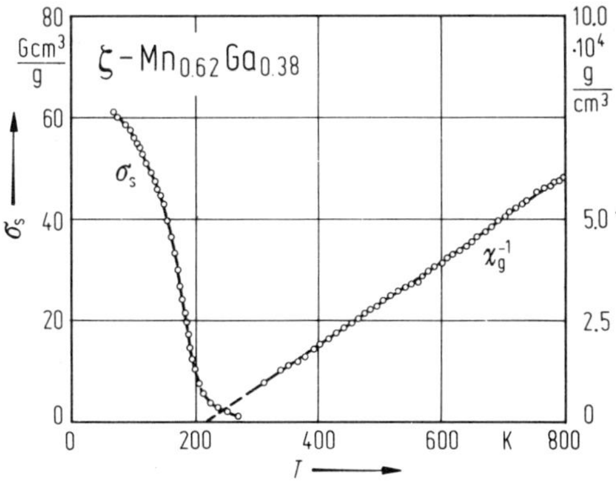

Fig. 52. ζ-$Mn_{0.62}Ga_{0.38}$. Temperature dependence of the saturation magnetization and inverse magnetic susceptibility. Samples quenched from 850 °C, annealed at 700 °C [63 T 1].

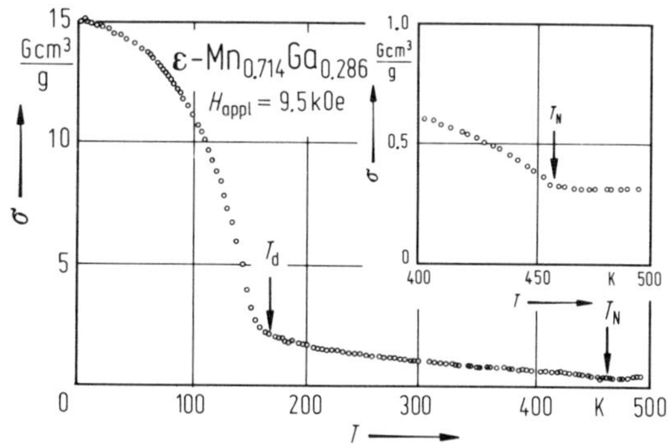

Fig. 53. ε-$Mn_{0.714}Ga_{0.286}$. Magnetization vs. temperature curve at an applied magnetic field of 9.5 kOe [83 N 1]. T_d is the lattice distortion temperature.

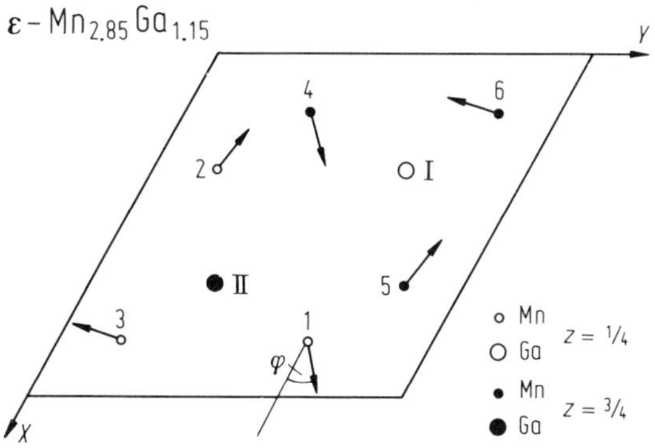

Fig. 54. ε-$Mn_{2.85}Ga_{1.15}$. Projection of the D019-type unit cell on the basal plane (001). The numbers indicate lattice sites. The triangular antiferromagnetic structure is illustrated [70 K 2].

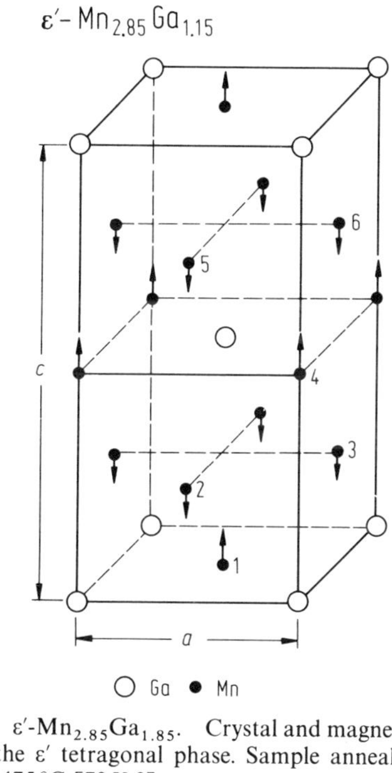

Fig. 55. ε'-$Mn_{2.85}Ga_{1.85}$. Crystal and magnetic structure of the ε' tetragonal phase. Sample annealed for 1 week at 475 °C [70 K 2].

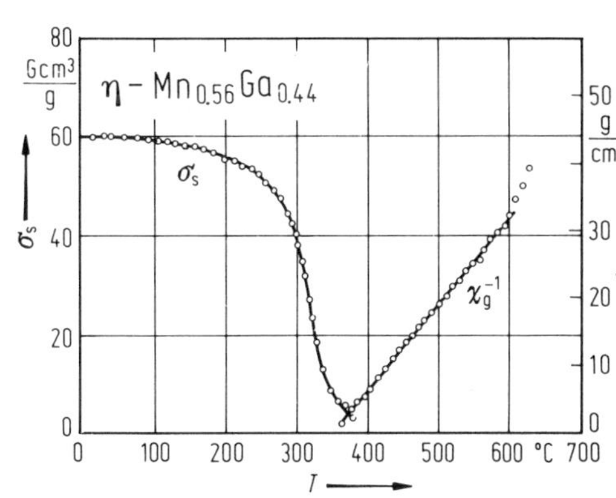

Fig. 56. η-$Mn_{0.56}Ga_{0.44}$. Temperature dependences of magnetization and reciprocal magnetic susceptibility [65 H 1].

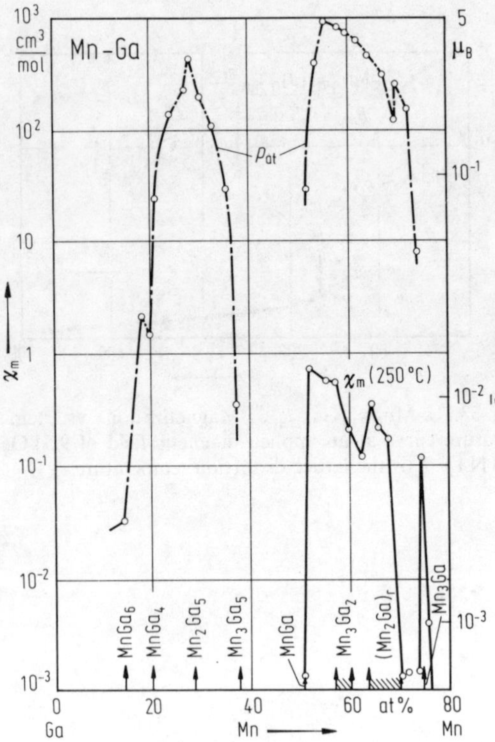

Fig. 57. Mn–Ga. Magnetic moment per atom, $\bar{p}_{at}$, and magnetic molar susceptibility χ_m at 250 °C vs. composition [65 W 1].

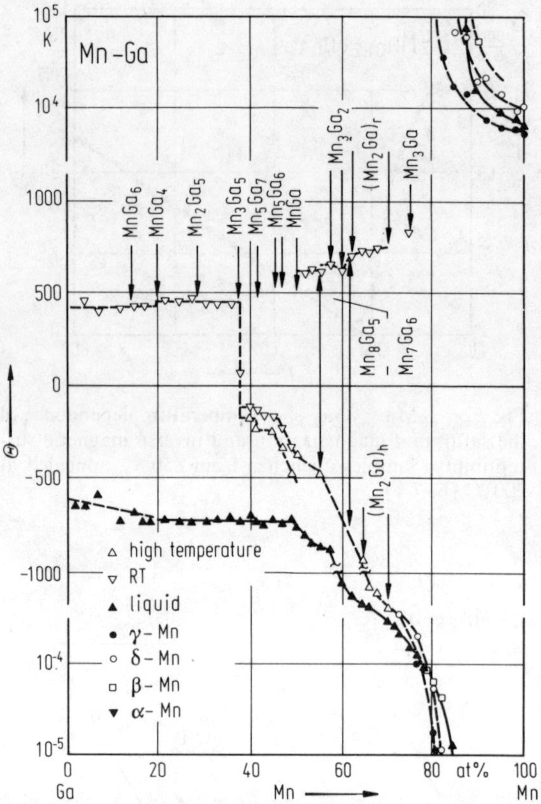

Fig. 58. Mn–Ga. Variation of Θ with composition [65 W 1].

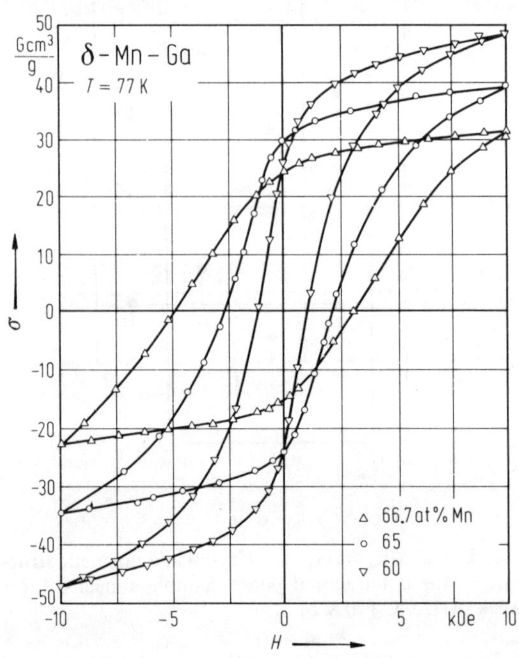

Fig. 59. δ-Mn–Ga. Hysteresis loops at 77 K for tetragonal Mn–Ga δ phase. Samples cooled from above Curie temperature in (+) magnetic field [65 B 1].

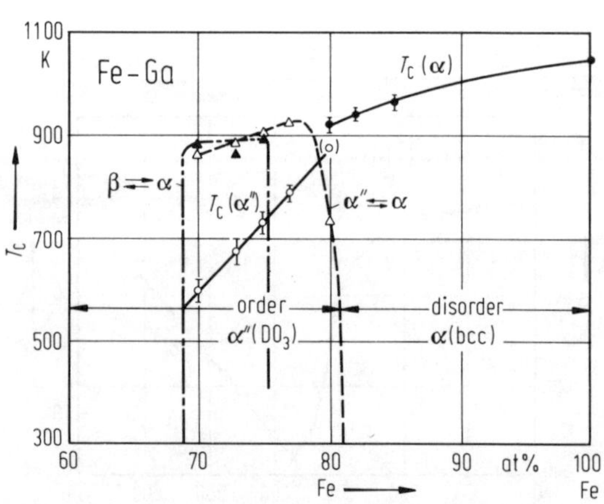

Fig. 60. Fe–Ga. Magnetic phase diagram, the dot-dash line indicates the boundary of the equilibrium fcc state, while the broken line and T_C (α″) are for the quenched bcc phase [72 K 1].

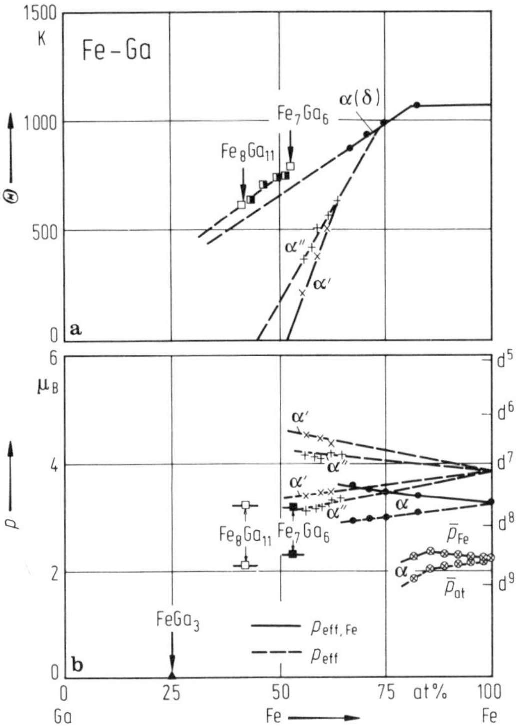

Fig. 61. Fe–Ga. (a) Variation with composition of the paramagnetic Curie temperature [67 W 1]. (b) Variation with composition of the effective magnetic moment per atom, p_{eff}, and per Fe atom, $p_{eff,Fe}$. The right-hand scale gives the corresponding 3d-band occupancies for an ionic model [67 W 1]. Also shown are the results from [66 A 2] for the magnetic moments per atom, $\bar{p}_{at}$, and per Fe atom, $\bar{p}_{Fe}$, from magnetization measurements.

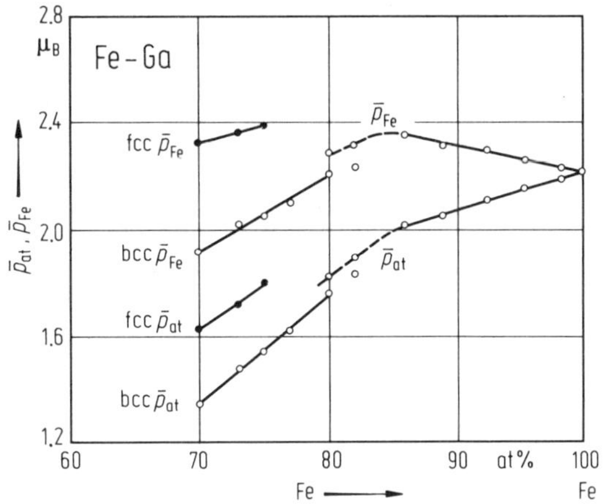

Fig. 62. Fe–Ga. Magnetic moment per atom, $\bar{p}_{at}$, and per Fe atom, $\bar{p}_{Fe}$ vs. composition for the bcc and fcc alloys [72 K 2, 66 A 2].

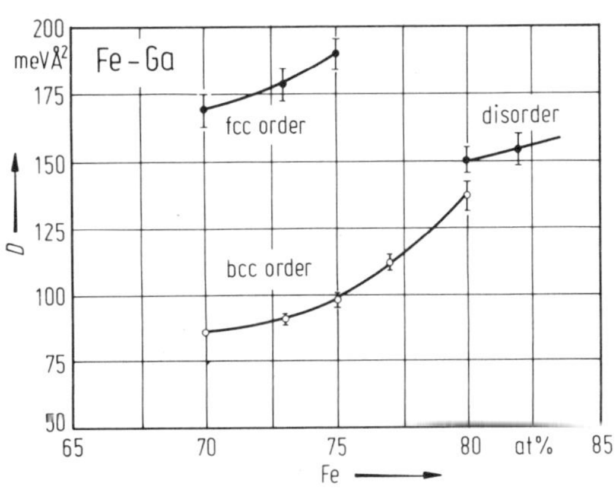

Fig. 63. Fe–Ga. Exchange stiffness constant D vs. composition [72 K 2].

Table 18. Fe_3Al, Fe_3Ga. Magnetic moments and hyperfine fields for ^{57}Fe on sites A and D [82 K 1].

		H_{hyp} [kOe]		$H_{hyp,D}/H_{hyp,A}$		p_{Fe} [μ_B]		$p_{Fe,D}/p_{Fe,A}$	
		0 K	RT	0 K	RT	0 K	RT	0 K	RT
Fe_3Al	A	234		1.427		1.50		1.45	
	D	334				2.18			
Fe_3Ga	A	230	204	1.565	1.559	1.71	1.50	1.46	1.45
	D	360	318			2.49	2.18		

Table 19. $Co_{0.503}Ga_{0.497}$. Magnetic susceptibility χ_m, X-ray and bulk densities, ϱ_X and ϱ_g, respectively, and concentrations of Co vacancies, $N(V_{Co})$, and of Co atoms on Ga sites, $N(Co_{Ga})$, relative to the number of atoms, for samples quenched from 900 °C into water and annealed for various times t_a at 560 °C [75 B 2].

t_a	χ_m	ϱ_X	ϱ_g	$N(V_{Co})$	$N(Co_{Ga})$
h	10^{-6} cm³ mol⁻¹	g cm⁻³		%	
0	304	8.99	8.61	4.5	2.5
1	294	8.99	8.61	4.4	2.5
2	265	8.97	8.64	3.8	2.2
4	204	8.92	8.68	2.7	1.7
8	191	8.91	8.71	2.3	1.5
16	177	8.91	8.72	2.1	1.4
[1])	163	8.91	8.75	1.8	1.2

[1]) Values after furnace-cooling.

Table 20. Molar Curie constant C_m, paramagnetic Curie temperature Θ, concentration of Co atoms on Ga sites, $N(Co_{Ga})$, relative to the number of atoms and effective magnetic moment p_{eff} of Co atoms on Ga sites for slow-cooled samples [75 B 2].

Co	C_m	Θ	$N(Co_{Ga})$	p_{eff}
at%	10^{-3} cm³ K mol⁻¹	K	%	μ_B/Co_{Ga}
45.2	–	–	–	–
48.0	4.2	−100	0.1	5.7
50.0	35	− 75	1.2	4.9
52.0	84	− 60	2.7	5.0
53.8	130	20	4.1	5.1
56.1	195	105	6.1	5.1

Table 21. Co–Ga. Curie temperatures, cluster sizes and spins as determined by magnetization measurements and neutron scattering measurements on quenched samples. Cluster magnetic moments $p_{cl} = 2\,\mu_B S$ [77 C 2].

Co	Magnetization measurements			Neutron scattering		
	T_C [1])	T_C	Cluster size	T_C	Cluster size	S
at%	K	K	Å	K	Å	
52	5	< 4.2	14(5)	< 4.2	–	–
53	12	< 4.2	12(5)	< 4.2	22(7)	27
56	56	70	33(5)	70	47(13)	32
57	84	80	26(5)	81	31(8)	39
60	200	200	25(5)	203	27(6)	58

[1]) [70 B 1].

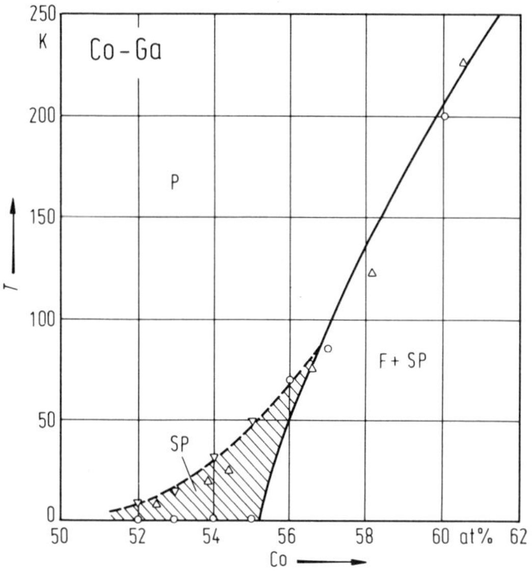

Fig. 64. Co–Ga. Magnetic phase diagram of quenched samples: Open circles: T_C from magnetization and neutron measurements; triangles, downward: superparamagnetic (SP) ordering temperatures from neutron measurements [77 C 2]. Triangles, upward: T_C from [70 B 1].

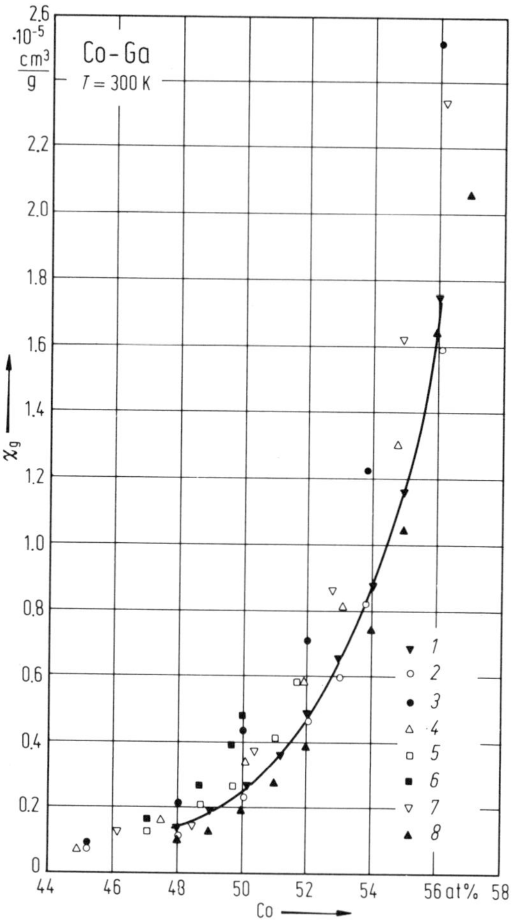

Fig. 65. Co–Ga. Dependence of the room-temperature magnetic susceptibility on composition and heat-treatment. The line through the data is only a guide. *1*: [82 M 1], annealing temperature T_a=950 °C, slowly cooled; *2*: [75 B 3], T_a = 900 °C, cooling rate 100 °C per day; *3*: [75 B 3], T_a = 900 °C, quenched; *4*: [73 W 1], T_a=950 °C, slowly cooled; *5*: [74 A 1], T_a=870 °C, slowly cooled; *6*: [74 A 1], T_a=870 °C, quenched; *7*: [69 G 1], T_a=900 °C, slowly cooled; *8*: [73 T 1], T_a=830/550 °C slowly cooled.

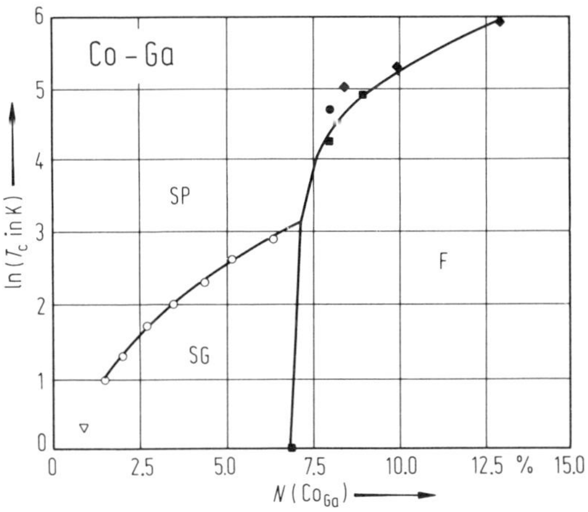

Fig. 66. Co Ga. Magnetic phase diagram for slowly cooled samples. The ordinate is the temperature of the susceptibility maximum. The abscissa is the magnetic defect concentration i.e. the number of Co atoms on Ga sites relative to the total number of atoms, $N(\mathrm{Co_{Ga}})$. Ferromagnetic boundary: lozenges [79 G 2], squares [73 T 1], solid circle [82 M 1]. Superparamagnetic (SP) to spin-glass (SG) transition: open circle [82 M 1]. Possible antiferromagnetic regime: triangle [82 M 1].

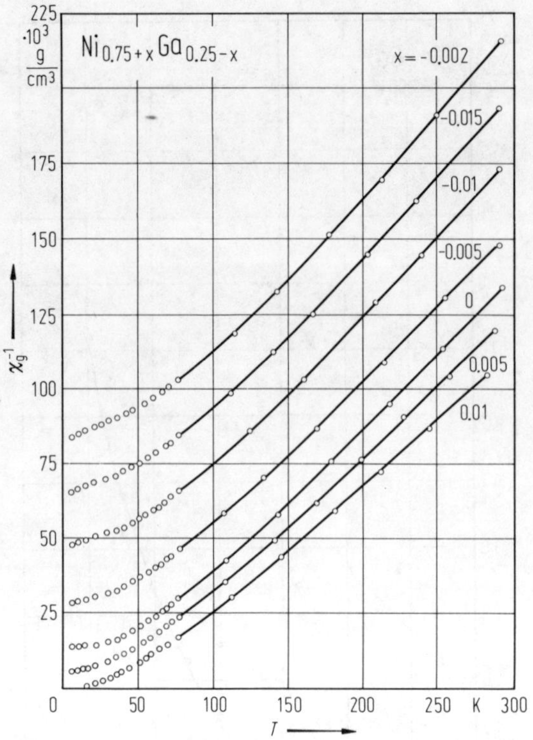

Fig. 67. $Ni_{0.75+x}Ga_{0.25-x}$. Reciprocal magnetic mass susceptibility vs. temperature [69 B 3].

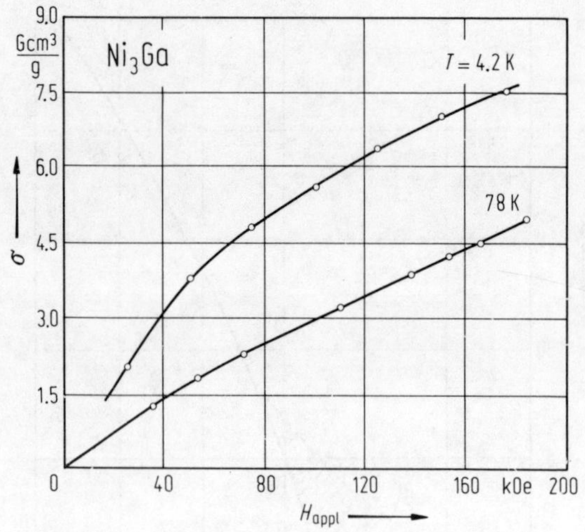

Fig. 68. Ni_3Ga. Magnetization as a function of applied magnetic field at 4.2 and 78 K of stoichiometric Ni_3Ga [67 B 2]. $\sigma = 6.8\,G\,cm^3\,g^{-1}$ corresponds to $0.1\,\mu_B$/Ni.

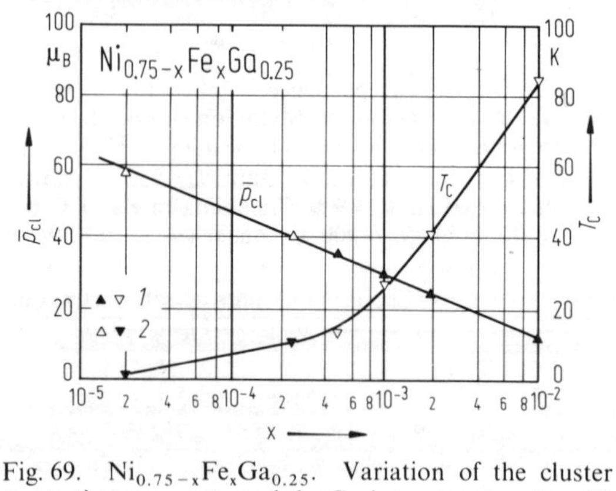

Fig. 69. $Ni_{0.75-x}Fe_xGa_{0.25}$. Variation of the cluster magnetic moment $\bar{p}_{cl}$ and the Curie temperature T_C with Fe content [72 M 1]. *1*: [68 S 1], *2*: [70 M 1].

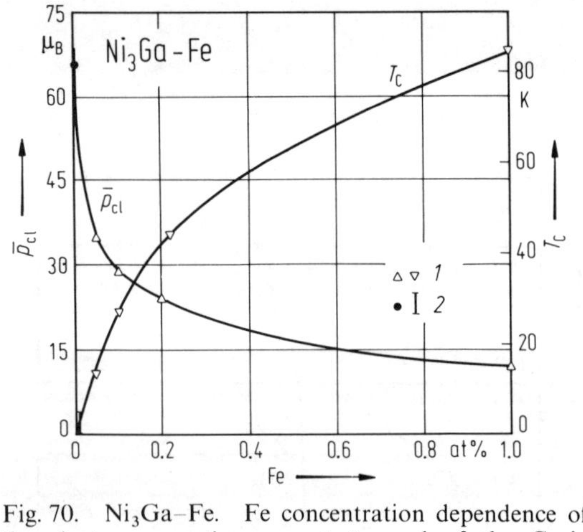

Fig. 70. Ni_3Ga–Fe. Fe concentration dependence of the cluster magnetic moment $\bar{p}_{cl}$ and of the Curie temperature T_C [70 M 1]. *1*: [68 S 1], *2*: [70 M 1].

5.3.5 References for 5.3

29 W 1 Weiss, P., Forrer, R.: Ann. Phys. (Paris) **12** (1929) 279.

37 B 1 Bradley, A.J., Jay, A.H.: Proc. R. Soc. London Ser. A **159** (1937) 56.

60 H 1 Höhl, M.: Z. Metallkd. **51** (1960) 85.

60 K 1 Koch, A.J.J., Hokkeling, P., Steeg, M., van dem, Vos, K.J., de: J. Appl. Phys. **31** (1960) 75 S.

60 K 2 Köster, W., Wachtel, E.: Z. Metallkd. **51** (1960) 271.

60 T 1 Tsuboya, I., Sugihara, M.: J. Phys. Soc. Jpn. **15** (1960) 1534.

61 L 1 Lawley, A., Cahn, R.W.: J. Phys. Chem. Solids **20** (1961) 204.

61 L 2 Lundquist, N., Meyer, H.: Ark. Fiz. **20** (1961) 463.

61 T 1 Taylor, M.A.: Proc. Phys. Soc. (London) **78** (1961) 1244.

61 T 2 Tsuboya, I.: J. Phys. Soc. Jpn. **16** (1961) 1875.

61 T 3 Tsuboya, I., Sugihara, M.: J. Phys. Soc. Jpn. **16** (1961) 1257.

61 T 4 Tsuboya, I., Sugihara, M.: Proc. Int. Conf. Magn. Crystallogr. (Kyoto, Sept. 1961). J. Phys. Soc. Jpn. **17** Suppl. B-1 (1962) 172.

62 C 1 Clogston, A.M., Gossard, A.C., Jaccarino, V., Yafet, Y.: Phys. Rev. Lett. **9** (1962) 262.

62 L 1 Lundquist, N., Myers, H.P., Weston, R.: Philos. Mag. **7** (1962) 1187.

63 C 1 Cooper, M.J.: Philos. Mag. **8** (1963) 805.

63 F 1 Fruchart, R.: C. R. Acad. Sci. **256** (1963) 3304.

63 S 1 Silver, A.H. Kushida, T.: J. Chem. Phys. **38** (1963) 865.

63 T 1 Tsuboya, I., Sugihara, M.: J. Phys. Soc. Jpn. **18** (1963) 1096.

64 O 1 Ostenburg, D.O., van, Lam, D.J., van, Tapp, H.D., Pracht, D.W., rowland, T.J.: Phys. Rev. A **135** (1964) 455.

65 B 1 Bither, T.A., Cloud, W.H.: J. Appl. Phys. **36** (1965) 1501.

65 C 1 Cadeville, M.C.: Thèse, Université de Strasbourg **1965**.

65 D 1 Daido, K., Tsuboya, I.: J. Phys. Soc. Jpn. **20** (1965) 1519.

65 H 1 Hasegawa, M., Tsuboya, I.: J. Phys. Soc. Jpn. **20** (1965) 464.

65 W 1 Wachtel, E., Nier, K.J.: Z. Metallkd. **56** (1965) 779.

66 A 2 Aldred, A.T.: J. Appl. Phys. **37** (1966) 1344.

66 A 1 Andersson, L., Dellby, B., Myers, H.P.: Solid State Commun. **4** (1966) 77.

66 C 1 Cadeville, M.C.: J. Phys. Chem. Solids **27** (1966) 667.

66 C 2 Cadeville, M.C., Daniel, E.: J. Phys. (Paris) **27** (1966) 449.

67 B 1 Boer, F.R., de, Biesterbos, J., Schinkel, C.J.: Phys. Lett. A **24** (1967) 355.

67 B 2 Boer, F.R., de, Schinkel, C.J., Biesterbos, J.: Phys. Lett. A **25** (1967) 606.

67 M 1 Magat, L.M., Shur, Ya.S., Kandaurova, G.S., Makarova, G.M. Gusel'nikova, N.I.: Fiz. Met. Metalloved. **23(2)** (1967) 226. Engl. Transl.: Phys. Met. Metallogr. **23(2)** (1967).

67 W 1 Wachtel, E., Maier, J.: Z. Metallkd. **58** (1967) 885.

68 S 1 Schinkel, C.J., Boer, F.R., de, Biesterbos, J.: Phys. Lett. A **26** (1968) 501.

69 B 1 Barnes, R.G., Creel, R.B.: Phys. Lett. A **29** (1969) 203.

69 B 3 Boer, F.R., de, Schinkel, C.J., Biesterbos, J., Proost, J.: J. Appl. Phys. **40** (1969) 1049.

69 B 2 Brodsky, M.B., Brittain, J.O.: J. Appl. Phys. **40** (1969) 3615.

69 C 1 Castaing, J., Caudron, R., Toupance, G., Costa, P.: Solid State Commun. **7** (1969) 1453.

69 G 1 Goto, T., Ohashi, M., Kamigaki, K.: J. Phys. Soc. Jpn. **26** (1969) 207.

69 K 1 Kallel, A.: C. R. Acad. Sci. B **268** (1969) 455.

69 R 1 Ravdel', M.P., Yevdokimova, O.I., Petrova, O.N., Yakimura, L.B.: Pretsizionuye Splavy Moscow, Metallurgiya **71** (1969) 93.

70 B 1 Booth, J.G., Marshall, J.D.: Phys. Lett. A **32** (1970) 149.

70 C 1 Cherchernikov, V.I., Nedel'ko, V.I., Vedyayev, A.V.: Fiz. Met. Metalloved. **29(6)** (1970) 1324. Engl. Transl.: Phys. Met. Metallogr. **29(6)** (1976) 223.

70 K 1 Keuntzler, R.: J. Appl. Phys. **41** (1970) 908.

70 K 2 Krén, E., Kádár, G.: Solid State Commun. **8** (1970) 1653.

70 M 1 Maletta, H., Mossbauer, R.L.: Solid State Commun. **8** (1970) 143.

70 P 1 Peshev, P., Etourneau, J., Naslain, R.: Mater. Res. Bull. **5** (1970) 319.

71 C 1 Carter, G.C., Swartz, J.C.: J. Phys. Chem. Solids **32** (1971) 2415.

71 C 2 Chakrabati, D.J., Beck, P.A.: J. Phys. Chem. Solids **32** (1971) 1609.

71 K 1 Kostetskii, I.I., L'vov, S.N., Kunitskii, Yu.A.: Ivz. Akad. Nuak. SSSR Neorg. Mater. **7** (1971) 951. Engl. Transl.: Inorg. Mater. USA **7** (1971) 839.

71 Y 1 Yoshida, K., Nagata, S.: Jpn. J. Appl. Phys. **10** (1971) 7.

72 B 1 Boydenko, V.S., Kandaurova, G.S., Magat, L.M., Makarova, G.M., Malyuk, V.P.: Phys. Met. Metalloved. **34(6)** (1972) 1295. Engl. Transl.: Phys. Met. Metallogr. **34(6)** (1972) 132.

72 K 1 Kawamiya, N., Adachi, K., Nakamura, Y.: J. Phys. Soc. Jpn. **33** (1972) 1318.

72 K 2 Kostetskii, I.I., L'vov, S.N.: Fiz. Met. Metalloved. **33(4)** (1972) 773. Engl. Transl.: Phys. Met. Metallogr. **33(4)** (1972) 95.

72 K 3 Kostetskii, I.I., L'vov, S.N.: Fiz. Met. Metalloved **34(1)** (1972) 475. Engl. Transl.: Phys. Met. Metallogr. **34(3)** (1972) 24.

72 M 1 Maletta, H.: Z. Phys. **250** (1972) 68.

72 R 1 Robbins, C.G., Claus, H.: AIP Conf. Proc. **5** (1972) 527.

72 S 1 Sellmyer, D.J., Caskey, G.R., Franz. J.: J. Phys. Chem. Solids **33** (1972) 561.

73 C 1 Caskey, G.R., Franz, J.M., Sellmyer, D.J.: J. Phys. Chem. Solids **34** (1973) 1179.

73 H 1 Horn, P.M., Lawrence, J., Parks, R.D., Thomas, G.A.: Phys. Rev. Lett. **31** (1973) 471.

73 M 2 Makarova, G.M., Magat, L.M., Boydenko, V.S., Deryagin, A.V., Kandaurova, G.S.: Fiz. Met. Metalloved. **35(1)** (1973) 209. Engl. Transl.: Phys. Met. Metallogr. **35(1)** (1973).

73 M 1 Miliayev, J., Besnus, M.J., Meyer, A.J.P.: Solid State Commun. **13** (1973) 1001.

73 S 1 Sidorenko, F.A., Kotov, A.P., Zelenin, L.P., Gel'd, P.V.: Fiz. Met. Metalloved. **35(1)** (1973) 218. Engl. Transl.: Phys. Met. Metallogr. **35(1)** (1973) 209.

73 T 1 Tamminga, Y.: Ph.D. Thesis, University of Amsterdam **1973**, unpublished.

73 W 1 Wachtel, E., Linse, V., Gerold, V.: J. Phys. Chem. Solids **34** (1973) 1461.

73 W 2 Webster, P.J., Ziebeck, K.R.A.: J. Phys. Chem. Solids **34** (1973) 1647.

74 A 1 Amamou, A., Gautier, F.: J. Phys. F **4** (1974) 563.

74 C 1 Coldea, M., Crisan, M., Néda, A. Pop, I.: J. Phys. Chem. Solids **35** (1974) 1095.

74 P 1 Prun, A.F., Chechernikov, V.I., Nedel'ko, V.I., Sidorenko, F.A., Kotov, A.P.: Fiz. Met. Metalloved. **38(2)** (1974) 431. Engl. Transl.: Phys. Met. Metallogr. **38(2)** (1974) 190.

74 V 1 Vintaykin, Ye.Z., Udovenko, V.A., Belyatskaya, I.S., Luarsabishvili, N.N., Makushev, S.Yu.: Fiz. Met. Metalloved. **38(2)** (1974) 398. Engl. Transl.: Phys. Met. Metallogr. **38(2)** (1974) 157.

75 B 1 Berner, D., Geibel, G., Gerold, V., Wachtel, E.: J. Phys. Chem. Solids **36** (1975) 221.

75 B 2 Besnus, A.J., Herr, A., Meyer, A.J.P.: J. Phys. F **5** (1975) 2138.

75 B 3 Booth, J.G., Pritchard, R.G.: J. Phys. F **5** (1975) 347.

75 T 1 Takacs, L., Cadeville, M.C., Vincze, I.: J. Phys. F **5** (1975) 800.

76 G 1 Grössinger, R., Steiner, W., Krec, K.: J. Mag. Magn. Mater. **2** (1976) 196.

76 J 1 Junod, A., Flükiger, R., Treyrand, A., Muller, J.: Solid State Commun. **19** (1976) 265.

76 S 1 Shull, R.D., Okamoto, H., Beck, P.A.: Solid State Commun. **20** (1976) 863.

76 T 1 Tanaka, T., Nozaki, H., Bannai, E., Ishizawa, Y., Kawai, S., Yamane, T.: J. Less-Common Met. **50** (1976) 15.

77 C 1 Cywinski, R., Booth, J.G.: J. Phys. F **7** (1977) 1531.

77 C 2 Cywinski, R., Booth, J.G., Rainford, B.D.: J. Phys. F **7** (1977) 2567.

78 A 1 Arajs, S., Anderson, E.E.: J. Appl. Phys. **49** (1978) 1433.

79 C 1 Cook, J.M., Pavlovic, A.S.: J. Appl. Phys. **50** (1979) 7710.

79 G 1 Goman'kov, V.I., Ravdel', M.P., Nogin, N.I., Petrova, O.N., Evdokimova, O.I.: Fiz. Met. Metalloved. **47(4)** (1979) 735. Engl. Transl.: Phys. Met. Metallogr. **47(4)** (1979) 50.

79 G 2 Grover, A.K., Malik, S.K., Radhakrishnamurty, C., Vijayaraghavan, R.: Solid State Commun. **32** (1979) 1323.

79 L 1 Leyarovska, L., Leyarovski, E., Popov, C., Midlarz, T.: J. Less-Common Met. **67** (1979) 389.

79 S 1 Suprunenko, P.A., Markiv, V.Ya., Storodyenko, A.I.: Ukr. Fiz. Zh. **24(1)** (1979) 114.

80 S 1 Shukla, P., Wortis, M.: Phys. Rev. B **21** (1980) 159.

81 K 1 Kanaizuka, T.: Mater. Res. Bull. **16** (1981) 1601.

81 V 1 Vlasova, N.I., Kandaurova, G.S., Shur, Ya.S., Bykhanova, N.N.: Fiz. Met. Metalloved. **51(6)** (1981) 1127. Engl. Transl.: Phys. Met. Metallogr. **51(6)** (1981) 1.

81 Y 1 Yoshida Y., Kawakami, M., Nakamichi, T.: J. Phys. Soc. Jpn. **50** (1981) 2203.

82 K 1 Kawamiya, N., Adachi, K.: Trans. Jpn. Inst. Met. **23** (1982) 296.

82 M 1 Meisel, M.W., Zhou, Wen Sheng, Owers-Bradley, J.R., Ochiai, Y., Brittain, J.O., Halperin, W.P.: J. Phys. F **12** (1982) 317.

82 S 1 Sigfusson, T.I., Lonzarich, G.G.: Phys. Scr. **25** (1982) 720.

83 K 1 Kawamiya, N., Adachi, K.: Proc. ICM 82 (Kyoto, 6–10 Sept. 1982), J. Mag. Magn. Mater. **31–34** (1983) 145.

83 M 1 Motoya, K., Shapiro, S.M., Muraoka, Y.: Phys. Rev. **B28** (1983) 6183.

83 N 1 Niida, H., Hori, T., Nakagawa, Y.: J. Phys. Soc. Jpn. **52** (1983) 1512.

83 T 1 Takahashi, S., Ikeda, K.: J. Phys. F **13** (1983) 2169.

85 B 1 Booth, J.G., Mankikar, R.M., Morris, P.: Solid State Commun. **60** (1986).

85 O 1 Okpalugo, D.E., Booth, J.G., Kepa, H., Hicks, T.J.: J. App. Phys. **61** (1987) 3997.

5.4 3d elements and C, Si, Ge, Sn or Pb

The magnetic properties of alloys between 3d elements and group 4B elements have been studied especially for Mn and Fe intermetallic compounds. They show almost all strong magnetic behavior such as ferromagnetism, ferrimagnetism, antiferromagnetism and helimagnetism. Many Ti and V intermetallic compounds with group 4B elements are superconductors whereas most of Cr, Co and Ni intermetallic compounds are paramagnets.

5.4.1 Ti and V alloys and compounds

Most of the Ti and V compounds are paramagnetic. Since V_3X compounds become superconductors at relatively high temperatures, their physical properties including magnetic properties have been studied in detail.

Table 1. Magnetic and related properties of TiC [59 B 1], Ti_5Si_3 [58 R 1], TiSi and $TiSi_2$ [70, R 1].

	TiC	Ti_5Si_3	TiSi	$TiSi_2$
Crystallographic structure	cubic B1(NaCl)	hexagonal $D8_8(Mn_5Si_3)$	orthorhombic B27(FeB)	orthorhombic C54
a [Å]	4.329	7.463	6.53	8.23
b [Å]			3.64	4.77
c [Å]		5.165	5.00	8.58
Magnetism	para	para	para	para
χ_g [10^{-6} cm^3/g]	0.10 at RT	2.50 at RT	0.20 at RT	1.19 at RT

Table 2. Magnetic and related properties of V_2C [70 C 1], V_3Si [61 C 1], VSi_2 [58 R 1], V_3Ge and V_3Sn [61 C 1]. In connection with superconducting properties, the Knight shifts of V_3X have been measured [61 C 1].

	V_2C	V_3Si	VSi_2	V_3Ge	V_3Sn
Crystallographic structure	hexagonal L'3(Fe_2N)	cubic A15(β-W)	hexagonal C40($CrSi_2$)	cubic A15(β-W)	cubic A15(β-W)
a [Å]	2.87	4.721	4.571	4.769	4.94
c [Å]	4.55		6.372		
c/a	1.59		1.394		
Magnetism	para	para	para	para	para
χ_g [10^{-6} cm^3/g]	1.55 at RT	5.6 at RT	1.53 at RT	4.3 at RT	3.4 at RT
T_s [1]) [K]		17.1		6.0	<1

[1]) Superconducting transition temperature.

5.4.2 Cr alloys and compounds

Cr-rich alloys especially with Si and Ge have extensively been investigated in connection with spin density waves (SDW) below their Néel temperatures. Cr compounds with Si and Ge studied so far show paramagnetism except for $Cr_{11}Ge_{19}$.

Table 3. Magnetic and related properties of Cr_3Si [79 M 1], CrSi [72 W 1], Cr_3Ge, Cr_5Ge_3, CrGe, $Cr_{11}Ge_{19}$ [80 K 1].

	Cr_3Si	CrSi	Cr_3Ge	Cr_5Ge_3	CrGe	$Cr_{11}Ge_{19}$
Crystallographic structure	cubic A15(β-W)	cubic B20(FeSi)	cubic A15(β-W)	tetragonal $D8_m(W_5Si_3)$ (T_1-phase)	cubic B20(FeSi)	tetragonal $Mn_{11}Si_{19}$
a [Å]	4.555	4.629	4.631	9.410	4.196	5.79
c [Å]				4.783		51.87
c/a				0.508		8.96
Magnetism	para	constant para	constant para	constant para	para	ferro
T_C [K]						91
Θ [K]						170
p_{eff} [μ_B/Cr]		3.01				2.13

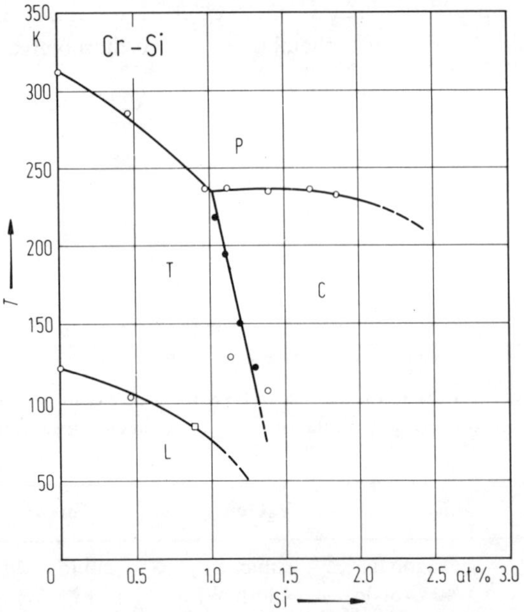

Fig. 1. Cr–Si. Magnetic phase diagram. Open circles: calorimetric and magnetic susceptibility measurements, solid circles: thermal expansion, square: neutron diffraction. Symbols P, C, T and L indicate the paramagnetic (P) state and the commensurate (C), incommensurate transverse (T) and incommensurate longitudinal (L) spin density wave states [82 B 1]. The P–C phase transition is of first order while those of P–T, T–L are of second order.

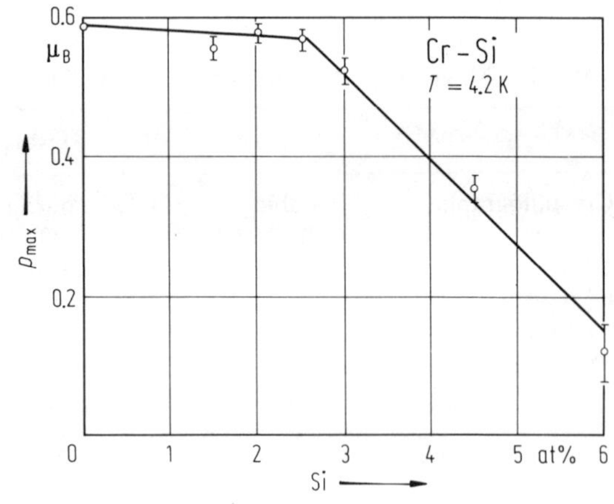

Fig. 2. Cr–Si. Variation of the amplitude of Cr spin density wave as a function of Si content as estimated from neutron powder diffraction at 4.2 K [80 C 1].

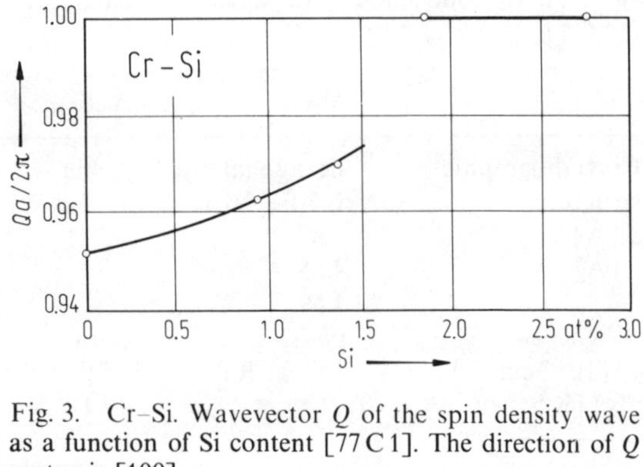

Fig. 3. Cr–Si. Wavevector Q of the spin density wave as a function of Si content [77 C 1]. The direction of Q vector is [100].

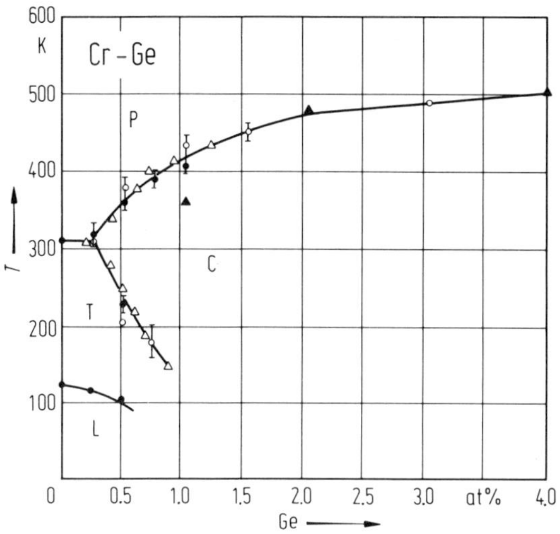

Fig. 4. Cr–Ge. Magnetic phase diagram. Solid circles: neutron diffraction, other symbols: thermal expansion measured by different authors. Symbols P, C, T and L indicate the paramagnetic (P) state and the commensurate (C), incommensurate transverse (T) and incommensurate longitudinal (L) spin density wave states [81 I 1].

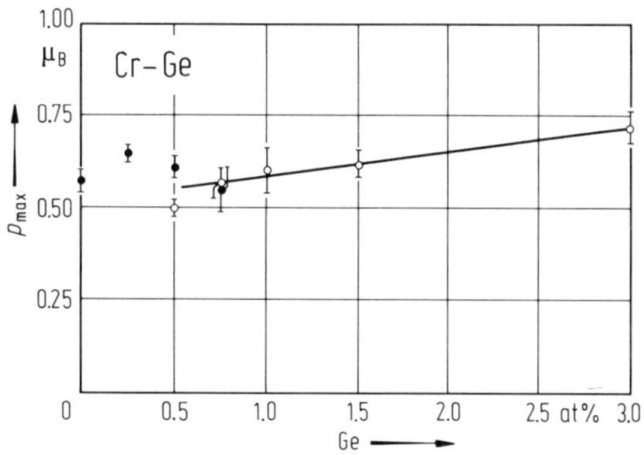

Fig. 5. Cr–Ge. Amplitude of the magnetic moment of the transverse incommensurate spin density wave (closed circles) just above the spin-flip temperature T_{sf} and of the commensurate spin density wave (open circles), just above the commensurate-incommensurate phase transition temperature T_{CI} for 0.5 and 0.75 at% Ge and at ≅77 K for ≧1 at% Ge, as a function of Ge concentration [81 I 1].

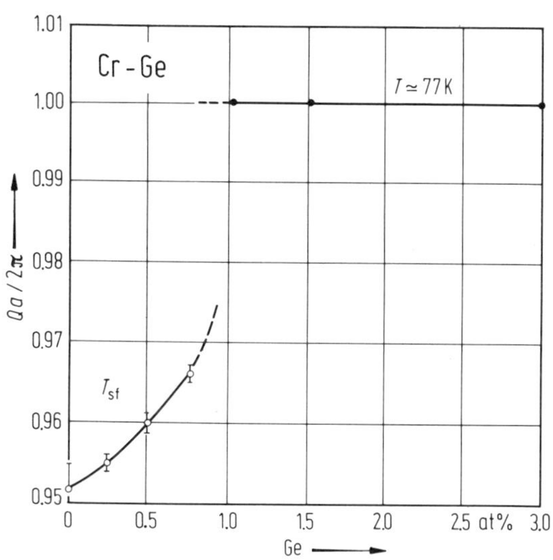

Fig. 6. Cr–Ge. Wavevector of the spin density wave, at ≅77 K for the commensurate phase and just above the spin-flip temperature T_{sf} for the incommensurate phase, as a function of Ge concentration [81 I 1].

5.4.3 Mn alloys and compounds

The solubilities of group 4B elements in Mn are a few percent at most. Therefore, the interest has been focused on intermetallic compounds. Mn–Si compounds show antiferromagnetism in MnSi which has extensively been studied because of its unusual magnetic behavior. Mn–Ge compounds show antiferromagnetism, ferrimagnetism and ferromagnetism depending upon compound. Mn–Sn system shows antiferromagnetism or ferrimagnetism.

Table 4. Magnetic and related properties of Mn_3Si [75 T 1], Mn_5Si_3 [84 B 2], MnSi [75 B 1] and $Mn_{11}Si_{19}$ [80 N 1]

	Mn_3Si	Mn_5Si_3	MnSi	$Mn_{11}Si_{19}$
Crystallographic structure	cubic $D0_3(AlFe_3)$	hexagonal $D8_8(Mn_5Si_3)$	cubic B20(FeSi)	tetragonal
a [Å]	5.722	6.912	4.558	5.53
c [Å]		4.812		117.9
c/a		0.696		
Magnetism	incommensurate antiferro	antiferro	weak itinerant ferro	antiferro
T_N [K]	25.8	98		42
T_C [K]			29	
Θ [K]				38
p_{Mn} [μ_B/Mn]	1.72(C) 0.19(B/D) at 4.2 K	0.4(4d) 1.2(6g) at 4.2 K [67 L 1]	0.4 at 1.4 K	0.056 at 4.2 K
p_{eff} [μ_B/Mn]		3.1 [62 K 1]	1.4 [66 W 1]	
structure	helical or sinusoidal. $Q \parallel$ [111] $p_{Mn} \perp$ [111]	$T < 64$ K: noncollinear, all p_{Mn} ordered $64 K < T < T_N$: collinear only $p_{Mn(4d)}$ ordered, $p_{Mn(4d)} \parallel a$	helical; $\lambda = 180$ Å at 4.2 K, $Q \parallel$ [111] $p_{Mn} \perp$ [111] [77 I 1]	helical; $\lambda = 163$ Å at 4.2 K, $Q \parallel c$ $p_{Mn} \perp c$

Table 5. Magnetic and related properties of Mn_3Ge [71 K 1], Mn_5Ge_2 [83 Y 1], Mn_5Ge_3 [63 C 1] and Mn_3Ge_2 [66 Z 1].

	Mn_3Ge		Mn_5Ge_2		Mn_5Ge_3	Mn_3Ge_2
concentration range x (Mn_xGe)	3.4	3.4	2.5···2.7 at 800 °C [61 O 1]	very narrow ≅2.5 [61 O 1]	–	–
Crystallographic structure	$T > 500$ °C hexagonal $D0_{19}(Ni_3Sn)$	$T < 500$ °C fct $D0_{22}$(AlTi)	$T > 630$ °C hexagonal $B8_2(Ni_2In)$	$T < 630$ °C undetermined $\varkappa$ phase	 hexagonal $D8_8(Mn_5Si_3)$	unknown
a [Å]	5.347	3.816	7.191		7.184	
c [Å]	4.374	3.631	39.19		5.58	
c/a	0.818	0.952	5.45 for x = 2.6		0.777 [62 F 1]	
Magnetism	antiferro	ferri	unusual antiferro	ferri	ferro	antiferro
T_N [K]	395		≅500 [61 O 1] 480 for x = 2.6			290 [62 F 1]
T_C [K]		920		710 [61 O 1]	304 [62 K 1]	
Θ [K]	−870···1010 [1])				335 [62 F 1]	339 [62 F 1]
dT_C/dp [K kbar^{-1}]					0.43 [62 B 1]	
p_{Mn} [μ_B/Mn]	2.4 at 77 K	3.4 (I), 1.9 (II) at RT or 3.3 (I), 1.7(II) at RT			3(4d), 2(6g) at 77 K [63 C 1] or 1.7(4d), 2.7(6g) at 77 K [65 F 1]	
p_s [μ_B/Mn]	0.02 at 77 K weak ferro	0.3 at RT		0.2 at 100 K [61 O 1]	2.49 at 0 K [62 K 1]	1.5 below 100 K
p_{eff} [μ_B/Mn]	3.7				3.3 [62 K 1]	2.35 [62 F 1]
K [erg/cm^3]					$1.2 \cdot 10^6$ at RT [62 S 1]	
structure	triangular, $p_{Mn} \perp c$	$p_{Mn} \parallel c$ (see Fig. 10)	$T < 120$ K: $p_{Mn} \parallel c$, $120\,K < T < 310$ K: $p_{Mn} \perp c$, $310\,K < T < T_N$: $p_{Mn} \parallel c$, for x = 2.6	N-type ferrimagnet; compensation point: 395 K [61 O 1]	$p_{Mn} \parallel c$	weak ferro below 160 K
$H_{hyp}(^{55}Mn)$ [kOe]					195(4d) 399(6g) at 4.2 K [65 J 1]	

[1]) x = 3.2···3.8 [67 K 1].

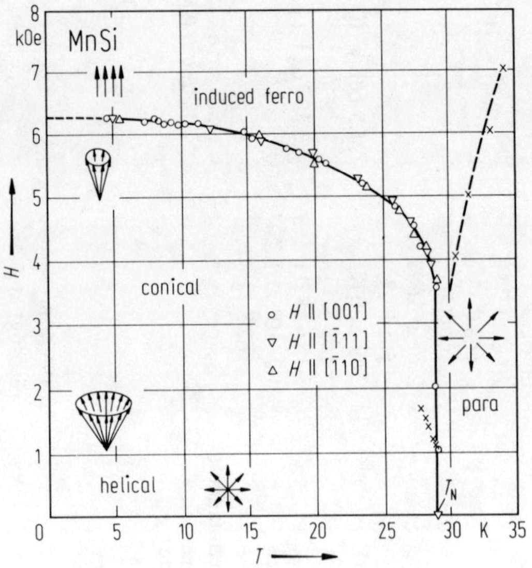

Fig. 7. MnSi. Magnetic phase diagram determined by ultrasonic study. Crosses indicate positions where an additional peak of the attenuation was observed [77 I 1].

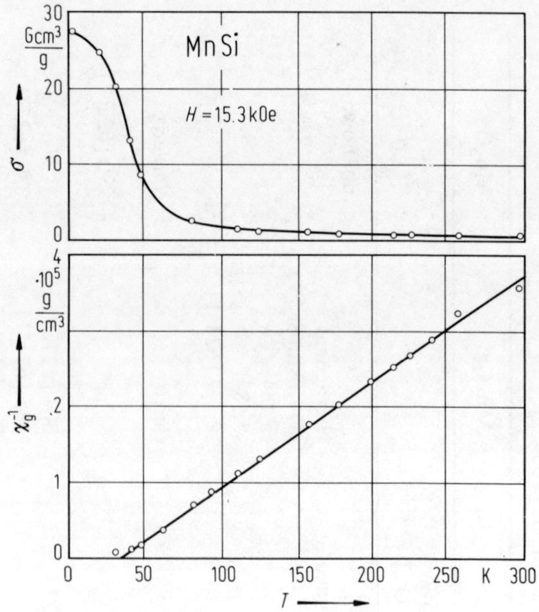

Fig. 8. MnSi. Inverse magnetic susceptibility and magnetization as functions of temperature [72 W 1].

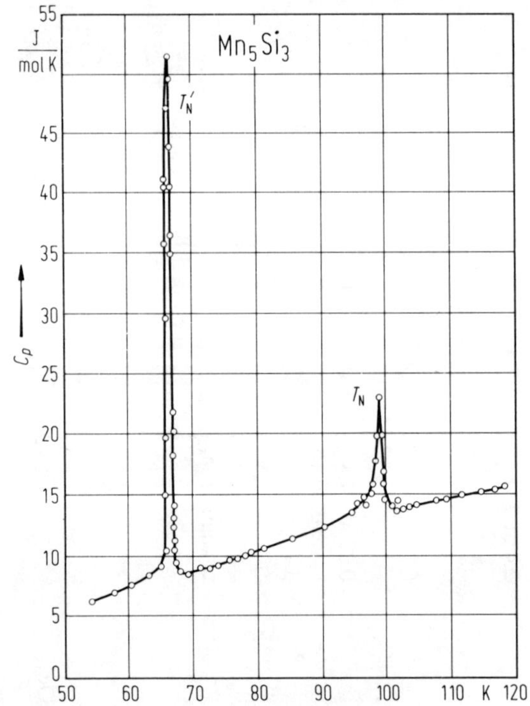

Fig. 9. Mn_5Si_3. Temperature dependence of the specific heat [72 G 1]. T_N and T_N' indicate the Néel temperature and the collinear-noncollinear antiferromagnetic transition temperature, respectively.

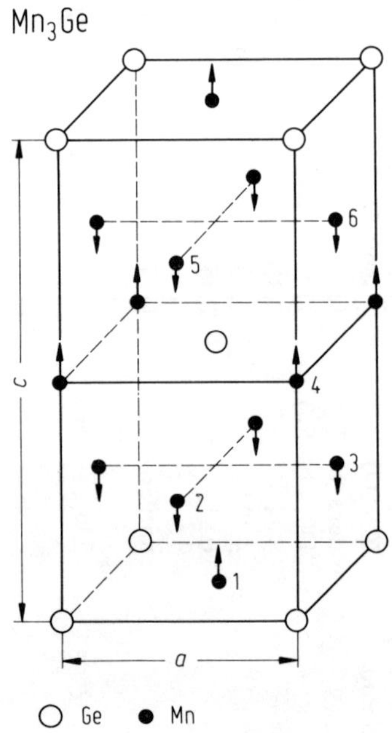

Fig. 10. Mn_3Ge. Magnetic structure of fct Mn_3Ge (low-temperature phase). Magnetic moments $p_{Mn(I)}$ at sites 1 and 4 are larger than and antiparallel to magnetic moments $p_{Mn(II)}$ at sites 2, 3, 5 and 6 [71 K 1].

Table 6. Magnetic and related properties of Mn_3Sn [67 K 1], Mn_2Sn [62 A 1, 62 Y 1] and $MnSn_2$ [68 C 1].

	Mn_3Sn	Mn_2Sn	$MnSn_2$
Concentration range x (Mn_xSn)	3.2···3.8	1.45···2.05	–
Crystallographic structure	hexagonal $D0_{19}(Ni_3Sn)$	hexagonal $B8_2(Ni_2In)$	tetragonal $C16(CuAl_2)$
a [Å]	5.64	4.40	6.659
c [Å]	4.52	5.52	5.436
c/a	0.80 for x = 3.8	1.25 for x = 1.77 [61 Y 1]	0.816
Magnetism	antiferro	ferri?	antiferro
T_N [K]	400···420		324 [61 Y 1]
T_C [K]		258···269 [1])	
Θ [K]	−500···−635		190 [61 Y 1] 175 [68 K 1]
p_{Mn} [μ_B/Mn]	2.4 at RT for x = 3.4 [72 Z 1]		2.33 at 4.2 K
p_s [μ_B/Mn]	0.003 at RT for x = 3.4 [72 Z 1]	1.7···0.8 [2]) at 0 K 1.58···0.98 [1]) at 0 K	0.5, $74\,K < T < T_N$
p_{eff} [μ_B/Mn]	4.1	3.4 for x = 1.77 [61 Y 1]	3.48 [61 Y 1] 3.75 [68 K 1]
structure	triangular [65 K 2]; $p_{Mn} \perp c$. $200\,K < T < 400\,K$: uncompensated weak ferro magnetism, for x = 3.4 [72 Z 1]	ferrimagnetism suggested from $\sigma(x)$ and $1/\chi_m(T)$.	$T < 74\,K$: (110) ferromagnetic sheets couple + + − −, $74\,K < T < T_N$: (110) ferromagnetic sheets couple + − + −, p_{Mn} sinusoidally modulated, $\lambda = 31$ Å, $p_{Mn}(max)$: 0.5 μ_B/Mn

[1]) x = 1.5···1.8 [62 A 1].
[2]) x = 1.45···2.05 [62 Y 1].

5.4.4 Fe alloys and compounds

Fe–C alloys have been studied for long time from the metallurgical view point. Fe–Si alloys are the most important soft magnetic materials which are treated in this subsection. Fe–Si compounds have been studied extensively. Fe_3Si and Fe_5Si_3 show ferromagnetism, while FeSi indicates unusual paramagnetism. Ternary intermetallic compounds Fe–M–Si have also been studied in detail. Fe–Ge intermetallic compound show ferromagnetism or antiferromagnetism. Fe–M–Ge ternary compounds have studied in detail. Fe–Sn intermetallics indicate ferromagnetism or antiferromagnetism.

Table 7. Magnetic and related properties of Fe_3C [59 H 1], Fe_5C_2 [67 B 1] and Fe_2C [59 H 1].

	Fe_3C (cementite)	Fe_5C_2 (χ carbide, Hegg's carbide)	Fe_2C (ε carbide)
Crystallographic structure	orthorhombic $D0_{11}$	monoclinic Mn_5C_2	hcp
a [Å]	4.525	11.56	6.27
b [Å]	5.087	4.573	
c [Å]	6.743	5.058	21.4
β		97°41′	
c/a			3.41
Magnetism	ferro	ferro	ferro
T_C [K]	483 [69 T 1]		653 [69 T 1]
Θ [K]	506		419
p_s [μ_B/Fe]	1.78 at 0 K	1.48 at RT	1.72 at 0 K
p_{eff} [μ_B/Fe]	3.89		
H_{hyp} (^{57}Fe) [kOe]	208 (I) 208 (II) at RT [67 B 1]	222 (I) 184 (II) 110 (III) at RT	
IS (^{57}Fe) [cm/s] relative to iron	0.29 at RT [67 B 1]	0.3 at RT	

Table 8. Magnetic and related properties of Fe_3Si [76 H 1], Fe_5Si_3 [68 J 1] and FeSi [81 T 1].

	Fe_3Si	Fe_5Si_3	FeSi
Crystallographic structure	cubic $D0_3(AlFe_3)$	hexagonal $D8_8(Mn_5Si_3)$	cubic B20(FeSi)
a [Å]	5.6554	6.7552	4.4891
c [Å]		4.7174	
c/a		0.6983	
Magnetism	ferro	ferro	unusual para
T_C [K]	839 [76 N 1]	376	
Θ [K]			−340
p_{Fe} [μ_B/Fe]	2.20(C) 1.35(B/D) at 0 K	1.05(4d) 1.55(6g) at 0 K	
p_{Si} [μ_B/Si]	−0.07 at 0 K		
p_s [μ_B/Fe]	1.61 at 6.5 K	0.97 at 0 K	
p_{eff} [μ_B/Fe]			2.65
K_1 [10^5 erg/cm^3]	1.08 at 0 K [83 G 1]		
H_{hyp} (^{57}Fe) [kOe]	−338(C) −218(B/D) at 1.4 K	243(4d) 132(6g) at 4.2 K [70 N 1]	
H_{hyp} (^{29}Si) [kOe]	37 at 1.4 K [83 N 1]		
ΔE_Q (^{57}Fe) [mm/s]	0 at 77 K [63 S 1]	−0.25(4d) 0.65(6g) at 77 K [63 S 1]	0.73 at 0 K [65 W 1]
IS [1]) (^{57}Fe) [mm/s]	0.2(C) 0.45(B/D) at 77 K [63 S 1]	0.3(4d) 0.2(6g) at 77 K [63 S 1]	

[1]) Source ^{57}Co electroplated onto Cu plate, IS (^{57}Fe) = 0.15 per iron.

Table 9, see pages 154 and 155.

Table 10. Magnetic and related properties of Fe_3Sn [70 T 1], Fe_5Sn_3 [70 T 1], Fe_3Sn_2 [78 M 1], FeSn [75 H 1] and $FeSn_2$ [85 C 1].

	Fe_3Sn	Fe_5Sn_3	Fe_3Sn_2	FeSn	$FeSn_2$
Crystallographic structure	hexagonal $D0_{19}$ stable 1750 °C < T < 850 °C	hexagonal $B8_2$ T < 840 °C vacancies ordered or T = 900 °C disordered vacancies	rhombohedral $R\bar{3}m$ [78 C 1]	hexagonal B35	tetragonal $C16(CuAl_2)$
a [Å]	4.230	4.217	5.344	5.30	6.539
c [Å]	5.208	5.245	19.845	4.45	5.325
c/a	1.231	1.244	3.714	0.84	0.814
Magnetism	ferro	ferro	ferro	antiferro	antiferro
T_C [K]	743	583	657		
T_N [K]				368 [70 T 1]	380
Θ [K]				−158 [67 Y 1]	−230 [60 K 1]
p_{Fe} [μ_B/Fe]	2.27 at 77 K	2.10 at 77 K	2.02 at 4.2 K	1.5 at 77 K [67 Y 1]	1.64 at 5 K
p_{eff} [μ_B/Fe]				4.5 [67 Y 1]	3.36 [60 K 1]
structure	$p_{Fe} \parallel c$		T < 60 K: $p_{Fe} \perp c$; 60 K < T < 250 K: p_{Fe} rotates to c axis: RT: $p_{Fe} \parallel c$	$p_{Fe} \parallel ab$ plane, forming ferromagnetic sheets [67 Y 1].	T < 93 K: noncollinear, p_{Fe} canted; T = 93 K: spin rotates to [110] direction; 93 K < T < T_N: ferromagnetic (100) planes couple antiferromagnetically along [100], p_{Fe} is in (001) plane near a axis

Table 9. Magnetic and related properties of Fe_3Ge [65 K 1], Fe_5Ge_3 [78 S 1], FeGe (hexagonal) [84 B 1], cubic [81 E 1], monoclinic [83 F 1] and $FeGe_2$ [64 F 1].

	Fe_3Ge *)		Fe_5Ge_3
Crystallographic structure	$T > 700\,°C$ hexagonal, ε $D0_{19}(Ni_3Sn)$	$T < 700\,°C$ cubic, $ε_1$ $L1_2(Cu_3Au)$	hexagonal $B8_2(Ni_2As)$
a [Å]	5.19	3.66	4.008
b [Å]			
c [Å]	4.22		5.009
β			
c/a	0.819		1.250
Magnetism	ferro	ferro	ferro
T_C [K]	665	755	480
T_N [K]			
p_{Fe} [μ_B/Fe]			1.3(2a) 2.4(2c) 1.9(2d) at 0 K
p_s [μ_B/Fe]	2.2 at 0 K	2.2 at 0 K	1.6 at 0 K
p_{eff} [μ_B/Fe]	3.6	3.6	
structure			
H_{hyp} (^{57}Fe) [kOe]	268 at 77 K [65 Y 1] 239 at RT [65 Y 1]	217 at 77 K [65 Y 1] 247 at RT [65 Y 1]	142(2a) 265(2c) 215(2d) at 4.2 K
ΔE_Q (^{57}Fe) [mm/s]	0 at 77 K and RT [65 Y 1]	0.26 at 77 K and RT [65 Y 1]	
$e^2qQ/2$ (^{57}Fe) [mm/s]			0.22(2a) −0.54(2c) −0.68(2d) at RT
IS (^{57}Fe) [mm/s]	0.45 [1]) at 77 K [65 Y 1] 0.30 [1]) at RT [65 Y 1]	0.50 [1]) at 77 K [65 Y 1] 0.30 [1]) at RT [65 Y 1]	0.13 [2]) (2a) 0.11 [2]) (2c) 0.13 [2]) (2d) at 520 K

*) [65 K 1]: $Fe_{1-x}Ge_x$. Hexagonal: $x = 0.238 \cdots 0.26$ at 710 °C; cubic: $x = 0.242 \cdots 0.258$ at 510 °C. Values listed are for $x = 0.25$.

[1]) Relative to ^{57}Fe in stainless steel.

[2]) Relative to ^{57}Fe in iron.

[3]) Average over the incommensurate modulation.

FeGe			$FeGe_2$
$T > 620\,°C$	$620\,°C < T < 740\,°C$	$T > 740\,°C$	
hexagonal	cubic	monoclinic	bct
B35(CoSn)	B20(FeSi)	C2/m(CoGe)	C16($CiAl_2$)
[78 F 1]	[78 F 1]		
5.002		11.838	2.48(d_1)
		3.937	4.18(d_2)
4.055		4.9336	4.86(d_3)
		103.514°	
0.811			
antiferro	conical	antiferro	antiferro
410	275 [84 B 1]	342 [84 B 1]	286 [84 F 1]
1.7$\parallel c$		0.98 3) (2a)	0.73 at 4.2 K
0.4$\perp c$ at 4.2 K		0.8 3) (2c)	
		1.85(4i) at 20 K	
			0.11 at 0 K [84 F 1]
			2.1
$T < 55$ K: double cone, $\lambda = 111$ Å; $55\,K < T < T_N$: collinear, $p_{Fe} \parallel c$	cone semi-angle with [111]: 55°, $\lambda = 700$ Å	$115\,K < T < T_N$: longitudinally polarized wave, $p_{Fe} \perp$ (1 0 1/2) plane, $T < 115$ K: further component, incommensurate modulation, $Q \parallel b$	$268\,K < T < T_N$. $p_{Fe} \perp c$ $T < 268$ K: collinear canted 30° to [100] at 100 K
151 at 20 K [70 W 1]	115 at 16 K [70 W 1]	95.5(2a) 120.2(2c) 201(4i) at 20 K [70 W 1]	148 at 77 K
		−0.03(2a) 0.05(2c) 0.09(4i) at 20 K [70 W 1]	
0.35 at RT [70 W 1]	0.16 at 80 K		−0.10 at 77 K
0.29 2) at RT [70 W 1]	0.61 2) at 80 K	≅0.25 2) at 342 K [70 W 1]	0.47 2) at 77 K

5.4.5 Co and Ni alloys and compounds

Most of the Co and Ni compounds are paramagnetic except Co_3Sn.

Table 11. Magnetic and related properties of CoSi, $CoSi_2$ [66 S 1], Co_3Ge, Co_2Ge [63 L 1], Co_3Sn [80 S 1], Ni_3Ge, Ni_2Ge [63 L 1] and Ni_3Sn, Ni_3Sn_2 [84 W 1].

	CoSi	$CoSi_2$	Co_3Ge	Co_2Ge	Co_3Sn
Crystallographic structure	cubic B20(FeSi)	cubic C1	hexagonal B8(NiAs)	hexagonal $B8_2(Ni_2In)$	cubic, bcc metastable high T phase
a [Å]	4.447	5.367	3.918		2.92
c [Å]			4.979		
c/a			1.271		
Magnetism	dia	pauli para	para	para	ferro
T_C [K]					≅ 500
Θ [K]					5200
χ_m [10^{-6} cm^3/mol]	−30 [67 W 1]	34.6 [67 W 1]			

	Ni_3Ge	Ni_2Ge	Ni_3Sn		Ni_3Sn
Crystallographic structure	cubic $L1_1(Cu_3Au)$·	hexagonal $B8_2(Ni_2In)$	$T<1223$ K cubic $D0_3(AlFe_3)$	$T>1223$ K hexagonal $D0_{19}$	hexagonal B8(NiAs)
a [Å]	3.556		5.98	5.293	4.145
c [Å]				4.244	5.213
c/a				0.802	1.258
Magnetism	para	para	para	para	para
T_C [K]					
Θ [K]					
χ_m [10^{-6} cm^3/mol]					

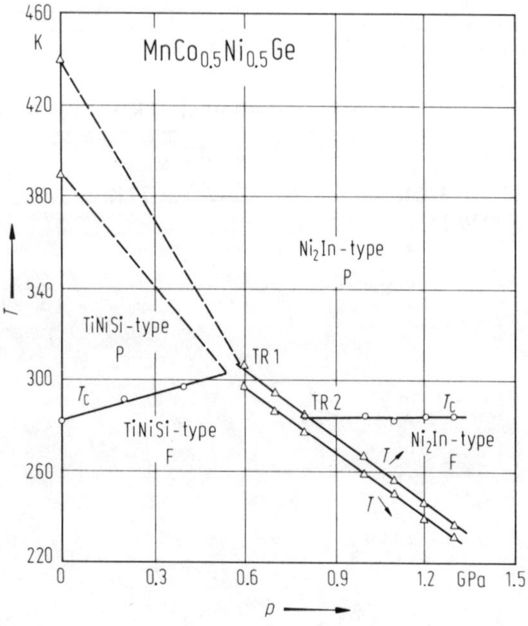

For Fig. 11, see next page.

Fig. 12. $MnCo_{0.5}Ni_{0.5}Ge$. $P-T$ phase diagram. Triple points (TR) are indicated [83 N 2].

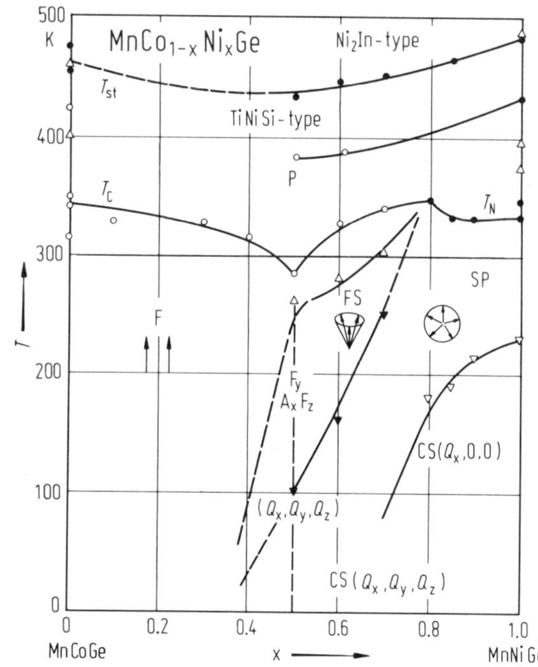

Fig. 11. $MnCo_{1-x}Ni_xGe$. Crystallographic and magnetic phase diagram. T_{st} is the temperature of the first-order phase transition, T_C and T_N denote the Curie temperature and Néel temperature, respectively. The magnetic structures are F (ferromagnetic (p_{Mn} and $p_{Co} \| c$)), F_y (ferromagnetic $p_{Mn} \| c$), A_xF_x (noncollinear p_{Co}), FS (ferromagnetic spiral), SP (simple spiral) and CS (cycloidal spiral). Q denotes the wavevector of the magnetic modulation [82 N 1].

5.4.6 References for 5.4

58 R 1 Robins, D.A.: Philos. Mag. **3** (1958) 318.
59 B 1 Bittner, H., Goretzki, H.: Monatsh. Chem. **91** (1960) 616.
59 H 1 Hoffer, L.J.E., Cohn, E.M.: J. Am. Chem. Soc. **81** (1959) 1576.
60 K 1 Kanematsu, K., Yasukochi, K., Ohoyama, T.: J. Phys. Soc. Jpn. **15** (1960) 2358.
61 C 1 Clogston, A.M., Jaccarino, V.: Phys. Rev. **121** (1961) 1357.
61 O 1 Ohoyama, T.: J. Phys. Soc. Jpn. **16** (1961) 1995.
61 Y 1 Yasukochi, K., Kanematsu, K., Ohoyama, T.: J. Phys. Soc. Jpn. **16** (1961) 1123.
62 A 1 Asanuma, M.: J. Phys. Soc. Jpn. **17** (1962) 300.
62 B 1 Boch, D., Pauthenet, R.: C. R. Acad. Sci. (Paris) **254** (1962) 1222.
62 F 1 Fontaine, R., Pauthenet, R.: C. R. Acad. Sci. (Paris) **254** (1962) 650.
62 K 1 Kanematsu, K.: J. Phys. Soc. Jpn. **17** (1962) 85.
62 S 1 Szceniowski, S.E., Wrzeciono, A.: J. Phys. Soc. Jpn. **17** (1962) Suppl. BI, p. 647.
62 Y 1 Yasukochi, K., Kanematsu, K., Ohoyama, T.: J. Phys. Soc. Jpn. **17** (1962) Suppl. BI, p. 165.
63 C 1 Ciszewski, R.: Phys. Status Solidi **3** (1963) 1999.
63 L 1 Lecoco, P.: Ann. Chim. **8** (1963) 85.
63 S 1 Shinjo, T., Nakamura, Y., Shikazono, N.: J. Phys. Soc. Jpn. **18** (1963) 797.
64 F 1 Forsyth, J.B., Johnson, C.E., Brown, P.J.: Philos. Mag. **10** (1964) 713.
65 F 1 Forsyth, J.B., Brown, P.J.: Proc. Int. Conf. on Magn., Inst. Phys. and Phys. Soc. London **1965**, p. 524.
65 J 1 Jackson, R.F., Scurlock, R.G., Utton, D.B., Wray, E.W.: Proc. Phys. Soc. **85** (1965) 127.
65 K 1 Kanematsu, K., Ohyama, T.: J. Phys. Soc. Jpn. **20** (1965) 236.
65 K 2 Kouvel, J.S., Kasper, J.S.: Proc. Int. Conf. on Magn., Inst. Phys. and Phys. Soc., London, **1965**, p. 169.
65 W 1 Wertheim, G.K., Jaccarino, V., Wernick, J.H., Seitchik, J.A., Williams, H.J., Sherwood, R.C.: Phys. Lett. **18** (1965) 89.
65 Y 1 Yamamoto, H.: J. Phys. Soc. Jpn. **20** (1965) 2166.
66 S 1 Shinoda, S., Asanabe, S.: J. Phys. Soc. Jpn. **21** (1966) 555.
66 W 1 Williams, H.J., Wernick, J.H., Sherwood, R.C., Wertheim, G.K.: J. Appl. Phys. **37** (1966) 1256.
66 Z 1 Zavadskii, E.A., Fakidov, I.G.: Zh. Eksper. Teor. Fiz. **51** (1966) 1317.

67 B 1 Bernas, H., Campbell, I.A., Fruchart, R.: J. Phys. Chem. Solids **28** (1967) 17.

67 K 1 Kanematsu, K.: Propriétés Thermodynamiques Physiques et Structurales des Derivés Semi-Métalliques, C.N.R.S., Paris **1967**, p. 135.

67 L 1 Lander, G.H., Brown, P.J., Forsyth, J.B.: Proc. Phys. Soc. London **91** (1967) 332.

67 Y 1 Yamaguchi, K., Watanabe, H.: J. Phys. Soc. Jpn. **22** (1967) 1210.

67 W 1 Walsted, R.E., Wernick, J.H., Jaccarino, V.: Phys. Rev. **162** (1967) 301.

68 C 1 Corliss, L.M., Hastings, J.M.: J. Appl. Phys. **39** (1968) 461.

68 J 1 Johnson, C.E., Forsyth, J.B., Lander, G.H., Brown, P.J.: J. Appl. Phys. **39** (1968) 465.

68 K 1 Kouvel, J.S., Jacobs, I.S.: J. Appl. Phys. **39** (1968) 467.

69 T 1 Tebble, R.S., Craik, D.J.: Magnetic materials, New York: John Wiley & Sons, **1969**, p. 90.

70 C 1 Caudron, R., Ducastelle, F., Costa, P.: J. Phys. Chem. Solids **31** (1970) 291.

70 N 1 Narashimhan, K.S.V.R., Reiff, W.M., Steinfink, H., Collins, R.L.: J. Phys. Chem. Solids **31** (1970) 1511.

70 R 1 Rabinovich, B.S., Radovskii, I.Z., Kozlov, F.N., Sidorenko, F.A., Gel'd, P.V.: Izv. Akad. Nauk SSSR Inorg. Mater. **6** (1970) 2202.

70 T 1 Trumphy, G., Both, E., Djéga-Mariadassou, D., Lecocq, P.: Phys. Rev. B **2** (1970) 3477.

70 W 1 Wäppling, R., Häggström, L., Karlsson, E.: Phys. Scr. **2** (1970) 233.

71 K 1 Kádár, G. Krén, E.: Int. J. Magn. **7** (1971) 143.

72 G 1 Gel'd, P.V., Mikhel'son, A.V., Kalishevich, G.I., Krentsis, P.P., Putintsev, Y.V., Sudakova, N.P.: Sov. Phys. Solid State **14** (1972) 798.

72 W 1 Wernick, J.H., Wertheim, G.K., Sherwood, R.C.: Mater. Res. Bull. **7** (1972) 1431.

72 Z 1 Zimmer, G.J., Krén, E.: AIP Conf. Proc. **5** (1972) 513.

75 B 1 Bloch, D., Voiron, J., Jaccarino, V., Wernick, J.H.: Phys. Lett. **51** A (1975) 259.

75 H 1 Häggström, L., Ericsson, T., Wäppling, R., Chandra, K.: Phys. Scr. **11** (1975) 47.

75 T 1 Tomiyoshi, S., Watanabe, H.: J. Phys. Soc. Jpn. **39** (1975) 295.

76 H 1 Hines, W.A., Menotti, A.H., Budnick, J.I., Burch, T.J., Litrenta, T., Niculescu, V., Raj, K.: Phys. Rev. B **13** (1976) 4060.

76 N 1 Niculescu, V., Raj, K., Budnick, J.I., Burch, T.J., Hines, W.A., Menotti, A.H.: Phys. Rev. B **14** (1976) 4160.

77 C 1 Cable, J.W.: J. Magn. Magn. Mater. **5** (1977) 112.

77 I 1 Ishikawa, Y., Komatsubara, T., Bloch, D.: Physica **86–88** B (1977) 401.

78 C 1 Le Caer, G., Malaman, B., Roques, B.: J. Phys. F **8** (1978) 323.

78 F 1 Forsyth, J.B., Wilkinson, C., Gardner, P.: J. Phys. F **8** (1978) 2195.

78 M 1 Malaman, B., Fruchart, D., Le Caër, G.: J. Phys. F **8** (1978) 2389.

78 S 1 Schurer, P.J., Hall, N.J.G., Morrish, A.H.: Phys. Rev. B **18** (1978) 4860.

79 M 1 Mikhaliov, I.G., Pan, V.M., Shevchenko, A.D., Bahr, H.A.: Sov. Phys. Solid State **21** (1979) 1609.

80 C 1 Costa, M.M.R., Ziebeck, K.R.A., Brown, P.J.: J. Phys. F **10** (1980) 2793.

80 K 1 Kolenda, M., Stoch, J., Szytula, A.: J. Magn. Magn. Mater. **20** (1980) 99.

80 N 1 Nakajima, T., Schelten, J.: J. Magn. Magn. Mater. **21** (1980) 157.

80 S 1 Schluckebier, G., Wachtel, E., Predel, B.: Z. Metallkd. **71** (1980) 456.

81 E 1 Ericsson, T., Karner, W., Häggström, L., Chandra, K.: Phys. Scr. **23** (1981) 1118.

81 I 1 Iida, S., Kawarazaki, S., Kunitomi, N.: J. Phys. Soc. Jpn. **50** (1981) 3612.

81 T 1 Takagi, S., Yasuoka, H., Ogawa, S., Wernick, J.H.: J. Phys. Soc. Jpn. **50** (1981) 2539.

82 B 1 Bendiktsson, G., Hedman, L., Åström, H.U., Rao, K.V.: J. Phys. F **12** (1982) 1439.

82 N 1 Niziol, S., Bombik, A., Bazela, W., Szytuta, A., Fruchart, D.: J. Magn. Magn. Mater. **27** (1982) 281.

83 F 1 Flecher, G.P., Jorgensen, J.D., Wäppling, R.: J. Phys. C **16** (1983) 6281.

83 G 1 Goto, M., Kamimori, T.: J. Phys. Soc. Jpn. **52** (1983) 3710.

83 N 1 Niculescu, V.A., Burch, T.J., Budnick, J.I.: J. Magn. Magn. Mater. **39** (1983) 223.

83 N 2 Niziot, S., Zieba, A., Zach, R., Baj, M., Dmowski, L.: J. Magn. Magn. Mater. **38** (1983) 205.

83 Y 1 Yamada, N., Ohoyama, T.: J. Magn. Magn. Mater. **31–34** (1983) 71.

84 B 2 Barmin, S.M., Mikhel'son, A.V., Sevast'yanov, A.A., Kortov, S.V., Gel'd, P.V.: Sov. Phys. Solid State **26** (1984) 1977).

84 B 1 Bernhard, J., Lebeck, B., Beckman, O.: J. Phys. F **14** (1984) 2379.

84 F 1 Franus-Muir, E., Fawcett, E., Pluzhnikov, V.: Solid State Commun. **52** (1984) 615.

84 W 1 Wachtel, E., Bayer, E.: Z. Metallkd. **75** (1984) 61.

85 C 1 Le Caër, G. Malaman, B., Venturini, G., Fruchart, D., Roques, B.: J. Phys. F **15** (1985) 1813.

6 Alloys and compounds between 4d or 5d elements and main group elements

6.1 Compounds of Zr

$ZrZn_2$ is a weak ferromagnet which is composed of two elements of nonmagnetic metals, Zr and Zn. The neutron diffraction experiments reveal that the magnetic moment density is spread out along the Zr–Zr bond [64 P 1, 84 B 1]; this is taken as evidence that the magnetism is due to itinerant electrons. Basic relations of the magnetization with a magnetic field *H* and temperature *T* for $ZrZn_2$ can be described by the Stoner model for weak itinerant ferromagnets; this weak ferromagnetism can be attributed to a high narrow peak of the density of states at the Fermi energy.

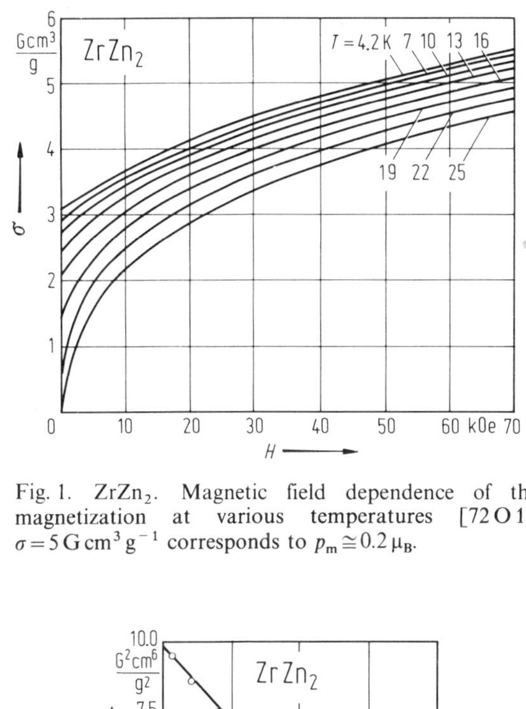

Fig. 1. $ZrZn_2$. Magnetic field dependence of the magnetization at various temperatures [72 O 1]. $\sigma = 5\,G\,cm^3\,g^{-1}$ corresponds to $p_m \cong 0.2\,\mu_B$.

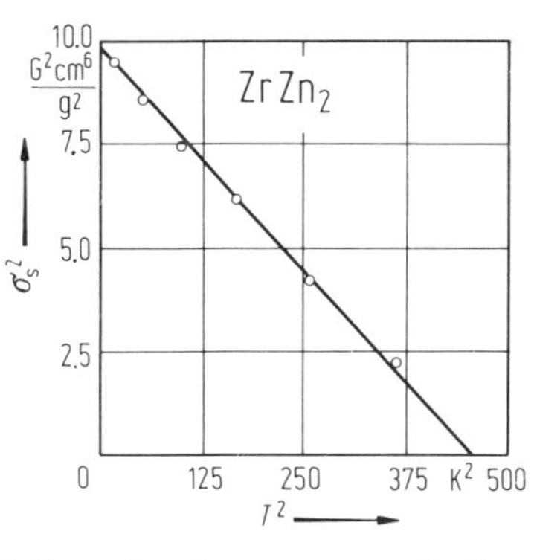

Fig. 2. $ZrZn_2$. Spontaneous magnetization squared plotted as a function of temperature squared [72 O 1].

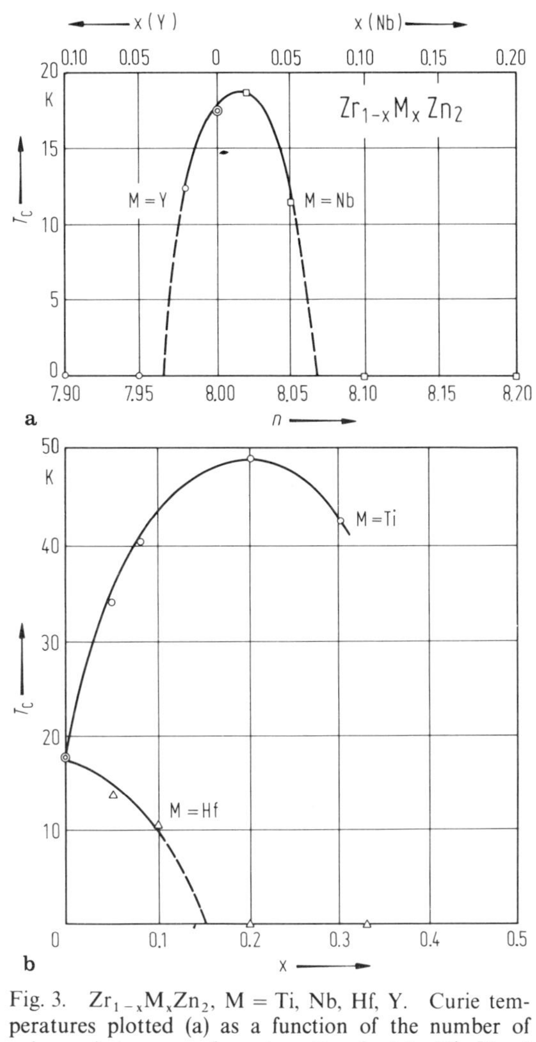

Fig. 3. $Zr_{1-x}M_xZn_2$, M = Ti, Nb, Hf, Y. Curie temperatures plotted (a) as a function of the number of valence electrons per formula unit, *n*, for M = Nb, Y and (b) as a function of concentration x for M = Ti, Hf [68 O 1].

Table 1. $ZrZn_2$. Magnetization data: saturation magnetic moment at 0 K, per unit mass, σ_s (0 K), and per formula unit, p_m, magnetic susceptibility at $T=0$ and $H=0$, χ_g (0 K), ferromagnetic and paramagnetic Curie temperatures, T_C and Θ, respectively, and effective magnetic moment per formula unit, p_{eff}.

Sample	Stoner-Wohlfarth				Curie-Weiss		Ref.
	σ_s (0 K) $Am^2 kg^{-1}$	p_m μ_B	T_C K	χ_g (0 K) $10^{-6} m^3 kg^{-1}$	Θ K	p_{eff} μ_B	
A	3.0	0.12	22	1.3			68 B 1
B	3.1	0.13	21	1.3			68 B 1
B	3.1	0.13	21	1.0	22		68 O 1
A	3.2(1)	0.130(4)	27.0(8)	0.98	32(2)	1.35	78 M 1
B	3.2(1)	0.130(4)	27.0(8)	0.98	32(2)	1.35	78 M 1

6.2 Alloys and compounds of Nb

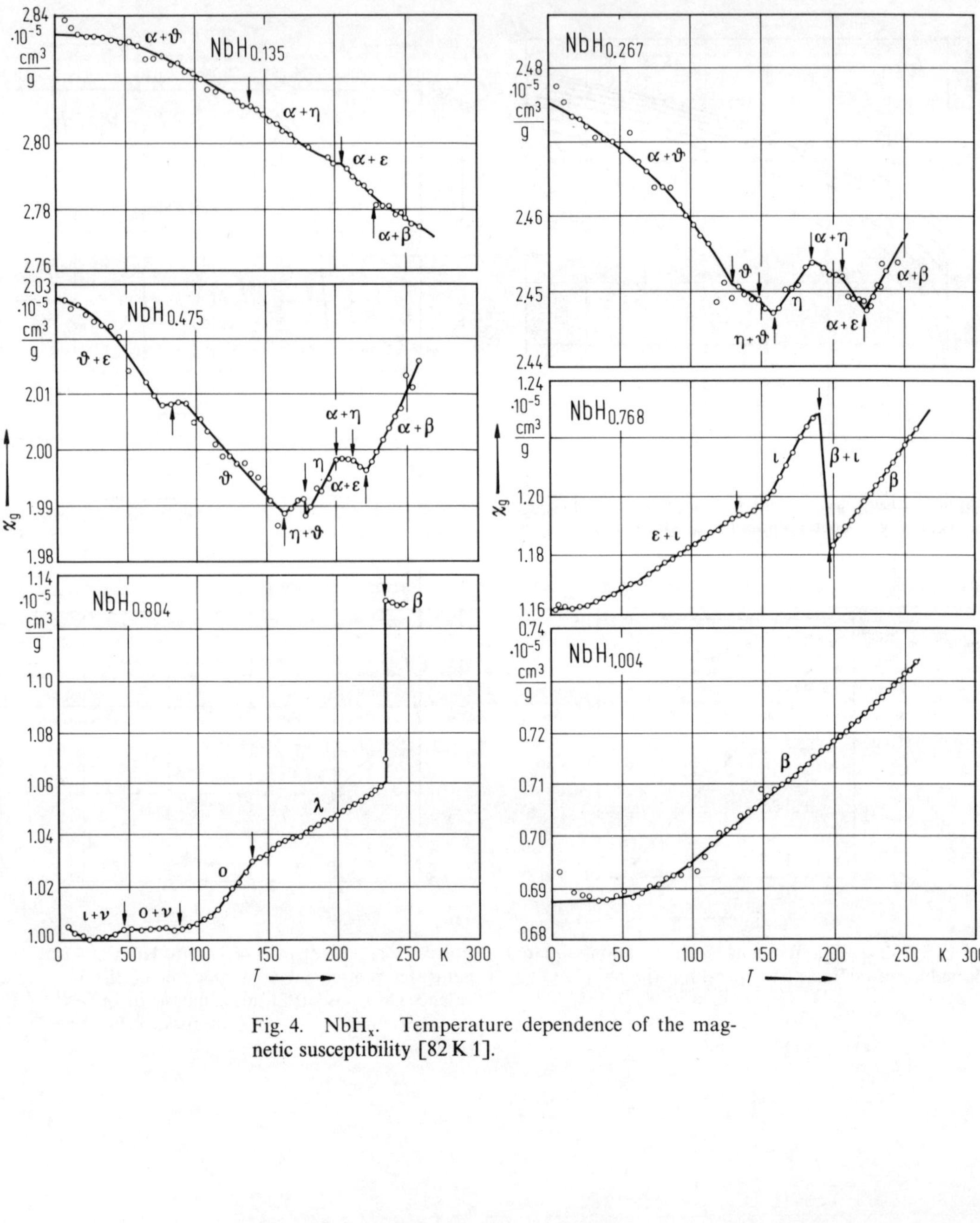

Fig. 4. NbH_x. Temperature dependence of the magnetic susceptibility [82 K 1].

Table 2. Nb–Al. Magnetic susceptibility, superconducting transition temperature and lattice parameters [77 L 1].

	χ_g (300 K)	T_s	a	c
	$10^{-6}\,cm^3\,g^{-1}$	K	Å	
Nb_3Al	2.14(11)	18.8(2)	5.198	
Nb_2Al	1.0(1)	0.74	9.945	5.174
$NbAl_3$	9.90(3)	0.64	3.841	8.641

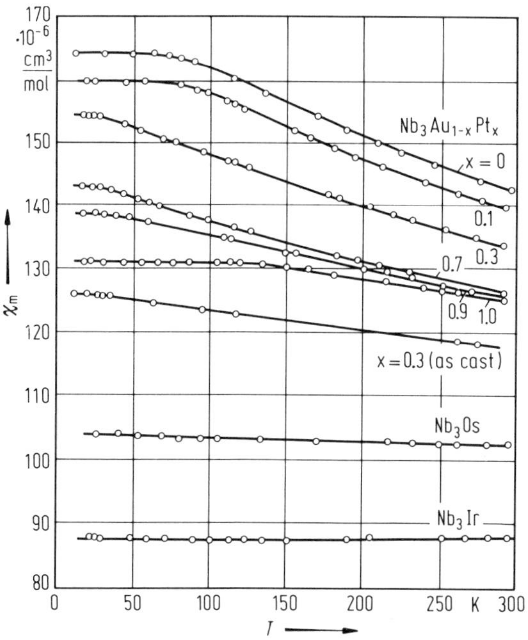

◀ Fig. 5. Nb_3Os, Nb_3Ir, $Nb_3Au_{1-x}Pt_x$. Temperature dependence of the magnetic susceptibility [70 B 1].

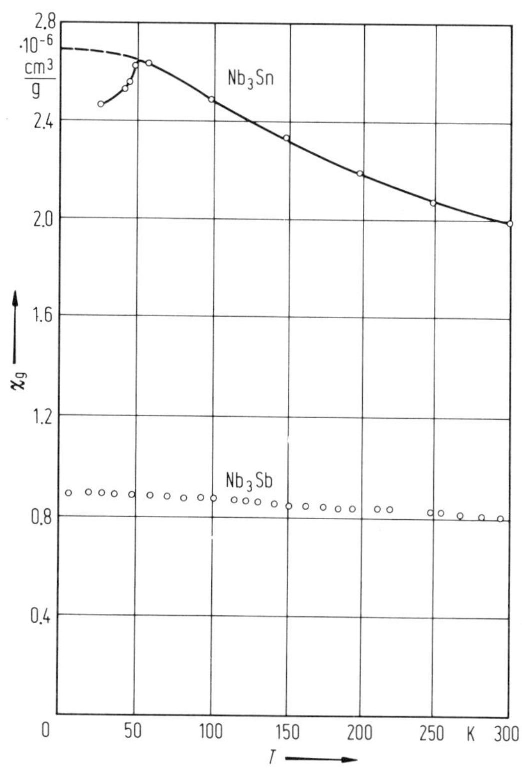

Fig. 6. Nb_3Sn, Nb_3Sb. Temperature dependence of the magnetic susceptibility. Solid line: [72 R 1]; dotted line: [79 W 1].

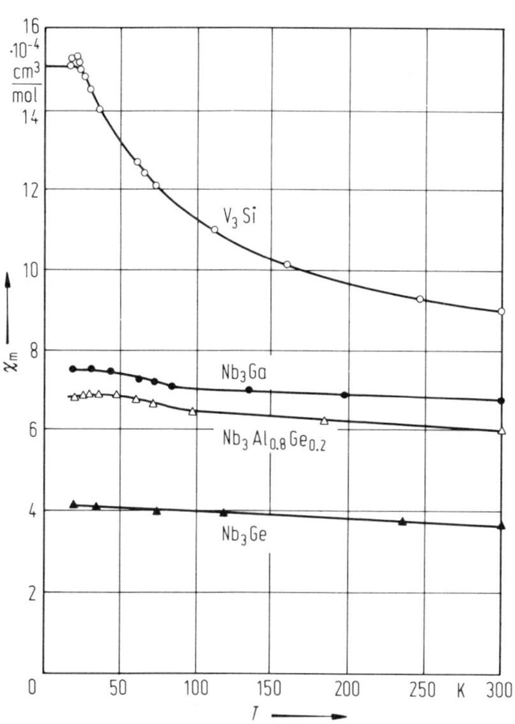

◀ Fig. 7. Nb_3M, V_3Si, M = Ga, Ge, $Al_{0.8}Ge_{0.2}$. Temperature dependence of the magnetic susceptibility of A15 phase alloys of Nb. For reference the data of V_3Si are added [71 E 1].

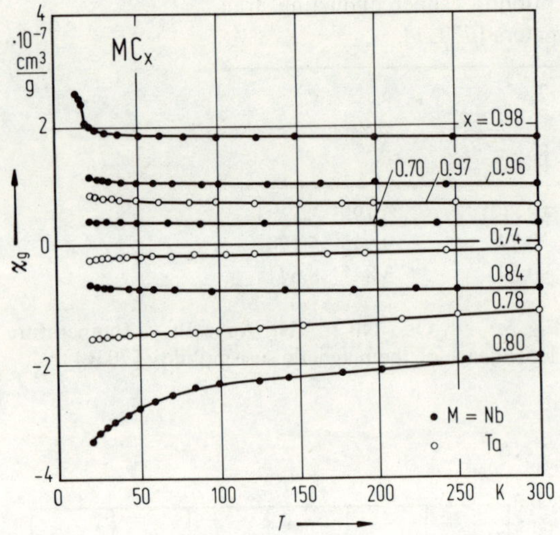

Fig. 8. MC_x, M = Nb, Ta. Temperature dependence of the magnetic susceptibility [69 B 1]. See also Fig. 10.

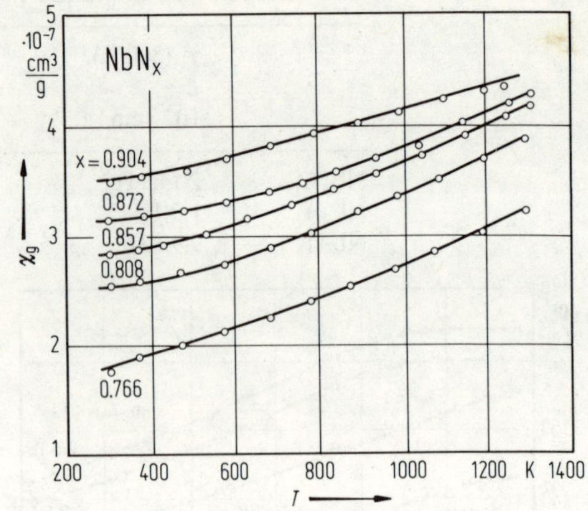

Fig. 9. NbN_x. Temperature dependence of the magnetic susceptibility [81 D 1].

6.3 Compounds of Mo

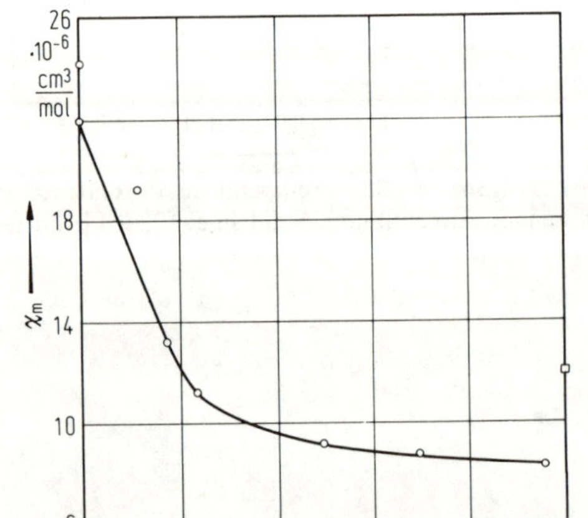

Fig. 10. $(NbC)_{1-x}(Mo_3C_2)_x$. Composition dependence of the magnetic susceptibility at 300 K [75 D 1].

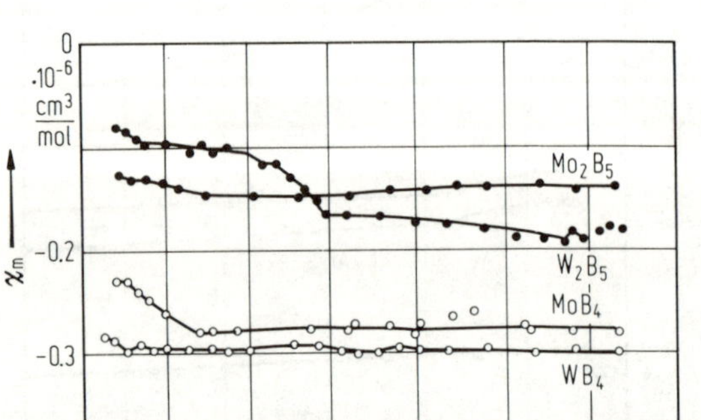

Fig. 11. Mo_2B_5, MoB_4, W_2B_5, WB_4. Temperature dependence of the magnetic susceptibility [79 K 2].

6.4 Alloys and compounds of Pd

On absorbing hydrogen at room temperature, the PdH_x system stays in the α-phase up to $x = 0.025 \cdots 0.030$, while the lattice parameter increases from 3.891 Å to 3.894 Å. When x exceeds 0.030, the system changes to the β-phase in which the lattice of Pd atoms still retains the fcc structure. At $x \approx 0.60$ there appears a first-order phase transition accompanied with a discontinuous change of the lattice parameter to 4.026 Å. Beyond $x \approx 0.60$ hydrogens occupy the octahedral interstitial sites in the lattice. For concentration range $0.03 \lesssim x \lesssim 0.60$, PdH_x is a mixture of the α- and the β-phase.

The magnetic susceptibility of PdH_x shows precisely a linear decrease with x from about $5.5 \times 10^{-4}\,cm^3\,mol^{-1}$ to essentially zero at the β-phase composition of $x = 0.60$, and remains at zero for $x > 0.60$ [78 M 2]. Since a similar magnetic susceptibility vs. concentration behavior has been observed in Pd–Ag alloys

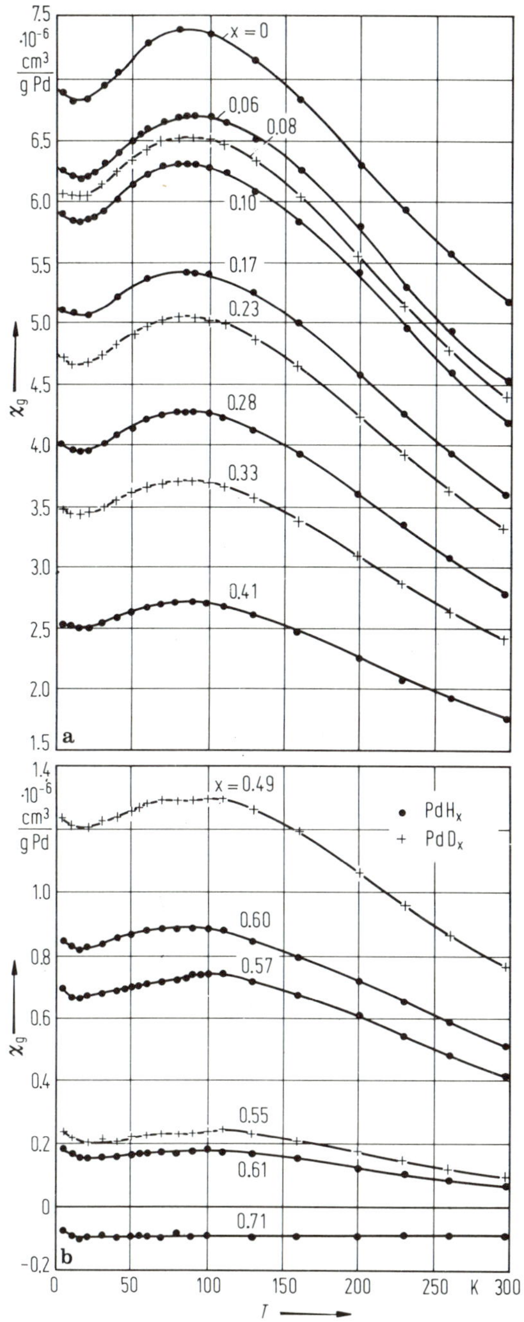

◄ Fig. 12. PdH_x, PdD_x. Temperature dependence of the magnetic susceptibility; (a) $x = 0 \cdots 0.41$, (b) $x = 0.49 \cdots 0.71$ [72 J 1].

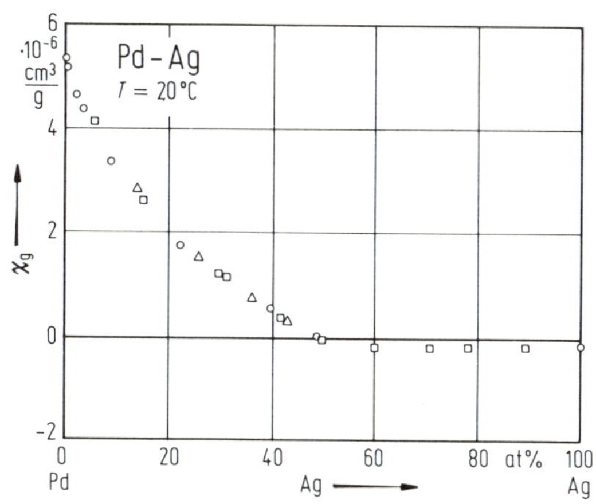

Fig. 13. Pd–Ag. Composition dependence of the magnetic susceptibility of Pd–Ag alloys at 20 °C. Squares: [32 S 1]; triangles: [50 W 1]; circles [53 H 1].

Table 3. Pd–Li, Pt–Li, $NaPd_2$, $NaPt_2$. Magnetic susceptibilities at room temperature [81 L 1].

	χ_m 10^{-6} cm³ mol⁻¹		χ_m 10^{-6} cm³ mol⁻¹
Pd	556	Pt	189
$LiPd_7$	245···188	$LiPt_7$	83···72
$LiPd_2$	90··· 16	$LiPt_2$	15···11
LiPd	4··· 0	LiPt	2(3)
Li_3Pd_2	0(2)		
Li_2Pd	0··· 24	Li_2Pt	2(2)
$Li_{15}Pd_4$	6(2)	$Li_{15}Pt_4$	7(3)
Li_5Pd	28(2)	Li_5Pt	11(4)
		Li_9Pt	16(3)
$NaPd_2$	≃240	$NaPt_2$	≃ 50

Table 4. Pd–In. Basic constants in the Curie-Weiss law of the magnetic susceptibility, $\chi_m = \chi_0 + C_m/(T-\Theta)$ [80 G 1].

Pd at%	χ_0 10^{-6} cm³ mol⁻¹	C_m 10^{-6} cm³ K mol⁻¹	Θ K	p_{eff} μ_B
50	−17.7	600	−11	–
52	−16.5	600	−10	–
54	−16.5	600	− 9	–
56	−16.0	9 700	0	1
58	−12	10 000	0	1

6.5 Compounds of Hf

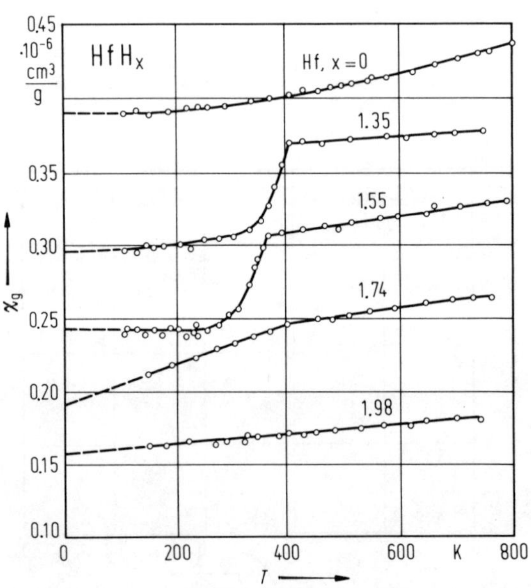

Fig. 14. HfH_x. Temperature dependence of the magnetic susceptibility [83 M 1].

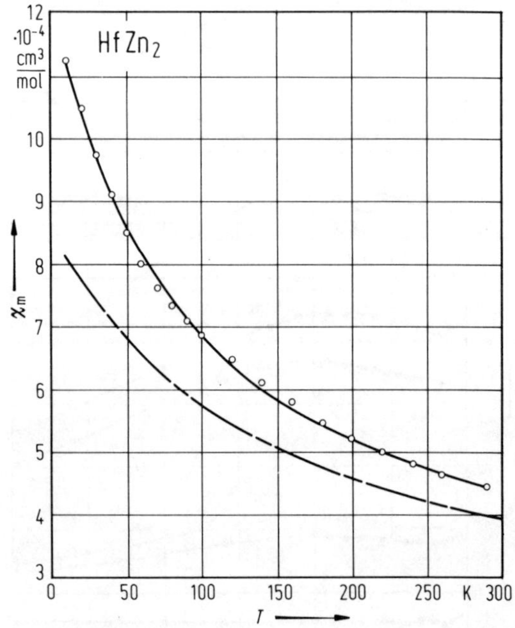

Fig. 15. $HfZn_2$. Temperature dependence of the magnetic susceptibility. Circles: [83 R 1]; dashed line: [71 K 1].

6.6 Compounds of Ta

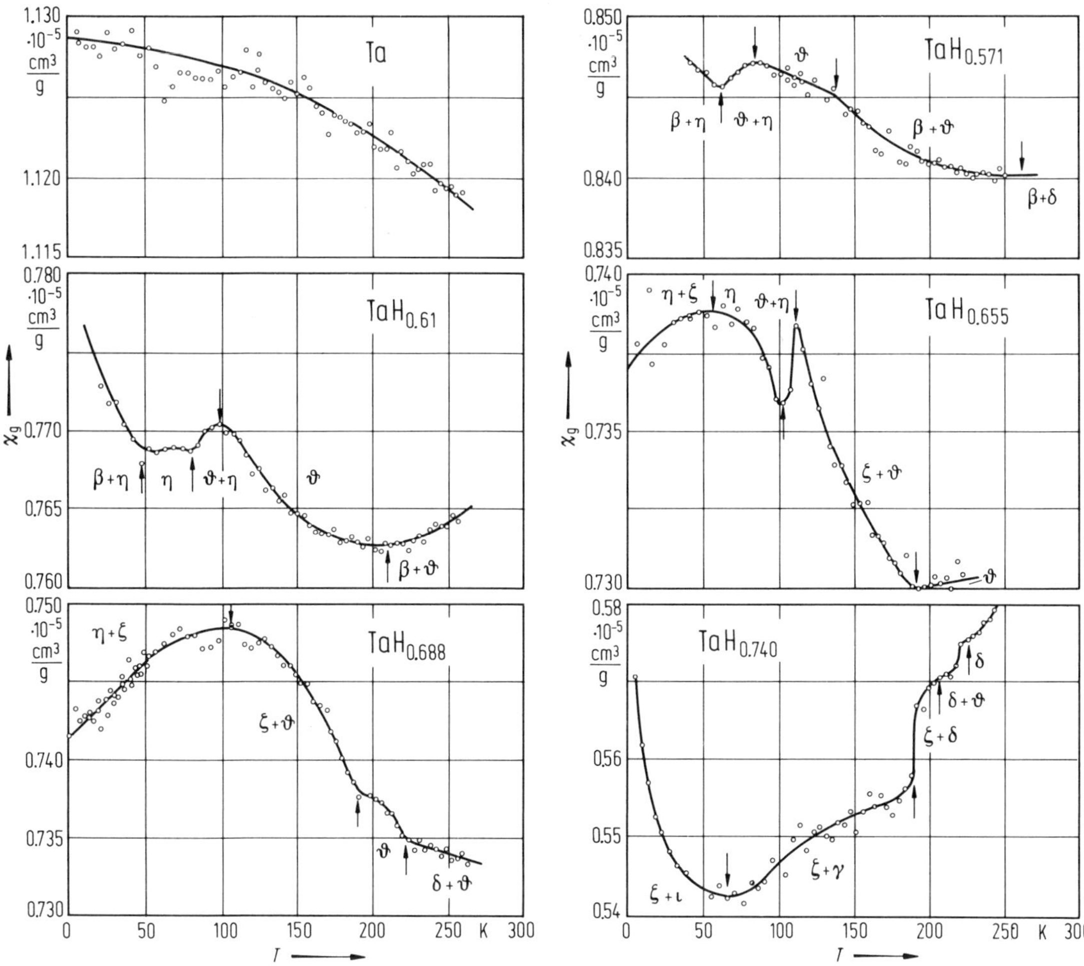

Fig. 16. TaH_x. Temperature dependence of the magnetic susceptibility [78 K 1].

See also Fig. 8.

6.7 Compounds of W

See Fig. 11.

The magnetic susceptibility of WP_4 is $1.73 \cdot 10^{-7}\ cm^3\ g^{-1}$ in the temperature range from liquid nitrogen to room temperature [83 K 1].

6.8 Alloys and compounds of Pt

See Table 3.

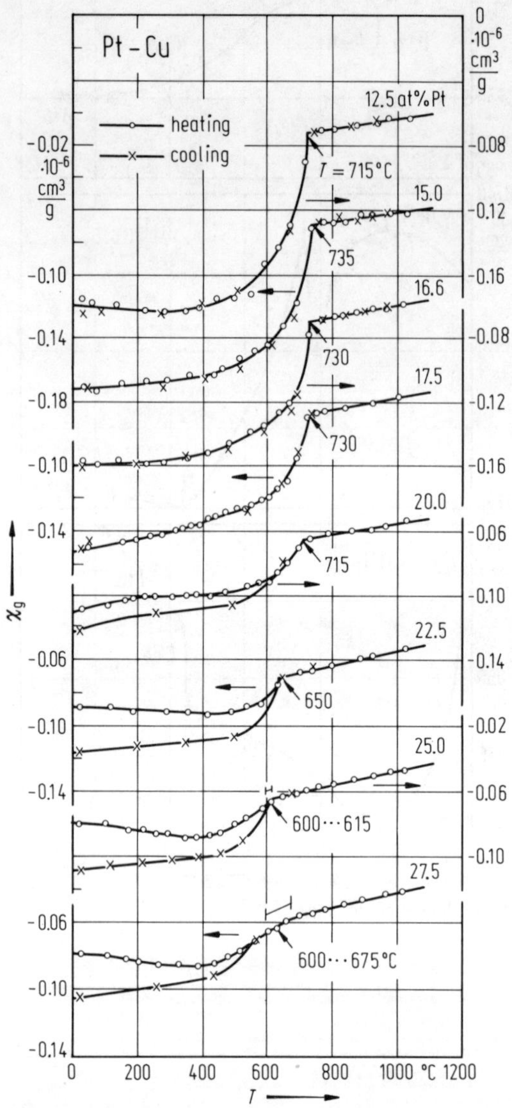

Fig. 17. Pt–Cu. Temperature dependence of the magnetic susceptibility for 12.5···27.5 at% Pt in heating-cooling cycles. Kinks in the magnetic susceptibility curves arise from the order-disorder transition. Prior to measurements, heat-treatments have been done below the transition temperature [76 C 1].

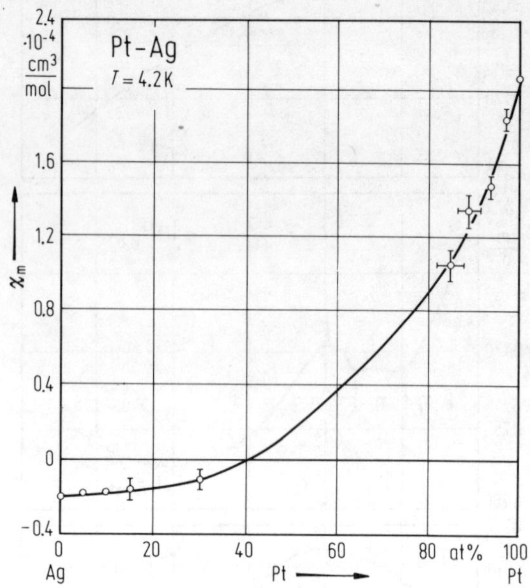

Fig. 18. Pt–Ag. Composition dependence of the magnetic susceptibility at 4.2 K [84 E 1].

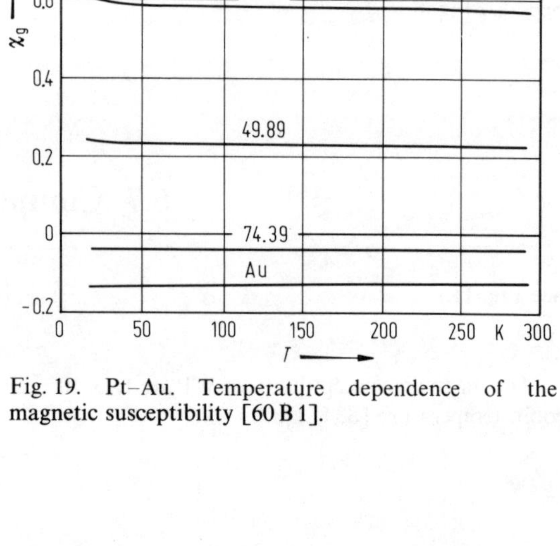

Fig. 19. Pt–Au. Temperature dependence of the magnetic susceptibility [60 B 1].

6.9 References for 6

32 S 1 Svensson, B.: Ann. Phys. (Leipzig) **14** (1932) 699.

50 W 1 Wucher, J.: Thesis, University of Strasbourg **1950**.

53 H 1 Hoare, F.E., Matthews, J.C., Walling, J.C.: Proc. R. Soc. London Ser. A **216** (1953) 502.

60 B 1 Budworth, D.W., Hoare, F.E., Preston, J.: Proc. R. Soc. London Ser. A **257** (1960) 250.

64 P 1 Pickart, S.J., Alperin, H.A., Shirane, G., Nathans, R.: Phys. Rev. Lett. **12** (1964) 444.

68 B 1 Blythe, H.J.: J. Phys. C **1** (1968) 1604.

68 O 1 Ogawa, S.: J. Phys. Soc. Jpn. **25** (1968) 109.

69 B 1 Borukhovich, A.S., Dubrovich, L.B., Dubrovskaya, L.B., Matveenko, I.I., Gel'd, P.V.: Sov. Phys. Solid State (English Transl.) **11** (1969) 681.

70 B 1 Bernasson, M., Descouts, P., Flükiger, R., Treyvaud, A.: Solid State Commun. **8** (1970) 837.

71 E 1 Ehrenfreund, E., Gossard, A.C., Gamble, F.R., Geballe, T.H.: J. Appl. Phys. **42** (1971) 1491.

71 K 1 Knapp, G.S., Veal, B.W., Culbert, H.V.: Int. J. Magn. **1** (1971) 93.

72 J 1 Jamieson, H.C., Manchester, F.D.: J. Phys. F **2** (1972) 323.

72 O 1 Ogawa, S.: Researches of the Electrotechnical Laboratory, Tokyo, Japan, No. 735, **1972**.

72 R 1 Rehwald,W., Rayl, M., Cohen, R.W., Cody, G.D.: Phys. Rev. B **6** (1972) 363.

75 D 1 Dubrovskaya, L.B.,Nazarova, S.Z., Ordanyan, S.S.: Phys. Met. Metallogr. USSR (English Transl.) **39** (1975) No. 4, 188.

76 C 1 Collings, E.W., Smith, R.D., Ho, J.C.: J. Less-Common Met. **46** (1976) 189.

77 L 1 Leyarovski, E., Leyarovska, L., Krasonopyorov, E., Kokot, L., Horyń, R., Mydlarz, T.: Z. Phys. B **27** (1977) 57.

78 K 1 Köbler, U., Schober, T.: J. Less-Common Met. **60** (1978) 101.

78 M 1 Mattocks, P.G., Melville, D.: J. Phys. F **8** (1978) 1291.

78 M 2 Miller, R.J., Brun, T.O., Satterthwaite, C.B.: Phys. Rev. B **18** (1978) 5054.

79 K 2 Koval'chenko, M.S., Bodrova, L.G., Nemchenko, V.F., Kolotun, V.F.: J. Less-Common Met. **67** (1979) 357.

79 W 1 Webb, G.W., Fisk, Z. Johnston, D.C.: Phys. Lett. **73A** (1979) 350.

80 G 1 Galoshina, E.V., Nomerovannaya, L.V., Kirillova, M.M., Shubina, T.S.: Phys. Met. Metallogr. USSR (English Transl.) **50** (1980) No. 1, 50.

81 D 1 Dubrovskaya, L.B., Zhikharev, V.M., Matyushina, L.N.: Phys. Met. Metallogr. USSR (English Transl.) **51** (1981) No. 3, 72.

81 L 1 Loebich, O., Jr., Raub, Ch.J.: Platinum Met. Rev. **25** (1981) 113.

82 K 1 Köbler, U., Welter, J.-M.: J. Less-Common Met. **84** (1982) 225.

83 K 1 Kinomura, N., Terao, K., Kikkawa, S., Koizumi, M.: J. Solid State Chem. **48** (1983) 306.

83 M 1 Markin, V.Ya., Savin, V.I.: Neorg. Mater. **19** (1983) 484.

83 R 1 Radousky, H.B., Knapp, G.S., Downey, J.W., Aldred, A.T., Freeman, A.J.: J. Magn. Magn. Mater. **40** (1983) 117.

84 B 1 Brown, P.J., Ziebeck, K.R.A., Mattocks, P. G.: J. Magn. Magn. Mater. **42** (1984) 12.

84 E 1 Ebert, H., Abart, J., Voitländer, J.: J. Phys. F **14** (1984) 749.

7 Heusler alloys

In 1903 Heusler [03 H 1] first reported that Mn–Cu bronze can be alloyed with Sn, Al, As, Sb, Bi or B to produce ferromagnetic alloys even though none of the constituent elements is itself ferromagnetic.

Heusler alloys are now defined as ternary intermetallic compounds, at the stoichiometric composition X_2YZ, which have the $L2_1$ structure.

In the Heusler, $L2_1$, and related structures the constituent atoms which may carry a magnetic moment can occupy simple cubic, body-centred cubic and face-centred cubic sublattices.

The majority of Heusler alloys order ferromagnetically and saturate in applied magnetic fields of less than 5 kOe at low temperatures. Some do order antiferromagnetically, particularly those compounds containing a group 3B element in which the magnetic moment is confined to the Y sites, usually occupied by Mn. If the magnetic moment is confined to the Mn sites, as it often is in the alloys X_2MnZ, a value close to $4\mu_B$ is usually observed. Although they are metals, these intermetallic compounds have localised magnetic properties and are ideal model systems for studying the effects of both atomic disorder and changes in the electron concentration on magnetic properties.

7.1 Ferromagnets X_2MnZ with X = d element

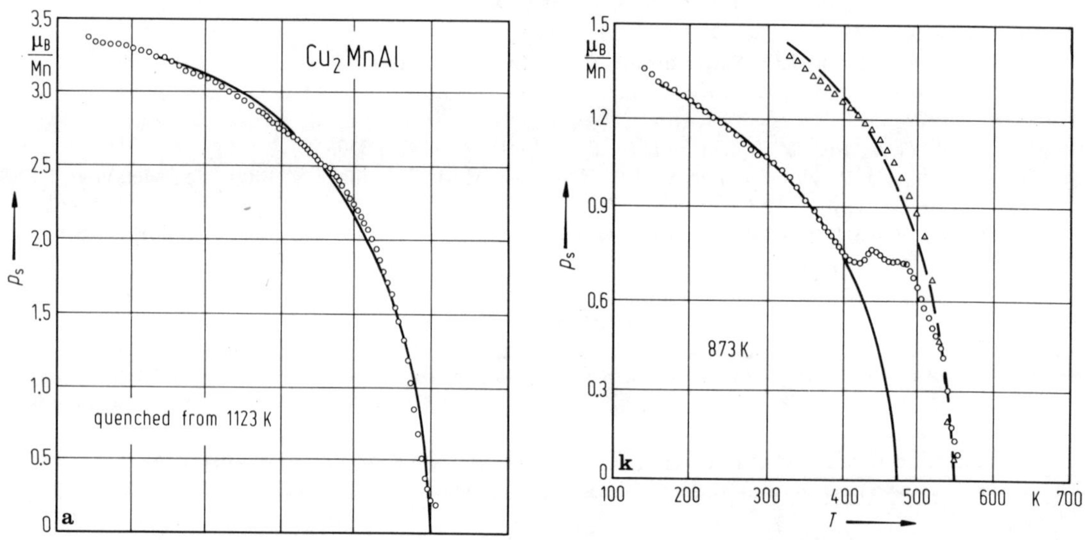

Fig. 1. Cu_2MnAl. Magnetisation vs. temperature. Circles and triangles denote the heating and cooling cycles, respectively, and the dashed and full curves represent a Brillouin function fit to the data [83 K 1]. (a) quenched from 1123 K, (k) followed by ageing at 873 K.

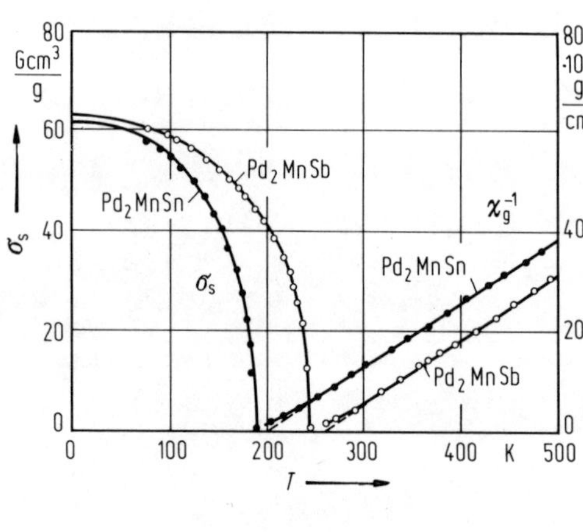

◀ Fig. 2. Pd_2MnSn, Pd_2MnSb. Spontaneous magnetisation and inverse static magnetic susceptibility vs. temperature [68 W 2].

Table 1. Structural and bulk magnetic properties of intermetallic compounds X_2MnZ. q: quenched, sc: slow-cooled.

	Structure	a (c)	T_C	T_N	Θ	p_s	p	p_{eff}	Ref.
		nm	K			μ_B/f.u.			
Cu_2MnAl	$L2_1$	0.5949	603			4.12			63 O 1
Cu_2MnIn	$L2_1$	0.6206	500			3.95			49 C 1
Cu_2MnSn	$L2_1$	0.6173	503			4.11			63 O 1
Ni_2MnAl (q)	$B2/L2_1$	0.5822		295	179				75 Z 1
Ni_2MnAl (sc)	$B2/L2_1$	0.5821		300	240				75 Z 1
Ni_2MnGa	$L2_1$	0.5825	374		381	4.17	3.86	4.75	84 W 1
Ni_2MnIn	$L2_1$	0.6069	314		318	4.43	4.24	5.14	69 W 1
Ni_2MnSn	$L2_1$	0.6053	360		357	4.22	4.18	5.07	69 W 1
Ni_2MnSb	$L2_1$	0.6004	365		408	3.52	3.18	4.06	84 W 2
Co_2MnAl	$B2/L2_1$	0.5756	693			4.01			71 W 1
Co_2MnSi	$L2_1$	0.5654	985			5.07			71 W 1
Co_2MnGa	$L2_1$	0.5770	694			4.05			71 W 1
Co_2MnGe	$L2_1$	0.5743	905			5.11			71 W 1
Co_2MnSn	$L2_1$	0.6000	829		856	5.08		5.29	71 W 1
$Co_{1.5}MnSb$	$L2_1/C1_b$	0.5929	600			4.90			71 W 1
Fe_2MnAl	$L2_1$	0.5816				1.58			81 B 1
Fe_2MnSi	$L2_1$	0.5663	214	69	238	1.76		4.56	76 Z 1
Pd_2MnAl	B2	0.6165		240	−215	4.4			68 W 2
Pd_2MnGe	$L2_1$ +	0.6174	170			3.2			68 W 1
Pd_2MnIn (q)	$B2/L2_1$	0.6373			50	4.3			67 W 1
Pd_2MnIn (sc)	$L2_1/B2$	0.6373		142	− 50	4.3			67 W 1
Pd_2MnSn	$L2_1$	0.6380	189		201	4.23			67 W 1
Pd_2MnSb	$L2_1$	0.6419	247		259	4.40			67 W 1
Rh_2MnAl	B2	0.6005	~ 95			>0.7	~3.3	~4.1	76 S 1
Rh_2MnGa	B2	0.6054	~ 80		60	>1.2	3.35	4.23	76 S 1
Rh_2MnGe	$L2_1$	0.5993	450		460	4.3	3.75	4.64	71 H 1
Rh_2MnIn	B2	0.6287	~105		81	>2.3	3.4	4.28	76 S 1
Rh_2MnSn	$L2_1$	0.6252	412		413	3.1	3.3	4.19	76 S 1
Rh_2MnSb	Tet	0.5898 (0.6897)	335		312				76 S 1
Rh_2MnTl	B2	0.6324			90	>2.2	3.86	4.75	76 S 1
Rh_2MnPb	$L2_1$	0.6332	338		348	4.12	3.6	4.49	76 S 1
Au_2MnAl	$L2_1$	0.636	233	64		4.2			67 B 1
Au_2MnGa	$L2_1$			AF					73 B 1
Au_2MnIn	$L2_1$			AF					73 B 1
Au_2MnZn	B2	0.636	253			4.6			73 B 1
Au_2MnCu	B2			AF					73 B 1
Pt_2MnAl	$L2_1+(L1_0)$	0.624		190					71 H 1
Pt_2MnGa	$L2_1$	0.616		75					71 H 1
Ir_2MnAl	$L2_1$								72 M 1

7.2 Antiferromagnets X_2MnZ with X = 3d element

Ni_2MnAl

As with many intermetallic compounds X_2MnAl considerable Mn–Al, B2 type, disorder occurs in Ni_2MnAl. The proportion of $L2_1/B2$ order depends on heat-treatment. Complete B2 disorder is obtained in samples quenched from 800 °C and approximately 68% B2 disorder in slow-cooled samples. Both slow-cooled and quenched samples show a maximum in susceptibility close to 300 K, indicating antiferromagnetic behaviour. Below 300 K the susceptibility shows field dependency suggesting a complex antiferromagnetic structure. This is confirmed by neutron diffraction measurements made at 4.2 K which reveal an incommensurate antiferromagnetic cone spiral structure.

Pd_2MnZ: Z = Al, In

The Pd-based Heusler alloys were the first in which antiferromagnetism was observed. Metallurgically the series is particularly tractable since single-phase compounds readily form and the degree of atomic order is sensitive to heat-treatment. Pd_2MnAl exhibits complete B2 disorder [68 W 2] whereas the atomic order in Pd_2MnIn is intermediate between B2 and $L2_1$ [67 W 1]. The type of atomic order existing strongly affects the type of antiferromagnetic structure observed and the ordering temperatures but does not affect the magnitude of the Mn moment (Table 1). For the B2 structure direct Mn–Mn interactions are believed to dominate whereas in the $L2_1$ structure the larger Mn–Mn separation necessitates an antiferromagnetic interaction via the Pd or In atoms [68 W 2] (Figs. 3, 5).

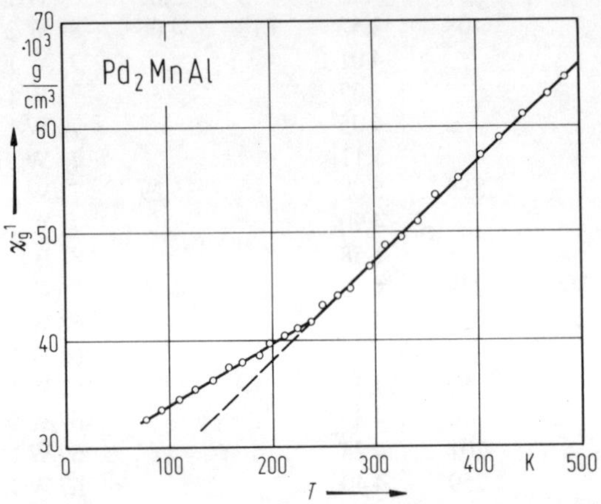

Fig. 3. Pd_2MnAl. Reciprocal static magnetic susceptibility vs. temperature. Pd_2MnAl has the B2 structure with Mn/Al disorder [68 W 2].

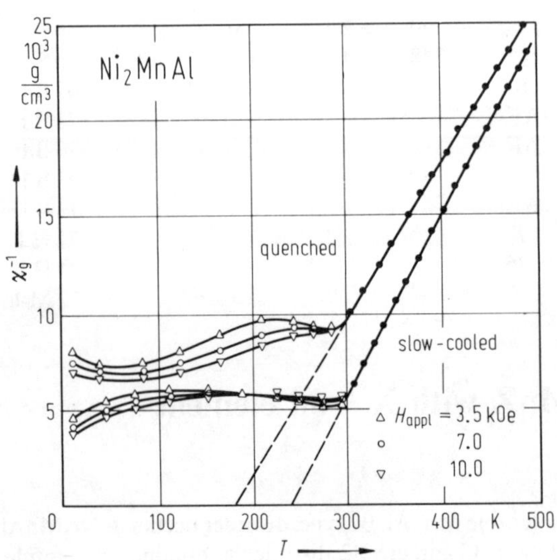

Fig. 4. Ni_2MnAl. Reciprocal static magnetic susceptibility vs. temperature for samples quenched and slow-cooled 800 °C [75 Z 1].

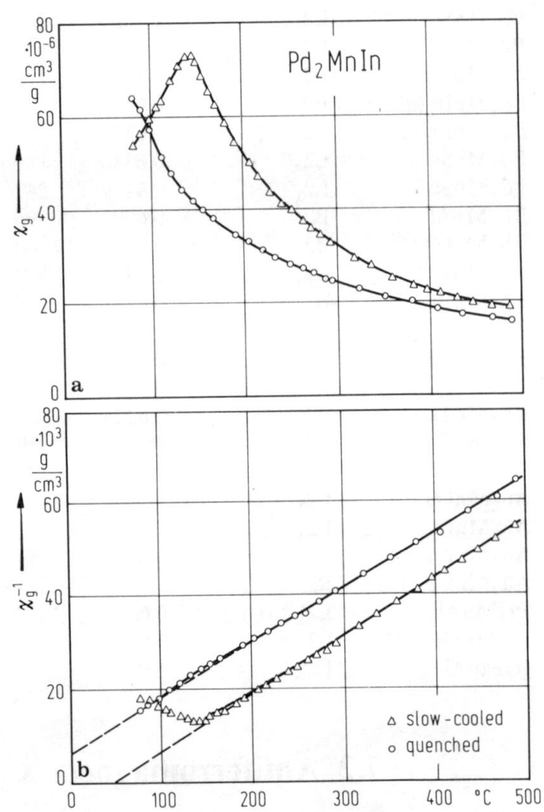

Fig. 5. Pd_2MnIn. (a) Static magnetic susceptibility and (b) reciprocal static magnetic susceptibility vs. temperature for samples slow-cooled and quenched from 800 °C. The slow-cooled sample is predominantly ordered in the $L2_1$ structure but the quenched sample is disordered into the B2 structure [67 W 1].

7.3 Ferromagnets X_2YZ for $Y \neq Mn$

Generally single-phase Heusler alloys form readily in these series, although it is not possible to form such alloys containing Si or Ge, and Co_2TiSb is two-phase as it is beyond the solubility level for Co. The majority of single-phase alloys that have been investigated were annealed at 800 °C and then quenched. In common with other series some of the alloys containing group 3B elements, particularly when Z = Al, exhibit preferential B2 type disorder, e.g. Ti–Al and V–Al. All the intermetallic compounds order ferromagnetically and saturate in fields of about 3 kOe. Neutron diffraction measurements indicate that the magnetic moments are confined to the Co sites [73 W 1, 74 Z 1]. Above the Curie temperature the thermal variation of the static susceptibility is consistent with Curie-Weiss behaviour [72 Z 1]. The results may be broadly classified into two groups, those with relatively large magnetic moments and higher Curie temperatures and those with smaller magnetic moments and correspondingly lower Curie temperatures. The intermetallic compounds containing group 4A and 4B elements or group 5A and 3B elements have the relatively large magnetic moments and Curie temperatures. It may be seen (Table 2) that the paramagnetic moment, p, is close to the ground state value. Compounds containing group 4A and 3B elements or group 5A and 4B elements have smaller magnetic moments and lower Curie temperatures but have paramagnetic moments greater than the ground state value. Maxima in magnetisation occur when the electronic contributions of the Y and Z atoms average to that expected from group 4 elements. Departures from this average lead to reductions in magnetisation.

In these compounds the magnetic atoms (Co) are separated by a distance $a/2$ so that direct magnetic exchange is expected to be important and the magnetic properties are expected to become more itinerant than in the classical Heusler alloys in which the magnetic moments are located on fcc Mn sites separated by $a/\sqrt{2}$. As many of these ferromagnetic Heusler alloys are formed from two binary, nonmagnetic, B2 compounds; e.g. CoTi and CoAl form Co_2TiAl, and the separation of the Co atoms remains essentially unchanged on combination, the onset of ferromagnetism must be due to the change in electronic environment of the Co sites. This change is two-fold. In Co_2TiAl, for example, the nearest neighbour Co environment becomes 4 Ti and 4 Al atoms, chemically ordered in the $L2_1$ array, instead of either 8 Ti or 8 Al atoms in the binary B2 compounds. Additionally there is also an important change in the electron per atom ratios, from those due to Ti or Al separately to the average due to the two combined. The onset of ferromagnetism in these B2–$L2_1$ systems has been investigated in two solid solutions, $Co_2Ti_{1-x}Al_x$ [77 E 1, 80 W 1] and $Co_2Ti_{1-x}Ga_x$ [75 B 1].

X-ray measurements indicate that Heusler alloys form at the compositions Fe_2NiAl, Fe_2NiGa, Fe_2CoGa, Fe_2CrAl, and Fe_2CrGa. Their bulk magnetic properties are listed in Table 2. In the series containing Ni and Cr there is a large difference between the magnetic moments observed in compounds containing Al and those containing Ga. It is also noticed that the total magnetic moment of Fe_2CoGa is close to that for Co_2FeGa [81 B 1].

The $D0_3$ compounds Fe_3Z may be regarded as Heusler alloys Fe_2FeZ as there are two distinct crystallographic and magnetic Fe sites. The three compounds that have been investigated form part of solid solutions extending from pure bcc Fe to the $D0_3$ composition and beyond. They are considered to be of particular interest since they enable a quantitative investigation of the environmental effects on the magnetic properties of elemental Fe to be made. In the $D0_3$ structure one of the Fe sites has a nearest neighbour environment of 8 Fe atoms with the other having 4 Fe and 4 Z atoms as nearest neighbours. Magnetic data are represented in Table 2. It is seen that the effect on the magnetic moment and Curie temperature of adding Si is more severe than the corresponding decrease observed when either Al or Ga is added. Furthermore, the $D0_3$ structure, which starts to form at 10 at% Si [58 P 1] is extremely stable and insensitive to heat-treatment. The phase diagram of the Fe_3Al series is, in contrast, complicated since the degree of atomic order is very sensitive to heat-treatment. Depending upon the composition and heat-treatment, the disordered bcc A2 structure (α), the B2 structure (α'), or the $D0_3$ structure (α''), or mixtures thereof, can be obtained in the vicinity of Fe_3Al stoichiometry. Since the system is metallurgically complex, it being difficult to attain homogeneity, short-range order and clustering substantially affect the observed magnetic properties. Diverse magnetic behaviour occurs in the series, including mictomagnetism, ferromagnetism and antiferromagnetism [76 S 2].

The Fe–Ga system is characterised by a large solubility of Ga in Fe at high temperatures, with some intermetallic compounds stable, or metastable, at room temperature. An ordered fcc (β) phase, Fe_3Ga, forms in the composition range 25 ··· 30 at% Ga with the $L1_2$, Cu_3Au type, structure. However, Fe–Ga solid solutions with up to 50 at% Ga may be retained at room temperature by rapid quenching from the molten state. These bcc alloys are found to form the $D0_3$ (α'') and B2 (α') structures in the composition ranges 25 ··· 30 at% and 34 ··· 50 at% Ga, respectively [72 K 1]. The magnetic properties are summarised in Table 2.

Since the Fe_3Si system forms the most stable and highly ordered arrangement it lends itself most readily to quantitative studies and interpretations and has been used as a host lattice for investigating the replacement of Fe by another transition element.

Compounds at the stoichiometric composition Mn_2VAl and Mn_2VGa have the Heusler $L2_1$ structure, with the Mn atoms occupying the X sites of symmetry m3m. Although the Mn–Mn separation is approximately 0.294 nm, and the direct exchange is expected to be antiferromagnetic, both the compounds are ferromagnets.

Table 2. Structural and bulk magnetic properties of intermetallic compounds, X_2YZ, Y $\neq$ Mn. Asterisk denotes total spin moment per formula unit, $S_C + 2S_{B/D}$.

	Structure	a	T_C	T_N	Θ	p_s	p	p_{eff}	Ref.
		nm	K			μ_B/f.u.			
Co_2TiAl	$L2_1$	0.5848	138		145	0.71	0.8	1.47	73 W 1
Co_2TiGa	$L2_1$	0.5848	130		134	0.75	0.76	1.44	73 W 1
Co_2TiSn	$L2_1$	0.6073	359		376	1.93	0.9	1.62	73 W 1
Co_2VAl	$L2_1$/B2	0.5770	310		320	1.65	0.92	1.64	74 Z 1
Co_2VGa	$L2_1$	0.5786	349		360	1.95	0.8	1.5	74 Z 1
Co_2VSn	$L2_1$	0.5980	70			0.56	0.8	1.5	72 F 1
Co_2ZrAl	$L2_1$	0.6082	185		190	0.61	0.66	1.33	74 Z 1
Co_2ZrSn	$L2_1$	0.6249	444		449	1.51	0.81	1.51	74 Z 1
Co_2NbAl	$L2_1$	0.5935	383			1.35			81 B 1
Co_2NbGa	$L2_1$	0.5950				1.39			81 B 1
Co_2NbSn	$L2_1$	0.6153	119		124	0.52	0.42	1.01	74 Z 1
Co_2HfAl	$L2_1$ +	0.6019	193		203	0.81	0.86	1.58	74 Z 1
Co_2HfGa	$L2_1$	0.6032	186		188	0.54	0.62	1.29	74 Z 1
Co_2HfSn	$L2_1$	0.6218	394		420	1.55	0.94	1.66	74 Z 1
Co_2TaAl	$L2_1$	0.5930	260			1.5			81 B 1
Co_2CrAl	$L2_1$	0.5887	334			1.55			81 B 1
Co_2CrGa	$L2_1$	0.5805							81 B 1
Co_2FeAl	$L2_1$	0.5730							81 B 1
Co_2FeSi	$L2_1$	0.5658	>980			5.9			81 B 1
Co_2FeGa	$L2_1$	0.5737							81 B 1
Fe_2CoGa	$L2_1$	0.5767				5.09			81 B 1
Fe_2VAl	$L2_1$	0.5762			4				83 W 1
Fe_2VSi	$L2_1$	0.5670			P				76 N 1
Fe_2VGa	$L2_1$	0.5776			16				83 W 1
Fe_2NiAl	$L2_1$	0.5758				4.25			81 B 1
Fe_2NiGa	$L2_1$	0.5780				3.21			81 B 1
Fe_2CrAl	$L2_1$	0.5805	246			1.67			81 B 1
Fe_2CrGa	$L2_1$	0.5824				2.6			81 B 1
Fe_3Si [1])	DO_3	0.5653	823			2.21 *)			68 S 1, 72 M 2
Fe_3Al [2])	DO_3	0.5792	713			2.59 *)			68 S 1
Mn_2VAl	$L2_1$	0.5932				1.82			83 N 1
Mn_2VGa	$L2_1$	0.6095				1.66			83 N 1
Ni_2TiSn	$L2_1$	0.609			P				72 F 1
Ni_2VSn	$L2_1$	0.604			P				72 F 1

[1]) See also chapter 5.4.4

[2]) See also chapter 5.3.3

7.4 Paramagnets X_2YZ with X, Y = 3d element

Both Ni_2TiSn and Ni_2VSn exhibit a temperature-independent paramagnetism with susceptibilities $1.56 \cdot 10^{-6}$ and $0.7 \cdot 10^{-6}$ $cm^3 g^{-1}$, respectively [72 F 1]. The static susceptibilities of both Fe_2VAl and Fe_2VGa are field-independent down to 4.2 K [83 W 1], with a non-Curie-Weiss temperature dependence. The inverse susceptibility gives rise to a smooth curve, bending towards the temperature axis.

7.5 References for 7

03 H 1 Heusler, F.: Verh. Dtsch. Phys. Ges. **5** (1903) 219.
49 C 1 Coles, B.R., Hume-Rothery, W., Myers, H.P.: Proc. R. Soc. London Ser. A **196** (1949) 125.
58 P 1 Parson, D., Sucksmith, W., Thomas, J.E.: Philos. Mag. **34** (1958) 1174.
63 O 1 Oxley, D.P., Tebble, R.S., Williams, K.C.: J. Appl. Phys. **34** (1963) 1362.
67 B 1 Bacon, G.E., Mason, E.W.: Proc. Phys. Soc. **92** (1967) 713.
67 W 1 Webster. P.J., Tebble, R.S.: Philos. Mag. **16** (1967) 347.
68 S 1 Stearns, M.B.: Phys. Rev. **168** (1968) 588.
68 W 1 Webster, P.J.: The magnetic and chemical structure of the Heusler alloys, Thesis, Sheffield University **1968**.
68 W 2 Webster, P.J., Tebble, R.S.: J. Appl. Phys. **39** (1968) 471.
69 W 1 Webster, P.J.: Contemp. Phys. **10** (1969) 559.
71 H 1 Hames, F.A., Crangle, J.: J. Appl. Phys. **42** (1971) 1336.
71 W 1 Webster, P.J.: J. Phys. Chem. Solids **32** (1971) 1221.
72 F 1 Fujita, Y., Endo, K., Terada, M., Kimura, R.: J. Phys. Chem. Solids **33** (1972) 1443.
72 K 1 Kawamiya, N., Adachi, K., Nakamura, Y.: J. Phys. Soc. Jpn. **33** (1972) 1318.
72 M 1 Masumoto, H., Watanabe, K.: J. Phys. Soc. Jpn. **32** (1972) 281.
72 M 2 Moss, J., Brown, P.J.: J. Phys. F **2** (1972) 358.
72 Z 1 Ziebeck, K.R.A.: Magnetic and chemical order in Heusler alloys, Thesis, Salford University **1972**.
73 B 1 Bacon, G.E., Plant, J.S.: J. Phys. F **3** (1973) 2003.
73 W 1 Webster, P.J., Ziebeck, K.R.A.: J. Phys. Chem. Solids **34** (1973) 1647.
74 Z 1 Ziebeck, K.R.A., Webster, P.J.: J. Phys. Chem. Solids **35** (1974) 1.
75 B 1 Booth, J.G., Pritchard, R.G.: J. Phys. F **5** (1975) 347.
75 Z 1 Ziebeck, K.R.A., Webster, P.J.: J. Phys. F **5** (1975) 1756.
76 N 1 Niculescu, V.A., Raj, K., Budnick, J.I., Burch, T.J., Hines, W.A., Menotti, A.H.: Phys. Rev. B **14** (1976) 4160.
76 S 1 Suits, J.C.: Phys. Rev. B **14** (1976) 4131.
76 S 2 Shull, R.D., Okamoto, H., Beck, P.A.: Solid State Commun. **20** (1976) 863.
76 Z 1 Ziebeck, K.R.A., Webster, P.J.: Philos. Mag. **34** (1976) 973.
77 E 1 Endo, K., An, I.C., Shinogi, A.: J. Phys. Lett. F **7** (1977) 99.
80 W 1 Waterman, E.H., Franse, J.J.M.: J. Phys. F **10** (1980) 947.
81 B 1 Buschow, K.H.J., van Engen, P.G.: J. Mag. Magn. Mater. **25** (1981) 90.
83 K 1 Kozubski, R., Soltys, J.: J. Mat. Science **18** (1983) 1689.
83 N 1 Nakamichi, T., Stager, C.V.: J. Mag. Magn. Mater. **31–34** (1983) 85.
83 W 1 Webster, P.J., Ziebeck, K.R.A.: Phys. Lett. A **98** (1983) 51.
84 W 1 Webster, P.J., Ziebeck, K.R.A., Town, S.L., Peak, M.S.: Philos. Mag. B **49** (1984) 295.
84 W 2 Webster, P.J., Mankikar, R.M.: J. Mag. Magn. Mater. **42** (1984) 300.

8 Metallic perovskites

8.1 Introduction

Earlier surveys of the magnetic properties of perovskites have been given in the Landolt-Börnstein volumes III/4a (subsect. 3.0.2) and III/12a (subsect. 2.0.3) by Goodenough and Longo [70 h 1], and by Nomura [78 h 1], respectively.

In spite of its very simple structure type (perovskite) the series of metallic compounds $M^cXM^f_3$ (X being mostly B, C or N) exhibits systematic anomalies in their structural and physical behaviours. Generally, at least one magnetic element is present in the structure, either on the M^f metal site (3d or rare earth metal) or on the M^c metal site (rare earth or actinide metal). The largest number of members in this family is found for M^f = Mn and M^c = 3d, 4d, 5d elements.

The symmetry classification for the perovskite structure $M^cXM^f_3$ is C = cubic (Pm3m) with tetragonal (T) and orthorhombic (O) distortions:

- T_1 = tetragonal (P4/mmm), either $c/a>1$ (T_1^+) or $c/a<1$ (T_1^-), the unit cell having the same volume as the chemical cell;
- T_2 = tetragonal with lattice parameters $a\sqrt{2}$, $a\sqrt{2}$, c relative to the chemical cell, the unit cell thus having twice the volume of the chemical cell [69 B 1];
- T_4 = tetragonal (I4/mcm), $c/a<1$, with alternate rotation of octahedra XM^f_6 along z-axis, the unit cell having 4 times the volume of the chemical cell [68 B 1];
- O_8 = orthorhombic (Fmmm), resulting from a combination of T_1^+ and T_4 along 2 orthogonal axes, the unit cell having 8 times the volume of the chemical cell [72 B 1].

The magnetic structures are generally complex noncollinear structures as the result of competiting direct and indirect interactions, long-range and local effects. They are dramatically modified at the boundaries of the phase diagrams. The family of metallic perovskites is a unique example of so a large series of metallic compounds, existing with so many different elements and exhibiting such a great number of lattice and phase diagram instabilities of electronic origin.

A survey of the relevant magnetic structures of the metallic perovskites is given in the Figs. 1a–k [76 F 1, 77 F 1, 78 F 1]. The analysis of the magnetic configurations is discussed in [80 M 1, 81 B 1].

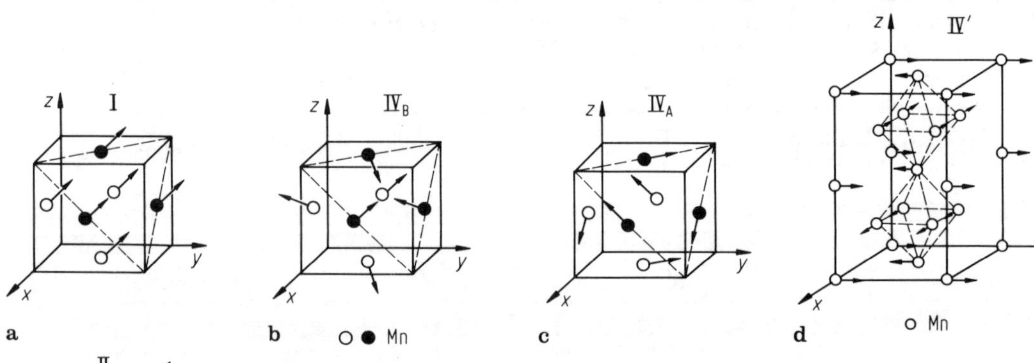

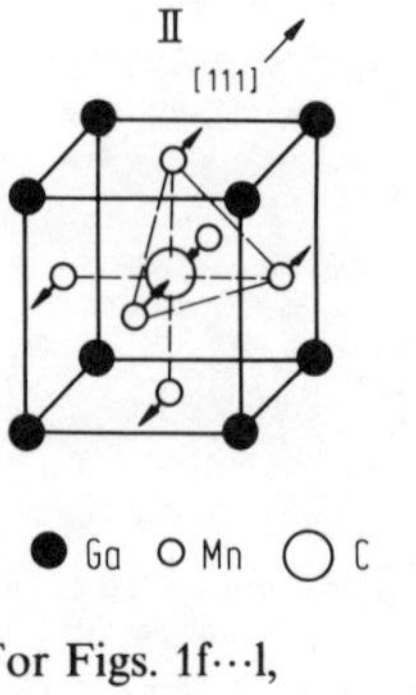

For Figs. 1f···l, see next page

Fig. 1. (a) Cubic and ferromagnetic structure (derived from the Γ_{4g} irreducible representation of Pm3m). It is observed in $ZnCMn_3$, $AlCMn_3$, $GaCMn_3$, $GaC_{0.935}Mn_3$ at high temperature. It is also the ferromagnetic mode associated with the antiferromagnetic one (b) encountered in $NiNMn_3$, $AgNMn_3$, $SnNMn_3$, when a rotation of spin occurs. Structure type designated I.

(b) Triangular antiferromagnetic mode (derived from the Γ_{4g} representation of Pm3m) as observed in $NiNMn_3$, $AgNMn_3$, $SnNMn_3$, associated with the collinear type of (a). A slight triangular antiferromagnetic component was also detected in Mn_4N on Mn^f(1, 2, 3) sites using polarized neutrons [76 F 1]. Structure type designated IV_B.

Fig. 1f···l.

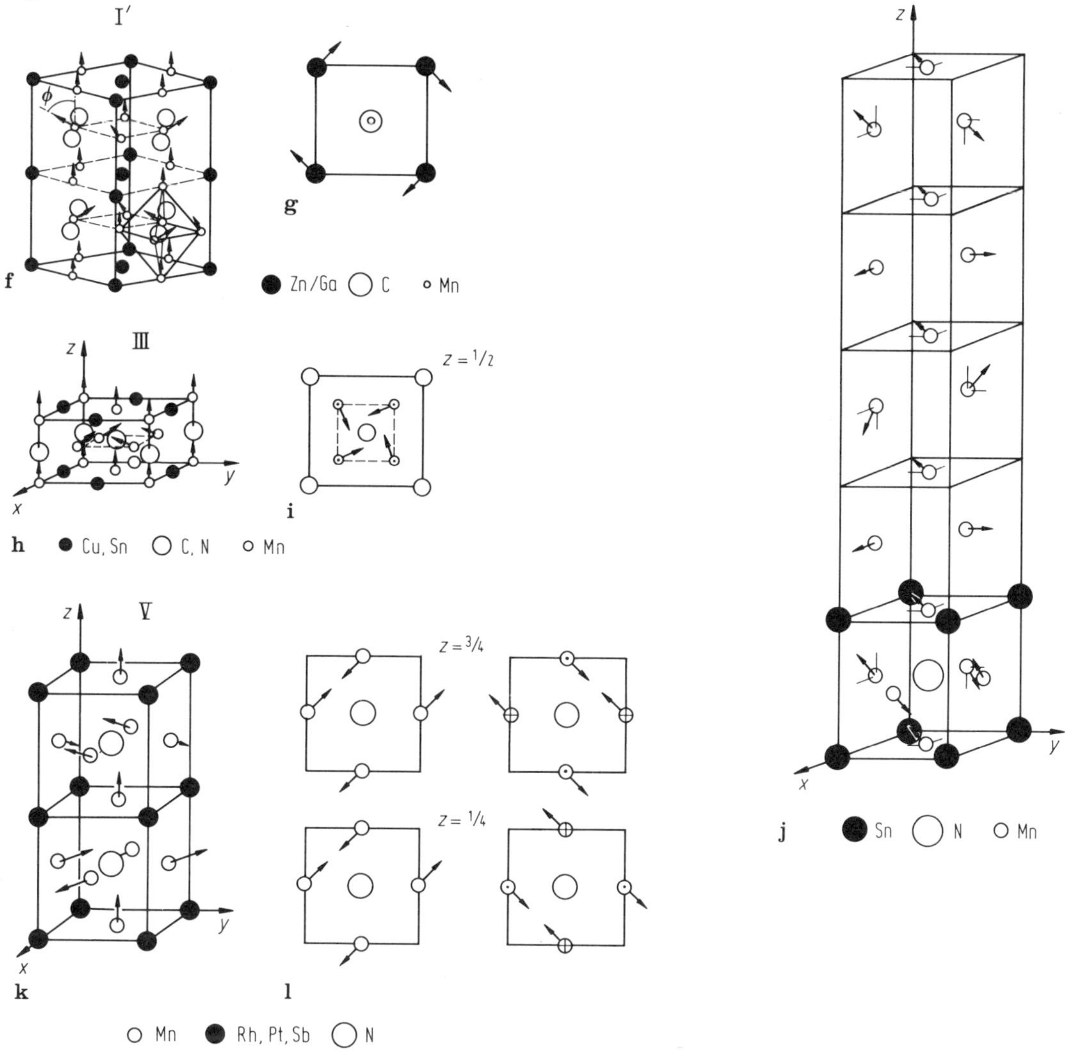

(c) Pure triangular antiferromagnetic mode as observed in $AgNMn_3$, $ZnNMn_3$, $GaNMn_3$, $SnNMn_3$ (high-temperature phases) described by the Γ_{5g} irreducible representation of Pm3m. Structure type designated IV_A.

(d) Magnetic structure of $ZnNMn_3$ ($T<140$ K). The lattice parameters of the magnetic cell are $a\sqrt{2}$, $a\sqrt{2}$, $2a$, with respect to the cubic one. The magnetic moments are all lying in the basal plane with $p_{Mn(1,2)}=0.61\ \mu_B$ and $p_{Mn(3)}=1.03\ \mu_B$, the easy axis of the two sublattices being orthogonal. Only the Mn atoms are represented. Structure type designated IV′.

(e) Collinear antiferromagnetic structure, observed in $GaCMn_3$ at low temperature. $p_{Mn}=1.8\ \mu_B$ antiferromagnetically coupled along [111] direction. Structure type designated II.

(f) Low-temperature magnetic structure of $ZnCMn_3$ ($T<233$ K) and $GaC_{0.935}Mn_3$ ($T<147$ K). The angle between $p_{Mn(1,2)}$ and the c-axis is 65° in the Zn compound and 20° in the Ga one. These magnetic moments from a square net as seen in (g). $p_{Mn(3)}$ is lying along the c-axis. The lattice parameters of the magnetic unit cell are $a\sqrt{2}$, $a\sqrt{2}$, $2a$. Structure designated I′.

(g) Arrangement of magnetic moments forming an octahedron in the type of structure depicted in (f).

(h) Magnetic cell of $SnCMn_3$ and $CuNMn_3$. Lattice parameters are $a\sqrt{2}$, $a\sqrt{2}$, a. Structure type designated III.

(i) Projection on the (001) plane of the $CuNMn_3$ model depicted in (h) showing the net of the Mn magnetic moments $p_{Mn(1)}$ and $p_{Mn(2)}$. $p_{Mn(3)}$ points along c-axis.

(j) Long-wave magnetic structure of $SnNMn_3$ at low temperature ($T<250$ K). The wavevector (00k) is $k\simeq 1/4$ at $T=4.2$ K and decreases continuously to $k\simeq 1/8$ at 237 K. $p_{Mn(1)}$ and $p_{Mn(2)}$ rotate in (100) and (010), respectively. $p_{Mn(3)}$ remains in the basal plane.

(k) Noncollinear magnetic structure of $RhNMn_3$, $PtN_{0.07}Mn_3$ (lattice parameters a, a, $\simeq 2a$). The $p_{Mn(3)}$ magnetic moments are lying along c-axis. Structure type designated V.

(l) Projections of $p_{Mn(1,2)}$ on the (001) plane of the type of structure depicted in (k) [76 F 1].

8.2 Magnetic and related data of $M^cXMn^f_3$ compounds

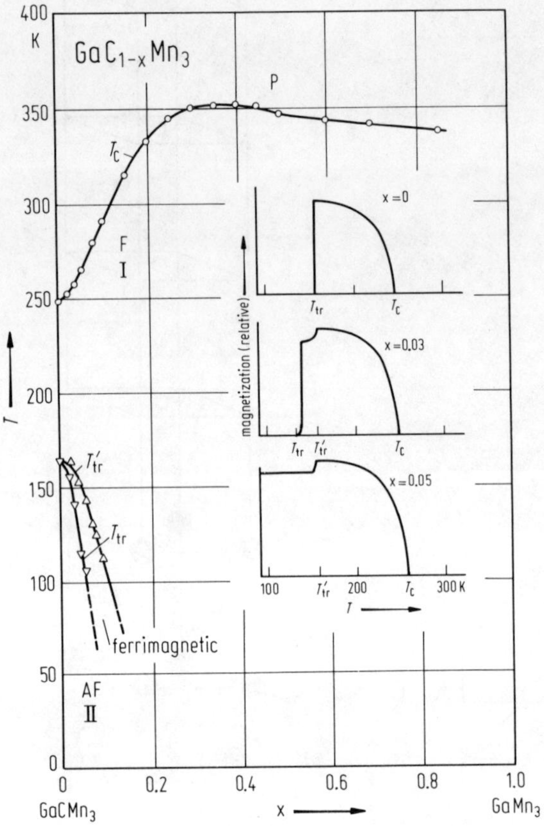

Fig. 2. $GaC_{1-x}Mn_3$. Magnetic phase diagram and magnetization vs. temperature. T_{tr} is a first-order antiferro- to ferromagnetic (or ferrimagnetic) transition temperature; T'_{tr} is a ferri- to ferromagnetic transition temperature. I: ferromagnetic, see Fig. 1a; II: antiferromagnetic, see Fig. 1e [65 B 1, 68 B 2, 70 H 1].

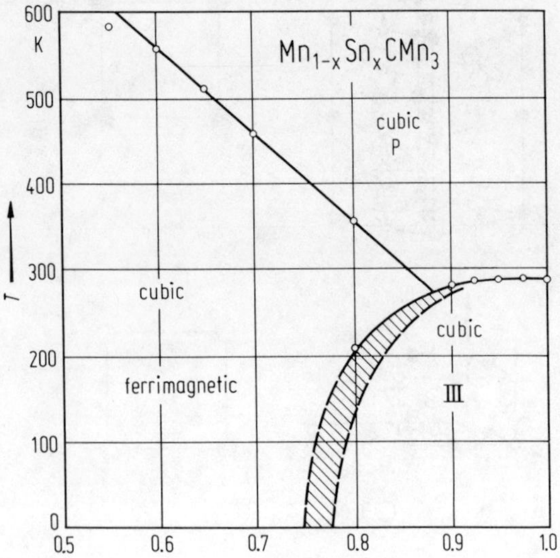

Fig. 3. $Mn_{1-x}Sn_xCMn_3$. Magnetic phase diagram. III: see Fig. 1h [80 L 1].

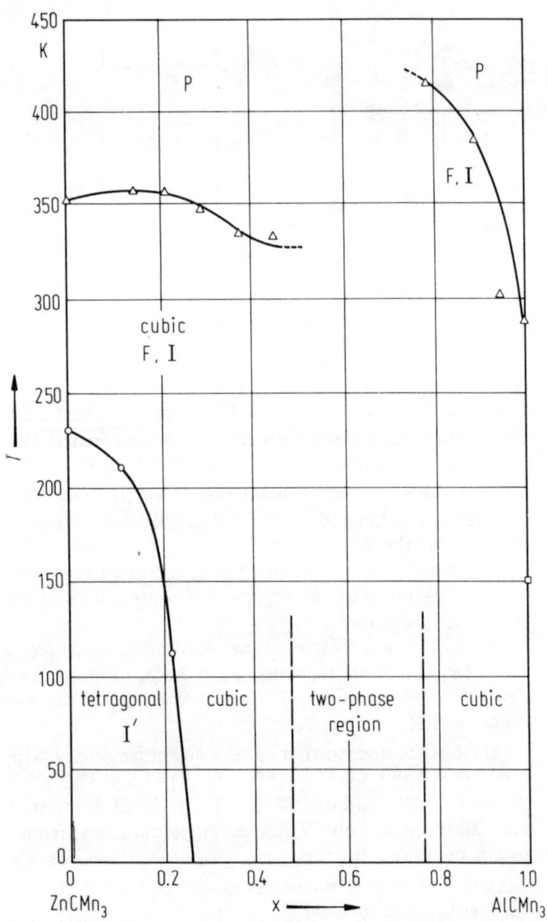

Fig. 4. $Zn_{1-x}Al_xCMn_3$. Magnetic phase diagram. I: ferromagnetic, see Fig. 1a; I′: see Fig. 1f [57 H 1].

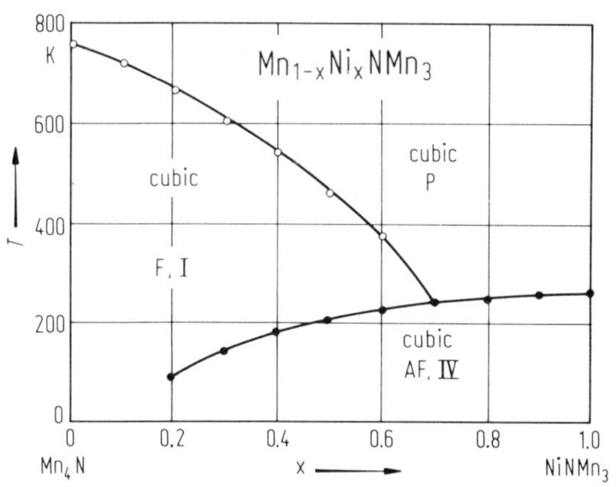

Fig. 8. $Mn_{1-x}Ni_xNMn_3$. Magnetic phase diagram. I: ferromagnetic, see Fig. 1a; IV: antiferromagnetic, see Figs. 1b, c [70 M 1].

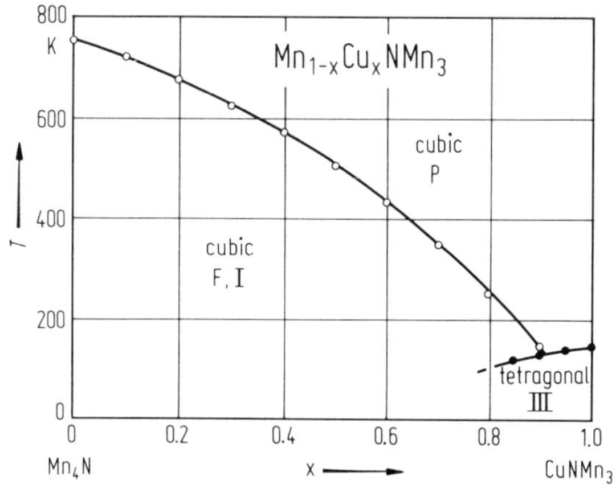

Fig. 9. $Mn_{1-x}Cu_xNMn_3$. Magnetic phase diagram. I: ferromagnetic, see Fig. 1a; III: see Fig. 1h [70 M 1].

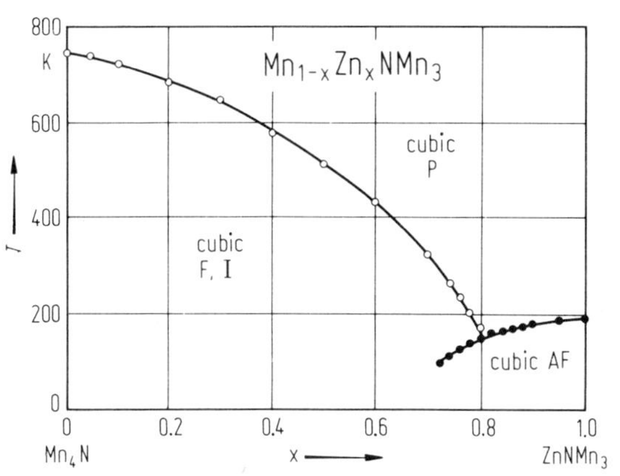

Fig. 10. $Mn_{1-x}Zn_xNMn_3$. Magnetic phase diagram. I: ferromagnetic, see Fig. 1a; AF: antiferromagnetic, see Figs. 1c, d [67 F 1].

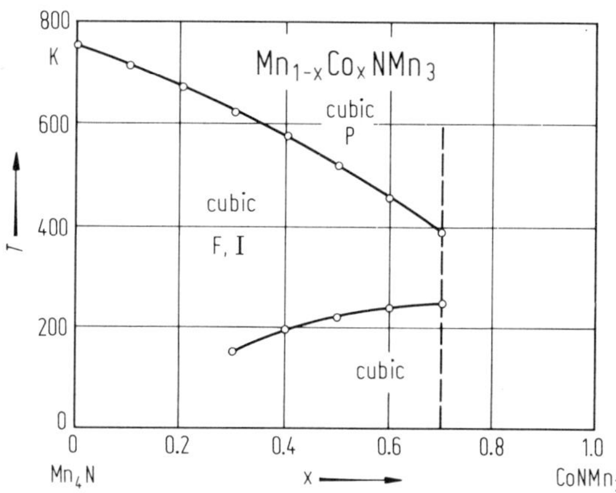

Fig. 11. $Mn_{1-x}Co_xNMn_3$. Magnetic phase diagram. I: ferromagnetic, see Fig. 1a [70 M 1].

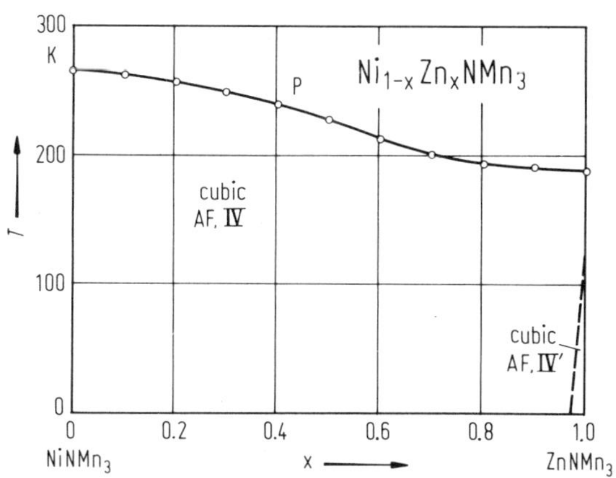

Fig. 12. $Ni_{1-x}Zn_xNMn_3$. Magnetic phase diagram. IV: antiferromagnetic, see Figs. 1b, 1c; IV′: antiferromagnetic, see Fig. 1d [70 M 1].

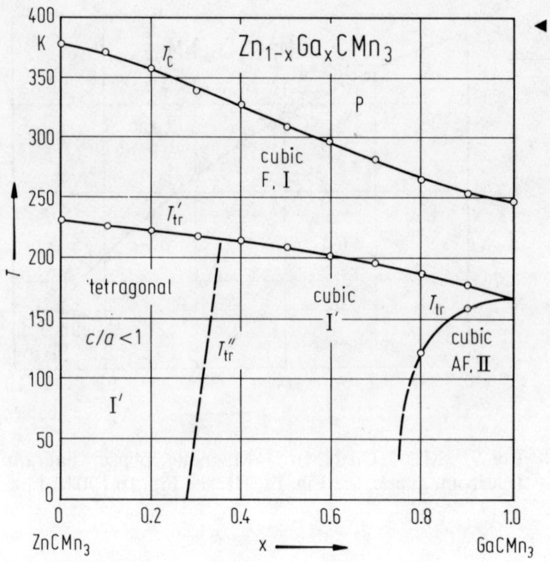

◄ Fig. 5. $Zn_{1-x}Ga_xCMn_3$. Magnetic phase diagram. I: ferromagnetic (see Fig. 1a); I′: see Fig. 1f; II: antiferromagnetic, see Fig. 1e [73 F 1].

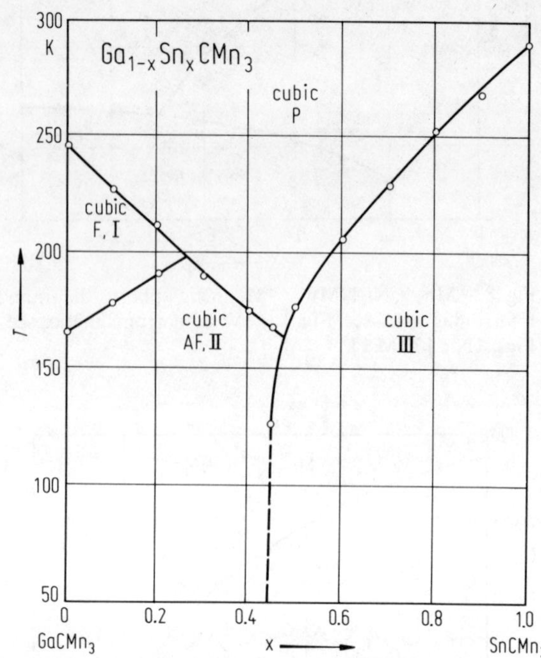

Fig. 6. $Ga_{1-x}Sn_xCMn_3$. Magnetic phase diagram. I: ferromagnetic, see Fig. 1a; II: antiferromagnetic, see Fig. 1e; III: see Fig 1h [80 L 1].

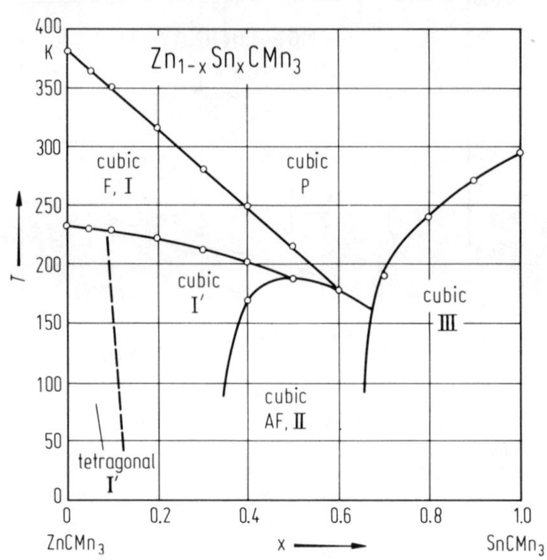

Fig. 7. $Zn_{1-x}Sn_xCMn_3$. Magnetic phase diagram. I: ferromagnetic, see Fig. 1a; I′: see Fig. 1f; II: antiferromagnetic, see Fig. 1e; III: see Fig. 1h [73 L 1].

Mn_4N

Magnetization [62 M 1, 62 T 1]
$T_C = 756$ K, $p_m(0\,K) = 1.2\mu_B$.

Neutron diffraction PW [60 T 1] [62 T 1, 66 M 1]
Collinear ferrimagnetic in first approximation.
$T = 77$ K: $p_{Mn^c} = 3.85\mu_B$, $p_{Mn^f} = -0.90\mu_B$.

Magnetic structure [79 F 1, 80 F 1]
Ferromagnetic component:
Direction of easy magnetization: [111] (checked for $T = 4.2 \cdots 500$ K).
$p_m(4.2\,k) = 1.17\mu_B$,
$p_m(300\,K) = 0.81\mu_B$: $p_{Mn^c} = 3.70(15)\mu_B$, $p_{Mn^f[111]} = -0.95\mu_B$.

Triangular antiferromagnetic component:
$p_{Mn^f(111)} = 0.36\mu_B$, nearly temperature-independent, turning in the (111) plane from a Γ_{4g}-cubic type (O_h^1) to a Γ_{5g}-cubic type (O_h^1) configuration, between 4.2 and 500 K, see Figs. 1a, b, c.

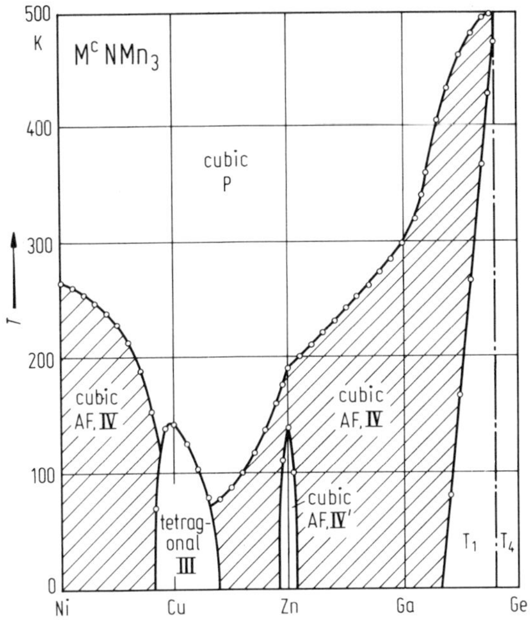

Fig. 13. M^cNMn_3 (M^c = Ni, Cu, Zn, Ga, Ge). Phase diagram. III: see Fig. 1h, IV: antiferromagnetic, see Figs. 1b, c; IV': antiferromagnetic, see Fig. 1d; T_1, T_4: tetragonal, see Introduction [70 M 1].

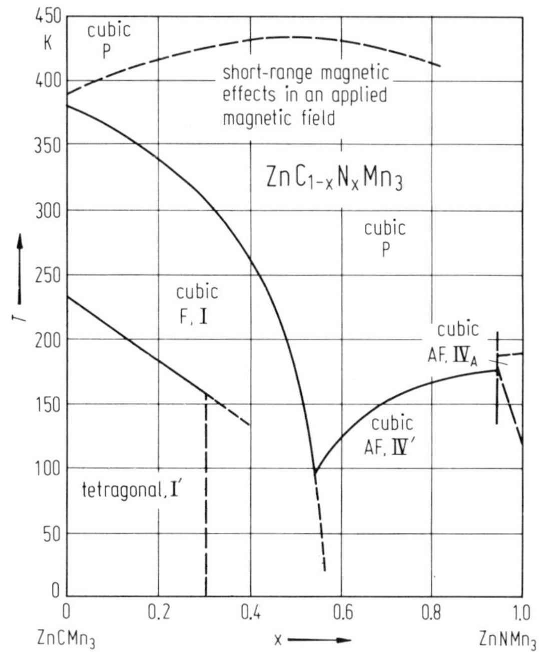

Fig. 14. $ZnC_{1-x}N_xMn_3$. Magnetic phase diagram. I: ferromagnetic, see Fig. 1a; I': see Fig. 1f, IV_A: antiferromagnetic, see Fig. 1c, IV': antiferromagnetic, see Fig. 1d [80 L 1].

8.3 References for 8

57 H 1 Howe, L., Myers, H.P.: Philos. Mag. **2** (1957) 554.

60 T 1 Takei, N.G., Shirane, G., Frazer, B.: Phys. Rev. **119** (1960) 122.

62 M 1 Mekata, M.: J. Phys. Soc. Jpn. **17** (1962) 796.

62 T 1 Takei, W.J., Heikes, R.R., Shirane, G.: Phys. Rev. **125** (1962) 1893.

65 B 1 Bouchaud, J.P., Fruchart, R.: Propriétés Thermodynamiques Physiques et Structurales des Dérivés Semi-Métalliques, Colloques Internationaux du C.N.R.S., N° 157, Orsay, Sept. **1965** (Editions C.N.R.S., Paris 1967) p. 117.

66 M 1 Mekata, M., Haruna, J., Takaki, H.: J. Phys. Soc. Jpn. **21** (1966) 2267.

67 F 1 Fruchart, R., Bouchaud, J.P., Fruchart, E., Lorthioir, R., Madar, R., Rouault, A.: Mater. Res. Bull. **2** (1967) 1009.

68 B 1 Boller, H.: Monatsh. Chem. **99** (1968) 2444.

68 B 2 Bouchaud, J.P.: Ann. Chim. **3** (1968) 81.

69 B 1 Boller, H.: Monatsh. Chem. **100** (1969) 1471.

70 H 1 Hellwege, K.-H. (ed.), Goodenough, J.B., Longo, J.M.: Landolt-Börnstein, Numerical Data and Functional Relationships in Science and Technology, New Series, Group III: Crystal and Solid State Physics, Volume 4, Magnetic and Other Properties of Oxides and Related Compounds, Part a, Springer: Berlin, Heidelberg, New York, **1970**.

70 M 1 Madar, R.: Thesis, University of Paris, **1970**; N° 686, Centre de Documentation du CNRS, Paris.

72 B 1 Barberon, M., Fruchart, E., Fruchart, R., Lorthioir, G., Madar, R., Nardin, M.: Mater. Res. Bull. **7** (1972) 109.

73 F 1 Fruchart, D., Bertaut, E.F., Clerc, B.L., Khoi, L.D., Veillet, P., Lorthioir, G., Fruchart, E., Fruchart, R.: J. Solid State Chem. **8** (1973) 182.

73 L 1 Lorthioir, G., Fruchart, E., Nardin, M., l'Héritier, P., Fruchart, R.: Mater. Res. Bull. **8** (1973) 1027.
76 F 1 Fruchart, D.: University of Grenoble (France) **1976**; No. 11922 at Centre de Documentation du CNRS, Paris.
77 F 1 Fruchart, D.: Physica **86–88** B (1977) 423.
78 F 1 Fruchart, D., Bertaut, E.F.: J. Phys. Soc. Jpn. **44** (1978) 781.
79 F 1 Fruchart, D., Givord, D., Convert, P., l'Héritier, Ph., Sénateur, J.P.: J. Phys. F: Met. Phys. **9** (1979) 2431.
80 F 1 Fruchart, D., l'Héritier. Ph., Fruchart, R.: Mater. Res. Bull. **15** (1980) 415.
80 L 1 l'Héritier, Ph.: Thesis, University of Grenoble (France) **1980**; deposited at Centre de Documentation du CNRS, Paris.
80 M 1 Michalski, K.J., Reich, W., Cofta, H.: Phys. Status Solidi (b) **97** (1980) 187.
81 B 1 Bertaut, E.F.: J. Magn. Magn. Mater. **24** (1981) 267.

Appendix: Contents of the volumes of the Landolt–Börnstein New Series dealing with magnetic properties of metals based on d elements

Group III Volume 19

Subvolume a

3d, 4d and 5d Elements, Alloys and Compounds

Subvolume b

Alloys and Compounds of d-Elements with Main Group Elements. Part 1

Subvolume c

Alloys and Compounds of d-Elements with Main Group Elements. Part 2

Subvolume a: Contents

Editor

H. P. J. Wijn

Institut für Werkstoffe der Elektrotechnik der Rheinisch-Westfälischen Technischen Hochschule Aachen, Templergraben 55, D-5100 Aachen

Contributors

K. Adachi

Department of Physics, Nagoya University, Chikusa-ku, Nagoya 464, Japan

D. Bonnenberg

Institut für Werkstoffe der Elektrotechnik der Rheinisch-Westfälischen Technischen Hochschule Aachen, Templergraben 55, D-5100 Aachen

J. J. M. Franse

Natuurkundig Laboratorium der Universiteit van Amsterdam, Valckenierstraat 65, 1018 XE Amsterdam, Nederland

R. Gersdorf

Natuurkundig Laboratorium der Universiteit van Amsterdam, Valckenierstraat 65, 1018 XE Amsterdam, Nederland

K. A. Hempel

Institut für Werkstoffe der Elektrotechnik der Rheinisch-Westfälischen Technischen Hochschule Aachen, Templergraben 55, D-5100 Aachen

K. Kanematsu

Department of Physics, Nihon University, Kanda-Surugadai, Chiyoda-ku, Tokyo 101, Japan

S. Misawa

Department of Physics, Nihon University, Kanda-Surugadai, Chiyoda-ku, Tokyo 101, Japan

M. Shiga

Department of Metal Science and Technology, Kyoto University, Sakyo-ku, Kyoto 606, Japan

M. B. Stearns

Department of Physics, Arizona State University, Tempe, Arizona, 85287, USA

H. P. J. Wijn

Institut für Werkstoffe der Elektrotechnik der Rheinisch-Westfälischen Technischen Hochschule Aachen, Templergraben 55, D-5100 Aachen

Subvolume b: Contents

Editor

H. P. J. Wijn

Institut für Werkstoffe der Elektrotechnik der Rheinisch-Westfälischen Technischen Hochschule Aachen, Templergraben 55, D-5100 Aachen

Contributors

J. G. Booth

Physics Department, University of Salford, Salford M5 4WT, United Kingdom

H.P.J. Wijn

Institut für Werkstoffe der Elektrotechnik der Rheinisch-Westfälischen Technischen Hochschule Aachen, Templergraben 55, D-5100 Aachen

G. Zibold

Fachhochschule Ravensburg-Weingarten, Postfach 1107, D-7967 Weingarten

Subvolume c: Contents

Editor

H. P. J. Wijn

Institut für Werkstoffe der Elektrotechnik der Rheinisch-Westfälischen Technischen Hochschule Aachen, Templergraben, D-5100 Aachen

Contributors

D. Fruchart

Laboratoire de Cristallographie du CNRS, 166X, F-38042 Grenoble Cedex

R. Fruchart

Laboratoire des Matériaux et du Génie Physique, UA 1109 CNRS, Ecole Nationale Supérieure de Physique de Grenoble, Domaine Universitaire, BP 46, F-38402 Saint-Martin-d'Hères

Ph. L'Héritier

Laboratoire des Matériaux et du Génie Physique, UA 1109 CNRS, Ecole Nationale Supérieure de Physique de Grenoble, Domaine Universitaire, BP 46, F-38402 Saint-Martin-d'Hères

K. Kanematsu

Physical Science Laboratories, Nihon University at Narashino, Funabashi, Japan

R. Madar

Laboratoire des Matériaux et du Génie Physique, UA 1109 CNRS, Ecole Nationale Supérieure de Physique de Grenoble, Domaine Universitaire, BP 46, F-38402 Saint-Martin-d'Hères

S. Misawa

Department of Physics, College of Science and Technology, Nihon University, Kanda Surugadai, Chiyoda-ku, Tokyo 101, Japan

Y. Nakamura

Department of Metal Science and Technology, Kyoto University, Sakyo-ku, Kyoto, Japan

P. J. Webster

Department of Civil Engineering, University of Salford, Salford M5 4WT, United Kingdom

K. R. A. Ziebeck

Department of Physics, Loughborough University of Technology, Loughborough LE II 3TU, United Kingdom